T0258159

Encyclopedia of Quantum Mechanics: Concepts and Principles

Volume II

Encyclopedia of Quantum Mechanics: Concepts and Principles
Volume II

Edited by **Ian Plummer**

New York

Published by NY Research Press,
23 West, 55th Street, Suite 816,
New York, NY 10019, USA
www.nyresearchpress.com

Encyclopedia of Quantum Mechanics: Concepts and Principles
Volume II
Edited by Ian Plummer

International Standard Book Number: 978-1-63238-157-6 (Hardback)

Printed in the United States of America.

Contents

Preface

Human understanding regarding the universe and nature driven forces have been thoroughly influenced by the Quantum theory as a scientific revolution. Possibly, the historical advancement of this concept mimics the history of human scientific endeavors from their birth. The book presents a valuable account of foundation, scientific history of the concept, distinct techniques to solve the Schrodinger equation, and relativistic quantum mechanics and field theory. It includes important topics such as Bohmian Trajectories and the Path Integral Paradigm, Approximate Solutions of the Dirac Equation and Theoretical Validation of the Computational Unified Field Theory. The aim of this book is to serve as an efficient reference for researchers and students.

After months of intensive research and writing, this book is the end result of all who devoted their time and efforts in the initiation and progress of this book. It will surely be a source of reference in enhancing the required knowledge of the new developments in the area. During the course of developing this book, certain measures such as accuracy, authenticity and research focused analytical studies were given preference in order to produce a comprehensive book in the area of study.

This book would not have been possible without the efforts of the authors and the publisher. I extend my sincere thanks to them. Secondly, I express my gratitude to my family and well-wishers. And most importantly, I thank my students for constantly expressing their willingness and curiosity in enhancing their knowledge in the field, which encourages me to take up further research projects for the advancement of the area.

Editor

Bohmian Trajectories and the Path Integral Paradigm – Complexified Lagrangian Mechanics

Valery I. Sbitnev

B. P. Konstantionv St.-Petersburg Nuclear Physics Institute,
Russ. Ac. Sci., Gatchina, Leningrad District
Russia

1. Introduction

All material objects perceivable by our sensations move in real 3D-space. In order to describe such movement in strict mathematical forms we need to realize, first, what does the space represent as a mathematical abstraction and how motion in it can be expressed? Isaac Newton had gave many cogitations with regard to categories of the space and time. Results of these cogitations have been devoted to formulating categories of absolute and relative space and time (Stanford Encyclopeia, 2004): (a) material body occupies some place in the space; (b) absolute, true, and mathematical space remains similar and immovable without relation to anything external; (c) relative spaces are measures of absolute space defined with reference to some system of bodies or another, and thus a relative space may, and likely will, be in motion; (d) absolute motion is the translation of a body from one absolute place to another; relative motion is the translation from one relative place to another.

Observe, that space coordinates of a body can be attributed to center of mass of the body, and its velocity is measured as a velocity of motion of this center. It means, that a classical body can be replaced ideally by a mathematical point situated in the center of mass of the body. Velocity of the point particle is determined from movement of the center of mass per unit of time. Both point particle coordinates and its velocity are measured exactly. Its behavior can be computed unambiguously from formulas of classical mechanics (Lanczos 1970).

Appearance of quantum mechanics in the early twentieth century brought into our comprehension of reality qualitative revisions (Bohm, 1951). One problem, for example, arises at attempt of simultaneous measurement of the particle coordinate and its velocity. There is no method that could propose such measurements. Quantum mechanics proclaims weighty, nay, unanswerable principle of uncertainty prohibiting such simultaneous measurements. Therefore we can measure these parameters only with some accuracy limited by the uncertainty principle. From here it follows, that formulas of classical mechanics meet with failure as soon as we reach small scales. On these scales the particles behave like waves. It is said, in that case, about the wave-particle duality (Nikolić, 2007).

It would be interesting to note here, that as far back as 5th century, B. C., ancient philosopher Democritus, (Stanford Encyclopeia, 2010), held that everything is composed of "atoms", which are physically indivisible smallest entities. Between atoms lies empty space. In such a view it means that the atoms move in the empty space. And only collisions of the atoms can effect on their future motions. One more standpoint on Nature, other than atomistic, originates from ancient philosopher Aristotle (Stanford Encyclopeia, 2008). Among his fifth elements (Fire, Earth, Air, Water, and Aether), composing the Nature, the last element, Aether, has a particular sense for explanation of wave processes. It provides a good basis for understanding and predicting the wave propagation through a medium.

By adopting wave processes underlying the Nature one can explain of interference phenomena of light. Huygens (Andresse, 2005) gave such an explanation. In contrast to Newtonian corpuscular explanation, Huygens proposed that every point to which a luminous wave reached becomes a source of a spherical wave, and the sum of these secondary waves determines the form of the wave at any subsequent time. His name was coined in the Huygens's wave principle, (Born & Wolf, 1999).

Such a competition of the two standpoints, corpuscular and wave, can provide more insight penetration into problems taking place in the quantum realm. Here we adopt these standpoints as a program for action (Sbitnev, 2009a). The article consists of five sections. Sec. 2 begins from a short review of the classical mechanics methods and ends by Dirac's proposition as the classical action can show itself in the quantum realm. Feynman's path integral is a summit of this understanding. The path integral technique is used in Sec. 3 for computing interference pattern from N-slit gratings. In Sec. 4 the path integral is analyzed in depth. The Schrödinger equation results from this consideration. And as a result we get the Bohmian decomposition of the Schrödinger equation to pair of coupled equations, modified the Hamilton-Jacobi equation and the continuity equation. Sec. 5 studies this coupled pair in depth. And concluding Sec. 6 gives remarks relating to sensing our 3D-space on the quantum level.

2. From classic realm to quantum

A path along which a classical particle moves, Fig. 1, obeys to variational principles of mechanics. A main principle is the principle of least action (Lanczos, 1970). The action S is a scalar function that is inner production of dynamical entities of the particle (its energy, momentum, etc.) to geometrical entities (time, length, etc.). For a particle's swarm moving through the space along some direction, the action is represented as a surface be pierced by their trajectories. Observe that adjoining surfaces are situated in parallel to each other and the trajectories pierce them perpendicularly.

The action S is the time integral of an energy function, that is the Lagrange function, along the path from A (starting from the moment t_0) to B (finishing at the moment t_1) :

$$S = \int_{t_0}^{t_1} L(\vec{q},\dot{\vec{q}};t)dt . \tag{1}$$

Here $L(\vec{q},\dot{\vec{q}};t)$ is the Lagrange function representing difference of kinetic and potential energies of the particle. And $\vec{q}$ and $\dot{\vec{q}}$ are its coordinate and velocity. Scientists proclaim that the action S remains constant along an optimal path of the moving particle. It is the principle

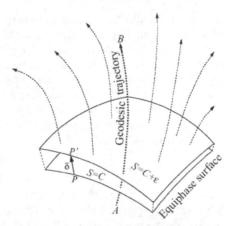

Fig. 1. Particle, at passing from A to B, moves along geodesic trajectory - the trajectory satisfying the principle of least action. All geodesic trajectories intersect equiphase surfaces, $S=C$, $S=C+\varepsilon$, perpendicularly (Lanczos, 1970).

of least action. According to this principle, finding of the optimal path adds up to solution of the extremum problem $\delta S = 0$. The solution leads to establishing the Lagrangian mechanics (Lanczos, 1970). We sum up the Lagrangian mechanics by presenting its main formulas via The Legendre's dual transformations as collected in Table 1:

Variables :		Variables :	
Coordinate:	$\vec{q}=(q_1,q_2,...,q_N)$	Coordinate:	$\vec{q}=(q_1,q_2,...,q_N)$
Momentum:	$\vec{p}=(p_1,p_2,...,p_N)$	Velocity:	$\dot{\vec{q}}=(\dot{q}_1,\dot{q}_2,...,\dot{q}_N)$
Hamiltonian function:		**Lagrangian function:**	
$H(\vec{q},\vec{p};t)=\sum_{n=1}^{N} p_n\dot{q}_n - L(\vec{q},\dot{\vec{q}};t)$		$L(\vec{q},\dot{\vec{q}};t)=\sum_{n=1}^{N} p_n\dot{q}_n - H(\vec{q},\vec{p};t)$	
$\dfrac{\partial H}{\partial p_n}=\dot{q}_n$		$\dfrac{\partial L}{\partial \dot{q}_n}=p_n$	
$\dfrac{\partial H}{\partial q_n}=-\dot{p}_n$		$\dfrac{\partial L}{\partial q_n}=\dot{p}_n$	

Table 1. The Legendre's dual transformations

The Hamilton-Jacobi equation (HJ-equation)

$$-\frac{\partial S}{\partial t}=H(\vec{q},\vec{p};t),\qquad\qquad(2)$$

describing behavior of the particle in $2N$-dimesional phase space is one of main equations of the classical mechanics. Let us glance on Fig. 1. Gradient of the action S can be seen as normal to the equiphase surface S = const. Consider two nearby surfaces $S = C$ and $S = C+\varepsilon$. Let us trace normal from an arbitrary point P of the first surface up to its intersection with the second surface at point P'. Next, make another shift of the surface that is 2ε distant away from the first surface, thereupon on 3ε, and so forth. Until all space will be filled with such secants. Normals drawn from P to P' thereupon from P' to P'', and so forth, disclose possible trajectory of the particle, since $\nabla S = \varepsilon / \delta$ represents a value of the gradient of S. When ε and δ tend to zero, this relation can be expressed in the vector form

$$\vec{p} = \nabla S. \qquad (3)$$

So far as the momentum $\vec{p} = m\vec{v}$ (m is a particle mass) has a direction tangent to the trajectory, then the following statement is true (Lanczos, 1970): *trajectory of a moving particle is perpendicular to the surface S* = const. Dotted curves in Fig. 1 show bundle of trajectories intersecting the surfaces S perpendicularly.

The particle's swarm moving through space can be dense enough. It is appropriate to mention therefore the Liouville theorem, that adds to the conservation law of energy one more a conservation law. Meaning of the law is that a trajectory density is conserved independently of deformations of the surface that encloses these trajectories. Mathematically, this law is expressed in a form of the continuity equation

$$\frac{\partial \rho}{\partial t} + \left(\vec{v} \cdot \nabla \rho \right). \qquad (4)$$

Here ρ is a density of moving mechanical points with the velocity $\vec{v} = \vec{p} / m$.

Thus we have two equations, the HJ-equation (2) and the continuity equation (4) that give mathematical description of moving classical particles undergoing no noise. Draw attention here, that the continuity equation depends on solutions of the HJ-equation via the term $\vec{v} = \nabla S / m$. On the other hand we see, that the HJ-equation does not depend on solutions of the continuity equation. This is essential moment at description of moving ensemble of the classical objects.

Starting from a particular role of the action, which it has in classical mechanics, Paul Dirac drew attention in 1933 (Dirac, 1933) that the action can play a crucial role in quantum mechanics also. The action can exhibit itself in expressions of type exp{ iS / ℏ}. It is appropriate to notice the following observation: the action here plays a role of a phase shift. According to the principle of least action, we can guess that the phase shift should be least along an optimal path of the particle. In 1945 Paul Dirac emphasize once again, that the classical and quantum mechanics have many general points of crossing (Dirac, 1945). In particular, he had written in this article: "We can use the formal probability to set up a quantum picture rather close to the classical picture in which the coordinates q of a dynamical system have definite values at any time. We take a number of times t_1, t_2, t_3, ... following closely one after another and set up the formal probability for the q's at each of these times lying within specified small ranges, this being permissible since the q's at any time all commute. We then get a formal probability for the trajectory of the system in quantum mechanics lying within certain limits. This enables us to speak of some trajectories being improbable and others being likely".

Next, Richard Feynman undertook successful search of acceptable mathematical apparatus (Feynman, 1948) for description of evolution of quantum particles traveling through an experimental device. The term

$$\exp\{iS/\hbar\} = \exp\{iL\delta t/\hbar\} \qquad (5)$$

plays a decisive role in this approach. Idea is that this term executes mapping of a wave function from one state to another spased on a small time interval δt. And L is Lagrangian describing current state of the quantum object.

Feynman's insight has resulted in understanding that the integral kernel (so called propagator) of the time-evolution operator can be expressed as a sum over all possible paths (not just over the classical one) connecting the outgoing and ingoing points, q_a and q_b, with the weight factor $\exp\{ iS(q_a, q_b;t)/\hbar \}$ (Grosche, 1993; MacKenzie, 2000) :

$$K(q_a, q_b) = \sum_{\text{all paths}} A\exp\{iS(q_a, q_b;T)/\hbar\}, \qquad (6)$$

where A is an normalization constant.

Observe that The Einstein-Smoluchowski equation which describes the Brownian motion of classical particles within some volume (Kac, 1957), served him as an example. As follows from idea of the path integral (6), there are many possible trajectories, that can be traced from a source to a detector. But only one trajectory, submitting to the principle of least action, may be real. The others cancel each other because of interference effects. Such an interpretation is extremely productive at generating intuitive imagination for more perfect understanding quantum mechanics.

It is instructive further to consider some quantum tasks by using the Feynman path integral. Here we will compute interference patterns as a result of incidence of particles on N-slit gratings.

3. Interference pattern from an N-slit gratin

Let a beam of coherent particles spreads through a grating. The grating shown in Fig. 2 has a set of narrow slits sliced in parallel. Width of the slits is sufficient in order that even large molecules could pass they through. Here we face with the uncertainty principle, $\Delta r \Delta p \geq \hbar/2$.

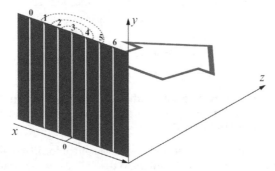

Fig. 2. Interference experiment in cylindrical geometry. Slit grating with $n=0,1, \dots ,N-1$ slits is situated in a plane (x,y). Propagation of particles occurs along axis z.

It means, if diameter of the molecule is close to width of the slit then direction of its escape from the slit is uncertain. One can draw, as commonly, cylindrical waves that are divergent from each slit, as shown in this illustration on the slit 3. They illustrate equally probable outcomes from slits in different directions. In other words, a particle may fly out in any direction with equal probability.

3.1 Passing through a slit

Before we will analyze interference on the N-slit grating, let us consider a particle passing through a single slit. The problem has been considered in detail in (Feynman & Hibbs, 1965). We will study migration of the free particle in transversal direction, let it be axis x, at passing along z with a constant velocity, see Fig. 3. Lagrangian is as follows

$$L = m\frac{\dot{x}^2}{2} + \text{const} . \tag{7}$$

Here m is mass of the particle and $\dot{x}$ is its transversal velocity. By translating a particle's position on a small value $\delta x = (x_b - x_a) \ll 1$, being performed for a small time $\delta t = (t_b - t_a) \ll 1$, we find that a weight factor, see (5), is as follows

$$e^{iL\delta t/\hbar} = \exp\left\{\frac{im(x_b - x_a)^2}{2\hbar(t_b - t_a)}\right\} . \tag{8}$$

Pay attention on the following situation: so far as argument of the exponent contains multiplication of the Lagrangian L by δt, as shown in Eq. (5), we obtain result $(x_b - x_a)^2$ divided by $(t_b - t_a)$. Next we will see, that the weight factor (8) plays an important role. By means of such small increments let us trace passing the particle from a source through the slit, Fig. 3.

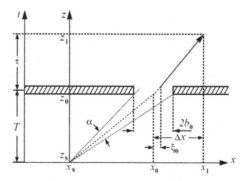

Fig. 3. A particle, being emitted from a source that is localized at a point (x_s, z_s) passes through a slit with width $2b_0$. It may undergo deflection from a straight direction at passing through the slit (Feynman & Hibbs, 1965).

We suppose, that at the time $t = 0$ the particle leaves a source localized at a point (x_s, z_s). Let we know, that after a time T the particle enters to the vicinity $x_0 \pm b$ of a point x_0, see Fig. 3. The question is: what is the probability to disclose the particle after a time τ at a point x_1 remote from the point x_0 at a distance $\Delta x = (x_1 - x_0)$? Let the particle outgoing from the point

x_s at the time $t=0$ passes a slit between the points $x_0 - b$ and x_0+b at the time $t=T$. Let us compute the probability of discovering the particle at some point x_1 after the time τ, i.e., at $t=T+\tau$. Because of existence of an opaque barrier a direct path to the point x_1 can be absent. In order to reach the point x_1 the particle should pass through the slit, maybe with some deflection from the direct path. In this connection, we partition the problem into two parts. Each part relates to movement of the free particle. In the first part we consider the particle which begins movement from the point x_s at the initial moment $t = 0$ and reaches to a point $x = x_0+\xi$, at the moment $t = T$, where $|\xi| \leq b$. In the second part we consider the same particle that after passing the point $x=x_0+\xi$ at the time $t=T$ moves to the point x_1 and reach it at the time $t=T+\tau$. A full probability amplitude is equal to integral convolution of two kernels, each describing movement of the free particle:

$$\psi(x_1,x_0,x_s) = \int_{-b}^{b} K(x_1,T+\tau;x_0+\xi,T)K(x_0+\xi,T;x_s,0)d\xi . \tag{9}$$

Here the kernel reads

$$K(x_b,t_b;x_a,t_a) = \left[\frac{2\pi i\hbar(t_b-t_a)}{m}\right]^{-1/2} \exp\left\{\frac{im(x_b-x_a)^2}{2\hbar(t_b-t_a)}\right\} . \tag{10}$$

It describes a transition amplitude from x_a to x_b for a time interval $(t_b - t_a)$ (Feynman & Hibbs, 1965). Consequently, the integral (9) computes the probability amplitude of transition from the source x_s to the point x_1 through the all possible intermediate points ξ situated within the interval $(x_0 - b, x_0 + b)$.

The expression (9) is written in accordance with a rule of summing amplitudes for successive events in time. The first event is the moving particle from the source to the slit. The second event is the movement of the particle from the slit to the point x_1. The slit has a finite width. Passage through the slit is conditioned by different alternative possibilities. For that reason, we need to integrate along all over the slit width in order to get a right result. All particles, moving through the slit, are free particles and their corresponding kernels are given by the expression (10). By substituting this kernel to the integral (9) we get the following detailed form

$$\psi(x_1,x_0,x_s) = \int_{-b}^{b} \left(\frac{2\pi i\hbar\tau}{m}\right)^{-1/2} \exp\left\{\frac{im(\Delta x-\xi)^2}{2\hbar\tau}\right\} \left(\frac{2\pi i\hbar T}{m}\right)^{-1/2} \exp\left\{\frac{im(x_0-x_s+\xi)^2}{2\hbar T}\right\}d\xi . \tag{11}$$

Integration here is fulfilled along the slit of a width $a=2b$, i.e., from $-b$ to $+b$.
Formally, range of the integration can be broadened from $-\infty$ to $+\infty$. But in this case, we need to introduce the step function $G(\xi)$ equal to unit in the interval $[-b,+b]$ and equal to zero outside this interval. In principle, we can approximate hard edged slits by series of the Gaussian functions, each with narrow halfwidth (Sbitnev, 2010). For sake of simplicity however, we confine themselves by a single Gaussian form-factor

$$G(\xi) = \exp\left\{-\xi^2/2b^2\right\} . \tag{12}$$

It simulates slits with fuzzy edges. Effective width of this curve is conditioned by a parameter b. For such a form-factor roughly two thirds of all its area is situated between the points $-b$ and $+b$. If the particles would move by classical way, then we can anticipate, that after the time τ a distribution of the particles will be similar to the distribution existing at T, see Fig. 4. New center x_1 of the distribution is shifted on a value Δx from the point x_0. Width b_1 of the new distribution is also broadened. The both parameters, x_1 and b_1, are determined from expressions

$$x_1 = x_0\left(1+\frac{\tau}{T}\right), \qquad b_1 = b\left(1+\frac{\tau}{T}\right). \tag{13}$$

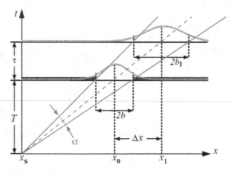

Fig. 4. Trajectories of particles passing through the Gaussian slit (Feynman & Hibbs, 1965), form a ray with an angle α of the divergent particle beam emanating from the source x_s.

Observe that quantum particles, in contrast to the classical ones, at scattering on the slit behave themselves like waves. The wavelike nature manifests itself via phase shifts of the moving particles in an observation point because of the de Broglie wavelength as innate character of quanta. According to the above stated remarks, Eq. (11) with inserted the form-factor of the slit, $G(\xi)$, now can be rewritten in the following form

$$\psi(x_1,x_0,x_s) = \int_{-\infty}^{\infty} \frac{mG(\xi)}{2\pi i\hbar\sqrt{T\tau}}\cdot\exp\left\{\frac{im}{2\hbar}\left[\frac{(x_1-x_0-\xi)^2}{\tau}+\frac{(x_0-x_s+\xi)^2}{T}\right]\right\}d\xi . \tag{14}$$

By substituting $G(\xi)$ from Eq. (12) to this expression and integrating it we obtain

$$\psi(x_1,x_0,x_s) = \sqrt{\frac{m}{2\pi i\hbar T}}\left(1+\frac{\tau}{T}+i\frac{\hbar\tau}{mb^2}\right)^{-1/2}$$
$$\exp\left\{\frac{im}{2\hbar}\left(\frac{(x_1-x_0)^2}{\tau}+\frac{(x_0-x_s)^2}{T}-\frac{\left((x_1-x_0)/\tau-(x_0-x_s)/T\right)^2\tau}{(1+\tau/T+i\hbar\tau/mb^2)}\right)\right\} . \tag{15}$$

At integrating Eq. (14) we use a standard integral

$$\int_{-\infty}^{\infty} e^{\alpha\xi^2+\beta\xi+\gamma}d\xi = \sqrt{\frac{\pi}{-\alpha}}e^{-\beta^2/4\alpha+\gamma} . \tag{16}$$

Before we will write out a final expression let us fulfill a series of replacements.

3.1.1 Series of replacements

First we define an effective slit's half-width $\sigma_0 = b/\sqrt{2}$. And further we define a complex time-dependent spreading

$$\sigma_\tau = \sigma_0 + i \frac{\hbar \tau}{2m\sigma_0(1 + \tau/T)}, \tag{17}$$

which has been defined in works (Sanz & Miret-Artês, 2007, 2008}. More one step is to replace flight times T and τ by flight distances $(z_0 - z_s)$ and $(z_1 - z_0)$, namely, $T = (z_0 - z_s)/v_z$ and $\tau = (z_1 - z_0)/v_z$. Here v_z is a particle velocity along the axis z. We note that $mv_z = p_z$ is z-component of the particle momentum. This component is not changed at passing through the grating. Next, we introduce the de Broglie wavelength $\lambda_{dB} = h/p_z$, where $h = 2\pi\hbar$ is the Planck constant. Rewrite in this view the complex time-dependent spreading (17) as the complex distance-dependent spreading

$$\sigma_{\tau \to z_1} = \sigma_0 + i \frac{\lambda_{dB}(z_1 - z_0)}{4\pi\sigma_0 \left(\dfrac{z_1 - z_s}{z_0 - z_s} \right)}. \tag{18}$$

Define now a dimensionless complex distance-dependent spreading as follows

$$\Sigma_{z_1} = \frac{z_1 - z_s}{z_0 - z_s} + i \frac{\lambda_{dB}(z_1 - z_0)}{4\pi\sigma_0^2} \tag{19}$$

and a dimensionless parameter characterizing remoteness of the source

$$\Xi_0 = 1 - \frac{(x_0 - x_s)\,(z_1 - z_0)}{(z_0 - z_s)\,(x_1 - x_0)}, \tag{20}$$

which tends to 1 as $z_s \to -\infty$.

Now we can use the above parameters, the dimensionless complex distance-dependent spreading Σ_{z_1} and the remoteness of the source Ξ_0, in order to write out the wave function behind the slit. By rewriting Eq. (15) via these parameters we obtain

$$\psi(x, x_0, x_s, z) = \sqrt{\frac{m}{2\pi i \hbar T \Sigma_z}} \exp\left\{ i\pi \left[\frac{(x - x_0)^2}{\lambda_{dB}(z - z_0)}\left(1 - \frac{\Xi_0^2}{\Sigma_z} \right) + \frac{(x_0 - x_s)^2}{\lambda_{dB}(z_0 - z_s)} \right] \right\}. \tag{21}$$

Here we have removed the subscript 1 at the variables, x, z, and Σ_z, since they relate to every points of the space behind the slit. In particular, at removing the source to infinity, $z_s \to -\infty$, the parameter Ξ_0 tends to 1 and the wave function reduces down to the paraxial approximation

$$\psi(x, x_0, z) = A \exp\left\{ i\pi \frac{(x - x_0)^2}{\lambda_{dB}(z - z_0)}\left(1 - \frac{1}{\Sigma_z} \right) \right\}. \tag{22}$$

Here a normalization factor A reads

$$A = \sqrt{\frac{m}{2\pi i \hbar T \Sigma_z}} . \tag{23}$$

One can see it vanishes at $T \to \infty$. It means, as the source moves away to infinity its intensity tends to zero. In the paraxial approximation we need to ignore this expression and consider the parameter A simply as a factor that normalizes the wave function.

Further, for the sake of simplicity, we will deal with the paraxial approximation.

3.2 Matter waves behind the grating

Let we have a screen, on which an incident monochromatic beam of the particles is scattered. It has N slits ($n=0,1,2, \ldots ,N-1$) located at equal distance from each other, as shown in Fig. 2. Origin of coordinates is placed in the center of the slit grating. In this frame of reference, n-th slit has a position $x_0 = (n - (N-1)/2)d$, where d is a spacing between slits. The spacing is measured in units multiple to the wavelength λ_{dB}.

We need now to compute contributions of all paths that pass from the source through all slits in the screen and farther to a point of observation (x,z). Per se, we should superpose in the observation point all wave functions (22) from all slits $n=0,1,2, \ldots ,N-1$. Such a superposition reads

$$|\Psi(x,z)\rangle = \frac{1}{N} \sum_{n=0}^{N-1} \psi\left(x,\left(n - \frac{N-1}{2}\right)d, z\right) \tag{24}$$

and probability density in the vicinity of the observation point (x,z) is

$$p(x,z) = \langle \Psi(x,z)|\Psi(x,z)\rangle . \tag{25}$$

3.2.1 Far-field diffraction

Before we will take up interference effects in the near-field region, let us consider an asymptotic limit of the formula (25) in the far-field region, Fig. 5. With this aim in mind, we replace the term $(n - (N-1)/2)d$ in Eq. (24) by kd, where k runs from $-(N-1)/2$ to $(N-1)/2$. Next, at summation we will neglect contribution of coefficients at $k^2 d^2$ emergent at decomposition $(x - kd)^2 = x^2 - 2xkd + k^2 d^2$. The point is that the terms with $k^2 d^2$ lead to phases muddled up on infinity. Because of it sum of all these exponents gives zero contribution. Other sums containing coefficients at x^2 and $2xkd$ can be easily computed. Next, at summation we use the mathematical equality

$$\sum_{k=-(N-1)/2}^{(N-1)/2} \exp\{ikx\} = \frac{\sin(Nx/2)}{\sin(x/2)} . \tag{26}$$

Intensity of the particle beam in the far-field region computed according to the above approximation is as follows

$$I(x,z) = I_0(x,z) \frac{\sin\left(\frac{N\zeta(x,z)}{2}\right)^2}{\sin\left(\frac{\zeta(x,z)}{2}\right)^2} . \tag{27}$$

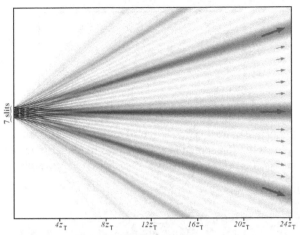

Fig. 5. Diffraction in the far-field zone at simulation of scattering thermal neutrons (λ_{dB}=0.5 nm) on N=7 slits grating. Width of slits a=2λ_{dB}, spacing d=10λ_{dB}, and the Talbot length z_T=2d^2/λ_{dB}=200λ_{dB}. Directions of principal and subsidiary maxima are pointed out by big red arrows and small blue arrows, respectively.

Here terms $\zeta(x,z)$ and $I_0(x,z)$ read

$$
\begin{cases}
\zeta(x,z) = \dfrac{xd\dfrac{z\lambda_{dB}}{4\pi\sigma^2}}{2\sigma_z^2}, \\[6mm]
I_0(x,z) = \dfrac{A^2}{N^2\sigma_z}\exp\left\{-\dfrac{x^2}{2\sigma_z^2}\right\}.
\end{cases}
\tag{28}
$$

The parameter A is the normalization factor, see Eq. (23), and σ_z has the following form

$$
\sigma_z = \sigma\sqrt{1+\left(\frac{z\lambda_{dB}}{4\pi\sigma^2}\right)}.
\tag{29}
$$

This parameter is equivalent to the instantaneous Gaussian width presented in (Sanz and Miret-Artês, 2007).

Fig. 6 shows diffraction in the far-field zone from the grating having N=7 slits. Distance to the observation screen is z=10^7 z_T =1 m, where z_T =2d^2/λ_{dB} = 200λ_{dB} is the Talbot length. It will be explained below. It is seen, that the principal maxima are partitioned from each other by N-2=5 subsidiary maxima.

3.2.2 Near-field interference

Above we have considered a coherent flow of thermal neutrons, λ_{dB}=0.5 nm. Radius of these particles is 10^{-15} m. It is much smaller the de Broglie wavelength λ_{dB} = $5\cdot10^{-10}$ m. For this reason, these particles can be considered as point particles, in contrast to enormous fullerene molecules shown in Fig 7.

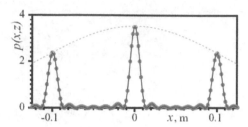

Fig. 6. Diffraction of thermal neutrons (λ_{dB}=0.5 nm) in the far-field zone from grating having N=7 slits. Distance to observation screen is z=1 m. Blue circles relate to the probability density calculated by Eqs. (24)--(25). Intensity (27) is drawn by red solid curve. Dotted green curve draws envelope $I_0(x,z)\,N^2$.

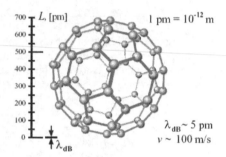

Fig. 7. The fullerene molecule C_{60} consists of 60 carbon atoms. Its radius is about 700 pm. At a flight velocity from a source v =100 m/s de Broglie wavelength of the fullerene molecule, λ_{dB}, is about 5 pm.

Here we consider interference phenomena in the near-field created by the fullerene molecules. Interest to such heavy molecules, having masses about 100 amu and more (Arndt et al., 2005; Brezger et al., 2002, 2005; Gerlich et al., 2011; Hackermüller et al., 2003, 2004; Nairz et al., 2003) is due to the fact that under ordinary circumstances they behave almost as classical objects. Indeed, diameter of the fullerene molecule C_{60}, see Fig. 7, is about 700 pm (Yanov & Leszczynski, 2004), but de Broglie wavelength is ~5 pm (Hackermüller et al., 2003; Juffmann et al., 2009). There is a problem to observe quantum interference for such large molecules having minuscule wavelengths.

At small distances from the grating we need in a acceptable scale in order to partition interference patterns on characteristic zones. Such a scale parameter is the Talbot length

$$z_T = 2\frac{d^2}{\lambda_{dB}}. \tag{30}$$

This length starts from Henry Fox Talbot who discovered in 1836 a beautiful interference pattern (Talbot, 1836), that carries his name. Here d is the spacing between slits and λ_{dB} is the de Broglie wavelength of particles under consideration. Figs. 8 and 9 show emergence of such interference patterns in the near-field.

Fig. 8 shows the density distribution function (25) in a transient region from near-field to far-field (it is shown in gray color). The Talbot length ranges from 0 to $8z_T$ = 0.8 m. The

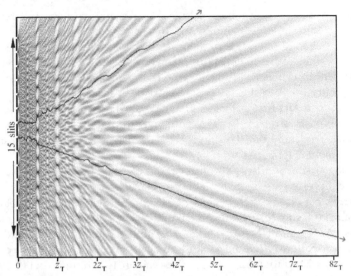

Fig. 8. Interference pattern of matter waves. The wave is presented by coherent fullerene molecule beam incident to a grating having N = 15 slits. De Broglie wavelength of the fullerene molecules is λ_{dB} = 5 pm. Spacing between slits d = 500 nm and slit width $a = 2b$ = 10 nm. The Talbot length z_T=0.1 m. Two Bohmian trajectories divergent from central area of the grating are shown in blue as examples.

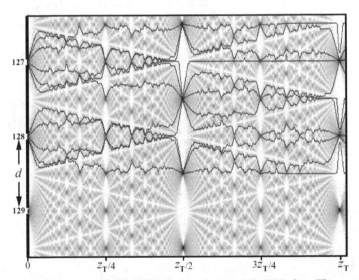

Fig. 9. Talbot carpet in the near-field of the grating having N = 255 slits. The pattern has been captured from central part of the grating. De Broglie wavelength of the fullerene molecules is λ_{dB} = 5 pm. Spacing between slits d = 500 nm and slit width $a = 2b$ = 10 nm. The Talbot length z_T=0.1 m. Some of the Bohmian trajectories passing by zigzag through spots with high density distribution are shown in blue as examples.

interference pattern emergent has been calculated for heavy particles, fullerene molecules, Fig. 7, incident on the grating containing $N = 15$ slits. Spacing between slits is $d = 500$ nm and slit width $a = 2b = 10$ nm. Mass of the fullerene molecules is about $m_{C60} \approx 1.2 \cdot 10^{-24}$ kg. And at average velocity about 100 m/s (Juffmann et al., 2009) the de Broglie wavelength is 5 pm. The Talbot length is about $z_T = 0.1$ m. One can easily evaluate that ratio of the Talbot length to the spacing between slits is equal to $2 \cdot 10^5$. So, a stripe between two slits extending from the grating up to the first Talbot length is extremely narrow. We can see that nearby the grating there exists a relatively perfect interference pattern. It decays with removing from the grating. And far from the grating characteristic rays divergent from it arise, as shown, for example, in Fig. 5

More fascinating picture arises at observation of the Talbot carpet as a peculiar manifestation of interference in near-field, see Fig. 9. The Talbot carpets arise if three conditions, Berry's conditions (Berry, 1996, 1997; Berry & Klein, 1996; Berry et al., 2001), are fulfilled: (a) paraxial beam; (b) arbitrary small ratio λ_{dB}/d; (c) arbitrary large number of slits. In a strict sense, in the limits $N \to \infty$ and $\lambda_{dB}/d \to 0$ the Talbot carpet should transform to fractal interference pattern. It would look like δ-peaks everywhere densely populating the probability density distribution function $p(x,z)$, as shown, for example, in Fig. 10.

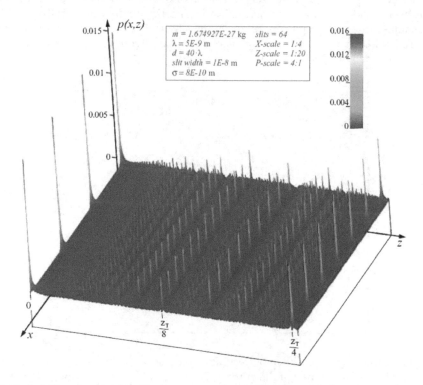

Fig. 10. Probability density distribution approaches infinite set of δ-functions as λ_{dB}/d tends to zero. Parameters here are as follows $N = 64$, $\lambda_{dB} = 0.5$ nm (thermal neutrons), $d = 20$ nm, z_T = 6400 nm (Sbitnev, 2009b).

In order to reach the Berry's conditions, we take number of slits as many as possible. The ratio λ_{dB}/d should be as small as possible, as well. Given λ_{dB} = 5 pm and d = 500 nm we have the ratio λ_{dB}/d = 10^{-5}. It is in good agreement with the condition (b). As for number of slits, as seen in Fig. 8 the interference patterns are washed away on first some Talbot lengths. It means, that number of the slits N = 15 is insufficient for observing of the Talbot carpet. Fig. 9 shows emergence of the Talbot carpet in the near-field in central part of the grating having 255 slits. As seen, N = 255 is sufficient number to get the Talbot carpet with perfect organization of alternation of high and low values of the density distribution.

3.3 Bohmian trajectories

How does particle pass through the slits? Answer to this problem is proposed in the Bohmian mechanics (Bohm, 1952a, 1952b; Bohm & Hiley, 1982; Hiley, 2002). In the next section we will consider this solution in detail. Here we show only some particular solutions. Two divergent Bohmian trajectories drawn in blue are shown in Fig. 8. They prefer to go along dark plots (high values of the density distribution) and avoid light-colored plots (low values of the density distribution). Fig. 9 also shows in blue a family of the Bohmian trajectories. In contrast to the trajectories shown in previous figure, here they demonstrate complex zigzag movements. The Bohmian trajectories result from solution of the guidance equation (Wyatt & Bittner, 2003; Nikolić, 2007; Sanz & Miret-Artês, 2007, 2008; Struyve & Valentini, 2009)

$$v_x = \dot{x} = \frac{\nabla S}{m} = \frac{\hbar}{m}\mathrm{Im}\left(|\Psi\rangle^{-1}\nabla|\Psi\rangle\right). \tag{31}$$

According to the equation (31), position (x,z) of the particle in 2D space is given as follows

$$x(t) = x_0 + \int_0^t v_x d\tau, \qquad z(t) = z_0 + v_z t. \tag{32}$$

Since we believe that longitudinal momentum, p_z, is constant in contrast to the transversal momentum p_x, the component z here is calculated by simple multiplication of v_z by t. In turn, velocity v_x, as seen from Eq. (31), is (a) proportional to gradient of the wave function; and (b) inversely proportional to the same wave function. It means: (a) a trajectory undergoes greatest variations in plots, where the wave function has slopes; and (b) the trajectory avoid areas, where the wave function tends to zero.

One could think that the Bohmian trajectories are physical artifacts, since they enter into a rough contradiction with the Heisenberg uncertainty principle, because of prediction in each time moment of exact values of coordinates and velocities of the particle (Bohm, 1952a, 1952b; Bohm & Hiley1982). However, there is no here contradiction so far as the uncertainty principle refers to the measurement problem. Whereas the Bohmian trajectories are simply geodesic trajectories. At drawing the density distribution function we could use an orthogonal grid represented by geodesic trajectories and surfaces of equal phases, see, for example, Fig. 1. In the absence of intervention in a particle's history by measuring its parameters, real particle prefer to move along a geodesic trajectory. However, as soon as we undertake measurement of the particle's parameters we destroy its history. For example, if we measure position of the particle, we destroy its future history. If we measure its momentum, then we lose its past history.

The Bohmian trajectories in Fig. 9 are seen to fulfill intricate zigzag dances. One can see, the trajectories pass through areas where the density distribution has high values and avoid areas with low its values. The particles one can guess should perform zigzag motions. However, as was noted above, the ratio z_T to d is about $2 \cdot 10^5$ and the observed pattern is within a very narrow strip. Consequently, these zigzags have very small curvatures. Vacuum fluctuations can provoke emergence of such deviations.

4. Variational computations

What could cause the particle to perform such a wavy and zigzag behaviors, as shown in the figures above? Possible answer could be as follows: a family of ordered slits in the screen poses itself as a quantum object that polarizes vacuum in the near-field region. The polarization, in turn, induces formation of a virtual particle's escort around of a flying real particle through the space. The escort corrects movement of the particle depending on the environment by interference of virtual particles with each other (Feynman & Hibbs, 1965).

4.1 Wave-particle duality, the Schrödinger equation

In contrast to classical mechanics where a single trajectory connecting the initial and final points submits to the principle of least action, in the quantum mechanics we need to consider all possible trajectories connecting these points in order to obtain clear answer. They pass through all intermediate points belonging to a transitional set $\mathcal{R}^3$. All these paths should be evaluated jointly. Such a description goes back to the integral Chapman-Kolmogorov equation (Ventzel, 1975):

$$p(x,z;t+\tau) = \int\limits_{R^n} p(x,y;t)p(y,z,\tau)dy \qquad (33)$$

which gives transitional probability densities of a Markovian sequence.

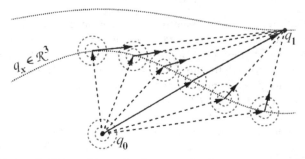

Fig. 11. Computation of all possible paths that pass from point q_0 to point q_1 through possible intermediate points $q_x \in \mathcal{R}^3$ represents a core of the path integral method. Pink circles conditionally represent radiation of Huygens waves.

Essential difference from the classical probability theory is that instead of the probabilities quantum mechanics deals with probability amplitudes containing imaginary terms. They bear information about phase shifts accumulated along paths. In that way, a transition from an initial state $\vec{q}_0$ to a final state $\vec{q}_1$ through all intermediate positions $\vec{q}_x$ given on a conditional set $\mathcal{R}^3$ (see Fig. 11) is represented by the following path integral

$$\psi(\vec{q}_1,\vec{q}_0,t+\delta t) = \int_{\mathcal{R}^3} K(\vec{q}_1,\vec{q}_x;t+\delta t,t)\,\psi(\vec{q}_x,\vec{q}_0,t)\,\mathcal{D}^3 q_x \tag{34}$$

in the limit $\delta t \rightarrow 0$ and $\vec{q}_1 \rightarrow \vec{q}_x$. Here symbol $\mathcal{D}^3 q_x$ represents a differential element of volume in the set $\mathcal{R}^3$.

Circular waves pictured by dotted circumferences in Fig. 11 illustrate working of the Huygens-Fresnel principle (Landsberg, 1957; Longhurst, 1970). The principle proclaims that each point $\vec{q}_x$ at an advanced wave front is, in fact, the center of a fresh disturbance and it is the source of a new wave radiation. The advancing wave as a whole may be regarded as the sum of all the secondary waves arising from points in the medium already traversed by the wave. All the secondary waves are coherent, since they are activated from the one source given in $\vec{q}_0$.

It is important to note, that all rays from such secondary sources represent virtual trajectories emanating from the source at $\vec{q}_0$ up to the point $\vec{q}_1$. Along with the other virtual trajectories generated by the other secondary sources, all together they create in the point $\vec{q}_1$ an averaged effect of contribution of these secondary sources. This averaged effect shows whether a real particle passes by this route and what probability of this event can be.

We suppose that the integral kernel

$$K(\vec{q}_1,\vec{q}_x;t+\delta t,t) = \frac{1}{A}\exp\left\{\frac{i}{\hbar}L\left(\vec{q}_x,\dot{\vec{q}}_x\right)\delta t\right\} \tag{35}$$

has a standard form of the Lagrangian (Feynman & Hibbs, 1965)

$$L\left(\vec{q}_x,\dot{\vec{q}}_x\right) = \frac{m}{2}\left(\frac{\vec{q}_1-\vec{q}_x}{\delta t}\right)^2 - U(\vec{q}_x). \tag{36}$$

Here $U(\vec{q}_x)$ is a potential energy of the particle localized at the point $\vec{q}_x \in \mathcal{R}^3$. And $(\vec{q}_1-\vec{q}_x)/\delta t$ is a velocity $\dot{\vec{q}}_x$ attached to the same point $\vec{q}_x$ and oriented in the direction of the point $\vec{q}_1$.

The next step is to expand terms, ingoing into the integral (34), into Taylor series. The wave function written at the left is expanded up to the first term

$$\psi(\vec{q}_1,\vec{q}_0,t+\delta t) \approx \psi(\vec{q}_1,\vec{q}_0,t) + \frac{\partial \psi}{\partial t}\delta t. \tag{37}$$

As for the terms under the integral, here we preliminarily make some transformations. We define a small increment

$$\vec{\xi} = \vec{q}_1 - \vec{q}_x \quad \Rightarrow \quad \mathcal{D}^3 q_x = -\mathcal{D}^3\xi. \tag{38}$$

The Lagrangian (36) is written as

$$L(\vec{q}_x,\dot{\vec{q}}_x) = \frac{m}{2}\frac{\xi^2}{\delta t^2} - U(\vec{q}_1-\vec{\xi}). \tag{39}$$

Here the potential energy $U(\vec{q}_1-\vec{\xi})$ is subjected to expansion into the Taylor series by the small parameter $\vec{\xi}$. The under integral wave function $\psi(\vec{q}_x,\vec{q}_0,t) = \psi(\vec{q}_1-\vec{\xi},\vec{q}_0,t)$ is subjected to expansion into the Taylor series up to the second terms of the expansion

$$\psi(\vec{q}_1 - \vec{\xi}, \vec{q}_0, t) \approx \psi(\vec{q}_1, \vec{q}_0, t) - (\nabla\psi \cdot \vec{\xi}) + \nabla^2\psi \cdot \xi^2/2 . \tag{40}$$

Taking into account the expressions (37)-(40) and substituting theirs into Eq. (34) we get

$$\psi(\vec{q}_1, \vec{q}_0, t) + \frac{\partial\psi}{\partial t}\delta t = -\frac{1}{A}\int_{R^3} \exp\left\{\frac{i}{\hbar}\frac{m}{2}\frac{\xi^2}{\delta t}\right\}\left(1 - \frac{i}{\hbar}\left(U(\vec{q}_1) - (\nabla U \cdot \vec{\xi}) + \Delta U \cdot \xi^2/2\right)\delta t\right)$$
$$\left(\psi(\vec{q}_1, \vec{q}_0, t) - (\nabla\psi \cdot \vec{\xi}) + \nabla^2\psi \cdot \xi^2/2\right)\mathcal{D}^3\xi \tag{41}$$

One can see that the term $\psi(\vec{q}_1, \vec{q}_0, t)$ is presented from both the left side and from the right side. These both term can remove each other, if the right part will satisfy the following condition

$$-\frac{1}{A}\int_{\mathcal{R}^3}\exp\left\{\frac{i}{\hbar}\frac{m}{2}\frac{\xi^2}{\delta t}\right\}\mathcal{D}^3\xi = -\frac{1}{A}\left(\frac{2\pi i\hbar\delta t}{m}\right)^{3/2} = 1 \tag{42}$$

From here it follows

$$A = -\left(\frac{2\pi i\hbar\delta t}{m}\right)^{3/2} . \tag{43}$$

The power 3 arises here because that the integration is fulfilled on the 3-dimensional set $\mathcal{R}^3$. It would be desirable also to integrate the terms $(\nabla\psi \cdot \vec{\xi})$ and $\nabla^2\psi \cdot \xi^2/2$ existing in the integral (41). With this aim in the mind, we mention the following two integrals (Feynman & Hibbs, 1965)

$$\frac{1}{A}\int_{\mathcal{R}^3}\exp\left\{\frac{i}{\hbar}\frac{m}{2}\frac{\xi^2}{\delta t}\right\}\vec{\xi}\mathcal{D}^3\xi = 0 \tag{44}$$

and

$$\frac{1}{A}\int_{\mathcal{R}^3}\exp\left\{\frac{i}{\hbar}\frac{m}{2}\frac{\xi^2}{\delta t}\right\}\xi^2\mathcal{D}^3\xi = \frac{i\hbar}{m}\delta t . \tag{45}$$

In accordance with the first integral, contributions of the terms $\nabla\psi$ and ∇U in the expression (41) disappear. Whereas, the terms with multiplier $\xi^2/2$ gains the factor $(i\hbar\delta t/m)/2$.

Taking into account the above stated expressions, let us rewrite Eq.(41)

$$\frac{\partial\psi}{\partial t}\delta t = i\frac{\hbar\delta t}{2m}\Delta\psi - i\frac{\delta t}{\hbar}U(\vec{q}_1)\psi(\vec{q}_1, \vec{q}_0, t) + \frac{\delta t^2}{2m}\Delta U(\vec{q}_1)\psi(\vec{q}_1, \vec{q}_0, t) . \tag{46}$$

The last term contains the factor δt^2 due to which contribution of this term to this equation is abolished in contrast with other terms as $\delta t \to 0$. By omitting this term, we come to the Schrödinger equation

$$i\hbar\frac{\partial\psi}{\partial t} = -\frac{\hbar^2}{2m}\Delta\psi + U(\vec{q}_1)\psi .\tag{47}$$

describing the function ψ , wave function, in the configuration space $\mathcal{R}^3$. The subscript 1 here can be dropped.

4.2 The Bohmian decomposition
Let us examine the Schrödinger equation (47) by substituting the wave function ψ in the following form:

$$\psi(\vec{q},\vec{p},t) = \sqrt{\rho(\vec{q},\vec{p},t)}\exp\{iS(\vec{q},\vec{p},t)/\hbar\} = R(\vec{q},\vec{p},t)\exp\{iS(\vec{q},\vec{p},t)/\hbar\} .\tag{48}$$

Here functions $S(\vec{q},\vec{p},t)$ and $\rho(\vec{q},\vec{p},t)$ are real functions of their variables $\vec{q}$, $\vec{p}$, and t. The first function is the action which was mentioned earlier. And the second function is the probability density distribution defined as follows

$$\rho(\vec{q},\vec{p},t) = |\psi|^2 = \psi^*\psi .\tag{49}$$

Here we will consider the decomposition in a general view, i.e., the variables $\vec{q} = (q_1,q_2,\cdots,q_N)$ and $\vec{p} = (p_1,p_2,\cdots,p_N)$ are those representing the quantum system in 2N-dimensional phase space. It means, in particular, that there are several particles which can be considered in this space as one generalized particle.
By substituting the wave function $\psi(\vec{q},\vec{p},t)$ into the Schrödinger equation (47) we obtain

$$\underbrace{-\frac{\partial S}{\partial t}\cdot\psi}_{(a)} + \underbrace{i\hbar\frac{1}{2\rho}\frac{\partial\rho}{\partial t}\cdot\psi}_{(b)} = \underbrace{\frac{1}{2m}(\nabla S)^2\cdot\psi + U(\vec{q})\cdot\psi}_{(a)}$$

$$\underbrace{-\frac{i\hbar}{2m}\nabla^2 S\cdot\psi - \frac{i\hbar}{2m}\left(\frac{1}{\rho}\nabla\rho\right)(\nabla S)\cdot\psi}_{(b)} \underbrace{-\frac{\hbar^2}{2m}\left(\frac{1}{2\rho}\nabla^2\rho\right)\cdot\psi + \frac{\hbar^2}{2m}\left(\frac{1}{2\rho}\nabla\rho\right)^2\cdot\psi}_{(c)}\tag{50}$$

Operators of gradient, ∇ , and laplacian, ∇^2 , read

$$\nabla = \left\{\frac{\partial}{\partial q_1}i_1,\frac{\partial}{\partial q_2}i_2,\cdots,\frac{\partial}{\partial q_N}i_N\right\}, \qquad \nabla^2 = \left\{\frac{\partial^2}{\partial q_1^2}+\frac{\partial^2}{\partial q_2^2}+\cdots+\frac{\partial^2}{\partial q_N^2}\right\}.\tag{51}$$

A set { $i_1,i_2, \ldots , i_N$ } represents orthonormal basis of N-dimensional state space $\mathcal{S}^N$. The orthonormality means that $i_k\cdot i_j = \delta_{k,j}$ for all k, j ranging 1 to N.
Collecting together real terms (a) and (c), and separately imaginary terms (b) in Eq. (50) we obtain two coupled equations for real functions $S(\vec{q},\vec{p},t)$ and $\rho(\vec{q},\vec{p},t)$

$$-\frac{\partial S}{\partial t} = \frac{1}{2m}(\nabla S)^2 + U(\vec{q}) \underbrace{-\frac{\hbar^2}{2m}\left[\frac{\nabla^2\rho}{2\rho}-\left(\frac{\nabla\rho}{2\rho}\right)^2\right]}_{(c)},\tag{52}$$

$$-\frac{\partial \rho}{\partial t} = \nabla \left(\rho \frac{\nabla S}{m} \right). \tag{53}$$

A term

$$Q = -\frac{\hbar^2}{2m} \left[\frac{\nabla^2 \rho}{2\rho} - \left(\frac{\nabla \rho}{2\upsilon} \right)^2 \right] = -\frac{\hbar^2}{2m} \frac{\nabla^2 R}{R} \tag{54}$$

enveloped by brace (c) in Eq. (52) is the quantum potential. It evaluates a measure of curvature of the N-dimensional state space induced by a prepared physical scene consisting of sources, detectors, and other experimental devices. Equations, (52) and (53), are seen to be the coupled pair of nonlinear partial differential equations. The first of the two equations, Eq. (52), is the Hamilton-Jacobi equation modified by the quantum potential $Q(\vec{q},t)$. The second equation, Eq. (53), is the continuity equation. In the above equations we define the following computations

$$\vec{p} = m\vec{v} = \nabla S, \tag{55}$$

and

$$\frac{1}{2m}(\nabla S)^2 = \frac{1}{2m} p^2. \tag{56}$$

Here $\vec{p}$ is momentum of the particle, $\vec{v}$ is its velocity, and the last equation represents kinetic energy of the particle.

Equation (52) states that total particle energy is the sum of the kinetic energy, potential energy, and the quantum potential (Hiley, 2002). Equation (53), in turn, is interpreted as the continuity equation for probability density $\rho(\vec{q},\vec{p},t)$. It says that all individual trajectories demonstrate collective behavior like a liquid flux (Madelung, 1926; Wyatt, 2005), perhaps, superconductive one. We shall see further, that the quantum potential $Q(\vec{q},\vec{p},t)$ introduces corrections both in the kinetic energy and in the potential energy of the particle.

4.2.1 The quantum potential as an information channel
According to the observation

$$\rho^{-1}\nabla\rho = \nabla \ln(\rho) \tag{57}$$

we can rewrite the quantum potential by the following way

$$Q(\vec{q},t) = \frac{\hbar^2}{2m} \left[\left(\frac{1}{2\rho}\nabla\rho \right)^2 - \frac{1}{2} \left(\frac{1}{\rho}\nabla \left(\rho \cdot \frac{1}{\rho}\nabla\rho \right) \right) \right] = -\frac{\hbar^2}{2m} \left(\frac{1}{2}\nabla\ln(\rho) \right)^2 - \frac{\hbar^2}{2m} \left(\frac{1}{2}\nabla^2\ln(\rho) \right). \tag{58}$$

Define a logarithmic function

$$S_Q(\vec{q},\vec{p},t) = -\frac{1}{2}\ln\left(\rho(\vec{q},\vec{p},t) \right) \tag{59}$$

to be called further *quantum entropy*. It is like to the Boltzmann entropy. However if the Boltzmann entropy characterizes degree of order and chaos of classical gases, the quantum entropy evaluates analogous quality of the quantum liquid mentioned above. To be more defined, one can imagine the quantum liquid as ensemble of partially ordered virtual vortices (particle-antiparticle pairs) within vacuum. For example, such virtual vortices may be presented by spinning electron-positron pairs.

Substituting (59) into Eq. (58) we find that the quantum potential can be expressed in terms of this function

$$Q(\vec{q},\vec{p},t) = \underbrace{-\frac{\hbar^2}{2m}(\nabla S_Q)^2}_{(a)} + \underbrace{\frac{\hbar^2}{2m}\nabla^2 S_Q}_{(b)}. \tag{60}$$

It should be noted, that the term $-S_Q$ (negative S_Q) is named C-amplitude in (Bittner, 2003; Wyatt, 2005; Wyatt & Bittner, 2003). Here the term enveloped by brace (a) is viewed as the quantum corrector of the kinetic energy. And the term enveloped by brace (b) corrects the potential energy. Namely, substituting into Eq. (52) we obtain

$$-\frac{\partial S}{\partial t} = \underbrace{\frac{1}{2m}(\nabla S)^2 - \frac{\hbar^2}{2m}(\nabla S_Q)^2}_{(a)} + \underbrace{U(\vec{q}) + \frac{\hbar^2}{2m}\nabla^2 S_Q}_{(b)}. \tag{61}$$

In this equation the terms enveloped by brace (a) relate to the kinetic energy of the particle, and those enveloped by brace (b) relate to its potential energy.

Substituting also S_Q in the continuity equation (53) instead of ρ we obtain the entropy balance equation

$$\frac{\partial S_Q}{\partial t} = -(\vec{v}\cdot\nabla S_Q) + \frac{1}{2}(\nabla\vec{v}). \tag{62}$$

Here $\vec{v} = \nabla S / m$ is a particle speed. The rightmost term, $(\nabla\vec{v})$, describes a rate of the entropy flow produced by spatial divergence of the speed due to curvature of the N-dimensional state space. This term is nonzero in regions where the particle changes direction of movement.

5. Beyond the Bohm's insight into QM

Pair of the equations, the modified HJ equation (61) and the entropy balance equation (62), describes behavior of the quantum particle, subject to influence of the quantum entropy. Let us now multiply Eq. (62) by the factor $-i\hbar$ and add the result to Eq. (61). We obtain

$$-\frac{\partial \mathbf{S}}{\partial t} = \underbrace{\frac{1}{2m}(\nabla S)^2 + i\hbar\frac{1}{m}(\nabla S\cdot\nabla S_Q) - \frac{\hbar^2}{2m}(\nabla S_Q)^2}_{(a)} + \underbrace{U(\vec{q}) - i\hbar\frac{1}{2}(\nabla\vec{v}) + \frac{\hbar^2}{2m}\nabla^2 S_Q}_{(b)}. \tag{63}$$

Here $\mathbf{S}$ is sum of the action S and the quantum entropy S_Q (*complexified action*)

$$\mathbf{S} = S + i\hbar S_Q. \tag{64}$$

Terms enveloped by brace (a) can be rewritten as gradient of the squared complexified action

$$\frac{1}{2m}(\nabla \mathbf{S})^2 = \frac{1}{2m}(\nabla S)^2 + i\hbar\frac{1}{m}(\nabla S \cdot \nabla S_Q) - \frac{\hbar^2}{2m}(\nabla S_Q)^2. \tag{65}$$

As for the terms enveloped by brace (b) they could stem from expansion into the Taylor's series of the potential energy extended previously to a complex space, like a complex extension, for example, in (Poirier, 2008). In our case, the potential function is extended in the complex space, which has a small broadening into imaginary sector. Let us expand into the Taylor's series the potential function that has a complex argument

$$U(\vec{q} + i\varepsilon) \approx U(\vec{q}) + i(\vec{\varepsilon} \cdot \nabla U(\vec{q})) - \frac{\varepsilon^2}{2}\nabla^2 U(\vec{q}) + \dots. \tag{66}$$

Now we will examine the last two terms. Here a small vector $\vec{\varepsilon}$ has dimensionality of length. But it should contain also the Planck constant, $\hbar$, in order to reproduce the second and third terms enveloped by brace (b) in Eq. (63). A minimal representation of this vector can be as follows

$$\vec{\varepsilon} = \frac{\hbar}{2m}s_B\vec{n}. \tag{67}$$

Here $\vec{n}$ is unit vector pointing direction of the small increment, m is the particle mass, and s_B is universal constant, "reverse velocity",

$$s_B = 4\pi\varepsilon_0 \frac{\hbar}{e^2} \approx 4.57 \times 10^{-7} \text{ [s/m]}. \tag{68}$$

Here $e \approx -1.6 \times 10^{-19}$ [C] is the elementary charge carried by a single electron and $\varepsilon_0 \approx 8.854 \times 10^{-12}$ [$C^2N^{-1}m^{-2}$] is the vacuum permittivity. The reverse velocity measures time required for traversing unit of a distance. Such a distance can be perimeter of orbit (Poluyan, 2005) at oscillating electron around. Observe that $r_B = s_B\hbar/m = 4\pi\varepsilon_0 \hbar^2/me^2$ is value of the electron radius under its travelling on first orbit around the nucleus (Dirac, 1982). In our case it can be an effective radius of electron-positron pair under their virtual revolution about the mass center on the first orbit. From the above it follows, that $v_B = 1/s_B \approx 2.188 \times 10^6$ m/s is the Bohr velocity of electron oscillating on the first orbit about the mass center, and $r_B = \hbar/mv_B \approx 0.529 \times 10^{-10}$ m is the Bohr radius of this orbit. Here mv_B is the electron momentum.

In light of these remarks, we can rewrite the expansion (66) in the following form

$$U(\vec{q} + i\varepsilon) \approx U(\vec{q}) + i\hbar\underbrace{\left(\vec{n} \cdot \left(\frac{s_B}{2m}\nabla U(\vec{q})\right)\right)}_{(b_1)} - \underbrace{\frac{\hbar^2}{2m}\left(\frac{s_B^2}{2m}\nabla^2 U(\vec{q})\right)}_{(b_2)}.$$

$$= U(\vec{q}) + \underbrace{\frac{i}{2}(\vec{n}r_B \cdot \nabla U(\vec{q}))}_{(b_1)} - \underbrace{\frac{1}{4}r_B^2\nabla^2 U(\vec{q})}_{(b_2)} \tag{69}$$

A term enveloped by brace (b_1) contains unit vector $\vec{n}$ that points out direction of the imaginary broadening. A force $\vec{F} = -\nabla U(\vec{q})$ multiplied by $\vec{n}r_B$ is elementary work performed at displacement on a length r_B along direction $\vec{n}$. The elementary work divided into $\hbar$ is a rate of variation of the particle velocity per unit length, i.e., it represents divergence of the velocity, $\nabla \vec{v}$. So, the term enveloped by brace (b_1) can be rewritten in the following form

$$(b_1): \quad \frac{1}{\hbar}\left(\vec{n}r_B \cdot \nabla U(\vec{q})\right) = -(\nabla \cdot \vec{v}).$$ (70)

As for the term $(s_B^2 / 2m) \cdot U(\vec{q}) = (1 / 2mv_B^2) \cdot U(\vec{q})$ which is placed over brace (b_2) in Eq. (69) it is dimensionless. Accurate to an additive dimensionless function $a\vec{q}^2 + (\vec{b}\vec{q}) + c$ this term is comparable with S_Q, i.e., with $\ln(\rho)$. Taking into account that $s_B = 1/v_B$ we proclaim

$$(b_2): \quad -\left(\frac{1}{2mv_B^2}\nabla^2 U(\vec{q})\right) = \frac{1}{2mv_B^2}(\nabla \vec{F}) = \nabla^2 S_Q.$$ (71)

Thus, a value of the Laplacian of $U(\vec{q})$ at the point $\vec{q}$ can be interpreted as the density of sources (sinks) of the potential vector field $\vec{F} = -\nabla U(\vec{q})$ at this point. Accurate to the denominator $2mv_B^2$, it is proportional to the Laplacian of the quantum entropy S_Q.

We have defined the corrections (70) and (71) by extending coordinates of the real 3D space into imaginary domain on the value ε. It is equal to about the Bohr radius of the first orbit of the electron-positron virtual pair, $r_B \approx 5.292 \times 10^{-11}$ m. Energy of this pair is much smaller of the energy creating two real particles from the vacuum. Therefore such a shift, $\varepsilon = r_B/2$, can be considered as a virtual small shift to the imaginary domain.

Now we can define complexified momentum

$$\vec{P} = m\dot{\vec{Q}} = \nabla \mathbf{S} = \nabla S + i\hbar \nabla S_Q$$ (72)

and complexified coordinate

$$\vec{Q} = \vec{q} + i\vec{\varepsilon}$$ (73)

as extended representations of the real vectors $\vec{p}$ and $\vec{q}$. The complexified momentum $\vec{P}$ differs from momentum $\vec{p}$ by additional imaginary term $\hbar \nabla S_Q$. And the complexified coordinate $\vec{Q}$ differs from real coordinate $\vec{q}$ by the small imaginary vector (67). Now we can rewrite Eq. (63) as complexified the Hamilton-Jacobi equation:

$$-\frac{\partial \mathbf{S}}{\partial t} = \frac{1}{2m}(\nabla \mathbf{S})^2 + U(\vec{q} + i\varepsilon) = \mathcal{H}\left(\vec{Q}, \vec{P}; t\right).$$ (74)

Here $H(\vec{Q}, \vec{P}; t)$ is a complexified Hamiltonian.

The total derivative of the complex action reads

$$\frac{d\mathbf{S}}{dt} = \frac{\partial \mathbf{S}}{\partial t} + \sum_{n=1}^{N} \frac{\partial \mathbf{S}}{\partial Q_n} \frac{dQ_n}{dt} = \frac{\partial \mathbf{S}}{\partial t} + \sum_{n=1}^{N} \mathcal{P}_n \dot{Q}_n$$ (75)

where complex derivative is (see Ch.2 in (Titchmarsh, 1976))

$$\frac{\partial \mathbf{S}}{\partial Q_n} = \frac{\partial S}{\partial q_n} + i\hbar \frac{\partial S_Q}{\partial q_n} = \mathcal{P}_n . \tag{76}$$

Combining Eq. (75) with (76) we come to the Legendre's dual transformation (Lanczos, 1970) that binds the Hamiltonian $\mathcal{H}$ and the Lagrangian $\mathcal{L}$, and conversely:

$$\frac{d\mathbf{S}}{dt} = -\mathcal{H}(\vec{Q},\vec{P};t) + \sum_{n=1}^{N} \mathcal{P}_n \dot{Q}_n = \mathcal{L}(\vec{Q},\dot{\vec{Q}};t) . \tag{77}$$

We summarize this section by collecting formulas of the complexified Hamiltonian and Lagrangian mechanics via the Legendre's dual transformations in Table 2:

Variables :	Variables :
Coordinate: $\quad \vec{Q} = \vec{q} + i\dfrac{\hbar}{2mv_B}\vec{n}$	Coordinate: $\quad \vec{Q} = \vec{q} + i\vec{\varepsilon} = \vec{q} + i\dfrac{\hbar}{2mv_B}\vec{n}$
Momentum: $\quad \vec{P} = \vec{p} + i\hbar\nabla S_Q$	Velocity: $\quad \dot{\vec{Q}} = \dot{\vec{q}} + i\dot{\vec{\varepsilon}} = \dot{\vec{q}} + i\dfrac{\hbar}{2mv_B}\vec{n}$
Hamiltonian function:	Lagrangian function:
$H(\vec{Q},\vec{P},t) = \displaystyle\sum_{n=1}^{N} \mathcal{P}_n\dot{Q}_n - L(\vec{Q},\dot{\vec{Q}},t)$	$L(\vec{Q},\dot{\vec{Q}},t) = \displaystyle\sum_{n=1}^{N} \mathcal{P}_n\dot{Q}_n - H(\vec{Q},\vec{P},t)$
$\dfrac{\partial H}{\partial P_n} = \dot{Q}_n$	$\dfrac{\partial L}{\partial \dot{Q}_n} = P_n$
$\dfrac{\partial H}{\partial Q_n} = -\dot{P}_n$	$\dfrac{\partial L}{\partial Q_n} = \dot{P}_n$

Table 2. The Legendre's dual transformations.

The Lagrangian equations of motions and the Legendre's transformations are invariant under the above fulfilled imaginary extension of the real momenta, p_n, and the real velocities, v_n, $n=1,2, \ldots N$. It should be noted, that the Hamiltonian function is quadratic in the momenta, $\mathcal{P}_n$, and the Lagrangian function is quadratic in the velocities, $\dot{Q}_n$. A conservation law in this case unifies conservation of energy represented by real part, $\mathrm{Re}\left[\mathcal{H}(\vec{Q},\vec{P};t)\right]$, and the entropy balance (62) represented by imaginary part, $\mathrm{Im}\left[\mathcal{H}(\vec{Q},\vec{P};t)\right]$. One can see from definition of the complexified velocity presented in this table, that tip of the small vector $\dot{\vec{\varepsilon}}$ performs rotating movements on the sphere of the Bohr radius $r_B = \hbar/2mv_B$. This radius is about 5.3×10^{-11} m for the electron-positron pair dancing on the first, virtual, orbit. Energy of this pair, $E = \hbar v_B/r_Be = 27$ V, lies much below energy of the electron-positron creation, $E = \hbar c / \lambda_{Ce} = \hbar(v_B\, a^{-1})/(r_B\, a)e = 511$ kV. Here e is the electric charge, c is the speed of light, λ_C is the Compton wavelength, and $\alpha \simeq 1/137$ is the fine structure constant. From here it is seen, that there is a wide scope of energies for correcting movement of real particle by virtual ones.

6. Concluding remarks

Classical mechanics supposes a principle possibility of simultaneous measuring both coordinates (x,y,z) of material objects and their relative velocities (v_x, v_y, v_z). In the beginning of 20th century scientists call in question such simultaneous measurements. Methods of the classical mechanics cease to give correct results on microscopic level. Instead of the classical equations describing behavior of a classical body, equations of quantum mechanics deal with wave functions that encompass behavior of any particle belonging to the same ensemble of coherent particles. The wave function bears information about distribution of particles that populate a space-time prepared by experimenter. It is said in that case, that it is a guidance function. It contains both the action S and a quantum entropy S_Q (logarithm of the density distribution with negative sign) in the following manner

$$\left| \Psi\left(\vec{Q}, \vec{P}, t \right) \right\rangle = \exp\left\{ \mathrm{i}\, S/\hbar - S_Q \right\}. \tag{78}$$

In contrast to the classical mechanics here the action traces all routes weighted with the factor proportional to the density distribution $\rho = \exp\{ -S_Q \}$.

Wave functions within the same physical scene (the scene is represented by particle sources, detectors, and different physical devices placed between them) obey to superposition principle. Namely, sum of the wave functions is again a wave function that bears information about organization of the physical scene. At measurements we detect interference effects that are conditioned by a specific physical scene. There is no collapse of a wave function at the moment of detecting particle. Information relating to the physical scene exists until destruction of the scene happens. It can be picked up by a new particle again as soon as the particle will be generated by the source. The physical scene prepared by experimenter defines a space-time volume in which particles emitted by sources evolve. The Schrödinger equation (Schrödinger, 1926) gives formulas that determine a probable evolution of the particles within the space-time predefined by boundary conditions of a task. Madelung (Madelung, 1926) and then Bohm (Bohm, 1952a, 1952a) have demonstrated that behind this new equation of quantum mechanics (Schrödinger, 1926), classical equations, Hamilton-Jacobi equations together with the continuity equation, can be discerned. In contrast with the classical equations here a new term emerges - the quantum potential. According to the Madelung's views, the wave function simulates laminar flow of a "fluid" along geodesic paths, named further the Bohmian trajectories. Equiphase surfaces, in turn, are represented by secant surfaces of the trajectory's bundles. Because of these findings we cannot nowadays consider the space-time with the same point of view how it was formulated by thinkers of 17th century. The quantum potential compels to expand the 3D coordinate space onto the imaginary sector by unification the action S and the quantum entropy S_Q, that is, by introducing a complex action $S + \mathrm{i}\, \hbar S_Q$. One way to envisage such a complex space is to imagine a hose-pipe. From a long distance it looks like a one dimensional line. But a closer inspection reveals that every point on the line is in fact a *circle*. It determines the unitary group $U(1)$, which generates the term $\exp\{ \mathrm{i}\, S/\hbar \}$ - a main term in the Feynman path integral.

7. Acknowledgement

Author thanks Miss Pipa (administrator of Quantum Portal) for preparing programs, that permitted to calculate and prepare Figs. 5, 8, and 9.

8. References

Andresse, C. D. (2005). *Huygens The Man Behind the Principle*, Cambridge University Press, Cambridge

Arndt, M., Hackermüller, L. & Reiger, E. (2005). Interferometry with large molecules: Exploration of coherence, decoherence and novel beam methods. *Brazilian Journal of Physics*, Vol. 35(No. 2A), 216–223

Berry, M. (1996). Quantum fractals in boxes. *J. Phys. A: Math. Gen.*, Vol. 29, 6617–6629

Berry, M. (1997). Quantum and optical arithmetic and fractals. In: *The Mathematical beauty of Physics*, J.-M. Drouffe & J.-B. Zuber, pp. 281–294, World Scientific, Singapore

Berry, M. & Klein, S. (1996) Integer, fractional and fractal Talbot effects, *Journal of Modern Optics*, Vol. 43(No. 10), 2139--2164

Berry, M., Marzoli, L. & Schleich, W. (2001). Quantum carpets, carpets of light. *Physics World*, (6), 39–44

Bittner, E. R. (2003). Quantum initial value representations using approximate Bohmian trajectories. http://arXiv.org/abs/quant-ph/0304012

Bohm, D. (1951). *Quantum theory*, Prentice Hall, N. Y.

Bohm, D. (1952a). A suggested interpretation of the quantum theory in terms of "hidden variables", I. *Physical Review*, Vol. 85, 166--179

Bohm, D. (1952b). A suggested interpretation of the quantum theory in terms of "hidden variables", II. *Physical Review*, Vol. 85, 180-193

Bohm, D. J. & Hiley, B. J. (1982). The de Broglie pilot wave theory and the further development of new insights arising out of it. *Foundations of Physies*, Vol. 12(No. 10), 1001–1016

Born, M. & Wolf, E. (1999). *Principles of Optics*, Cambridge University Press, Cambridge

Brezger, B., Hackermüller, L., Uttenthaler, S., Petschinka, J., Arndt, M. & Zeilinger, A. (2002). Matter-wave interferometer for large molecules. *Phys. Rev. Lett.*, Vol. 88, 100404

Brezger, B., Arndt, M. & Zeilinger, A. (2003). Concepts for near-field interferometers with large molecules. *J. Opt. B: Quantum Semiclass. Opt.* , Vol. 5(No. 2), S82--S89

Dirac, P. A. M. (1933). The Lagrangian in quantum mechanics, *Physikalische Zeitschrift der Sowjetunion*, Vol. 3, 64-72

Dirac, P. A. M. (1945). On the analogy between classical and quantum mechanics, *Phys. Rev. Lett.*, Vol. 17(Nos. 2 & 3),   195-199

Dirac, P. A. M. (1982). *Principles of Quantum Mechanics*. Oxford University Press, Oxford, 4nd edition

Feynman, R. P. (1948). Space-time approach to non-relativistic quantum mechanics, *Rev. Mod. Phys..*, Vol. 20, 367-387

Feynman, R. P. & Hibbs, A. (1965). *Quantum mechanics and path integrals*, McGraw Hill, N. Y.

Gerlich, S., Eibenberger, S., Tomandl, M., Nimmrichter, S., Hornberger, K., Fagan, P. J., Tuxen, J., Mayor, M. & Arndt, M. (2011). Quantum interference of large organic molecules. *Nature Communications*, Vol. 2 (No. 263)

Grosche, C. (1993). *An introduction into the Feynman path integral.* http://arXiv.org/abs/hep-th/9302097

Hackermüller, L., Hornberger, K., Brezger, B., Zeilinger, A. & Arndt, M. (2003). Decoherence in a Talbot Lau interferometer: the influence of molecular scattering. *Appl. Phys. B,* Vol. 77, 781–787

Hackermüller, L., Hornberger, K., Brezger, B., Zeilinger, A. & Arndt, M. (2004). Decoherence of matter waves by thermal emission of radiation. *Nature,* Vol. 427, 711–714

Hiley, B. J. (2002). From the Heisenberg picture to Bohm: a new perspective on active information and its relation to Shannon information. In: *Quantum Theory: reconsideration of foundations, Proc. Int. Conf.,* A. Khrennikov, pp. 1--24, Växjö University Press, June 2001, Sweden

Juffmann, T., Truppe, S., Geyer, P., Major, A. G., Deachapunya, S., Ulbricht, H. & Arndt, M. (2009). Wave and particle in molecular interference lithography. *Phys. Rev. Lett.,* Vol. 103, 263601.

Kac, M. (1957). On some connections between probability theory and differential and integral equations, *Proc. Second   Berkeley Symposium on Mathematical Statistics and Probability,* pp .189-215, Berkeley, USA

Lanczos, C. (1970). *The variational principles of mechanics,* Dover Publ., Inc., N. Y.

Landsberg, G. S. (1957). *Optika,* Moscow-Leningrad, 4nd edition

Longhurst, R. S. (1970). *Geometrical and Physical Optics.* Longmans, London, 2nd edition

MacKenzie, R. (2000). *Path integral methods and applications,* http://arXiv.org/abs/quant-ph/0004090

Madelung, E. (1927). Quantumtheorie in hydrodynamische form. *Zts. f. Phys.,* Vol. 40, 322–326

Nairz, O., Arndt, M. & Zeilinger, A. (2003). Quantum interference experiments with large molecules. *Am. J. Phys.,*   Vol. 71(No. 4), 319–325

Nikolić, H. (2007). Quantum mechanics: Myths and facts, *Found. Phys.,* Vol. 37(No. 11), 1563-1611

Poirier, B. (2008). On flux continuity and probability conservation in complexified Bohmian mechanics. http://arXiv.org/abs/0803.0193

Poluyan, P. V. (2005). Nonclassical ontology and nonclassical movement. Kvantovaya Magiya, Vol. 2(No. 3), 3119--3134:

Sanz, A. S. & Miret-Artês, S. (2007). A causal look into the quantum Talbot effect, *J. Chem. Phys.,* Vol. 126, 234106

Sanz, A. S. & Miret-Artês, S. (2008). A trajectory-based understanding of quantum interference, *J. Phys. A: Math. Gen.,* Vol. 41, 435303

Sbitnev, V. I. (2009a). Bohmian trajectories and the Path Integral Paradigm. Complexified Lagrangian Mechanics. *Int. J. Bifurcation & Chaos,* Vol. 19(No. 9), 2335-2346

Sbitnev, V. I. (2009b). N-slit interference: fractals in near-field region, Bohmian trajectories. http://arxiv.org/abs/0907.4638

Sbitnev, V. I. (2010). Matter waves in the Talbot-Lau interferometry, http://arxiv.org/abs/1005.0890

Schrödinger, E. (1926). An undulatory theory of the mechanics of atoms and molecules, *Phys. Rev.,* Vol. 28(No. 6), 1049-1070

Stanford Encyclopeia of Philosophy, (2004). Newton's Views on Space, Time, and Motion, http://plato.stanford.edu/entries/newton-stm/

Stanford Encyclopeia of Philosophy, (2008). Aristotle,
      http://plato.stanford.edu/entries/aristotle/
Stanford Encyclopeia of Philosophy, (2010). Democritus,
      http://plato.stanford.edu/entries/democritus/
Struyve, W. & Valentini, A. (2009). De Broglie-Bohm guidance equations for arbitrary
      Hamiltonians. *J. Phys. A: Math.   Theor.*, Vol. 42, 035301
Talbot, H. F. (1836). Facts relating to optical science. *Philos. Mag.*, (No. 9), 401 − 407
Titchmarsh, E. C. (1976). *The Theory of Functions*, Oxford Science Publ., Oxford
Ventzel, A. D. (1975). *The course of stochastic processes theory*, Nauka, Moskow, in Russian
Wyatt, R. E. (2005). *Quantum Dynamics with Trajectories: Introduction to Quantum
      Hydrodynamics*, Springer, N. Y.
Wyatt, R. E. & Bittner, E. R. (2003). Quantum mechanics with trajectories: Quantum
      trajectories and adaptive grids.
      http://arXiv.org/abs/quant-ph/0302088
Yanov, I. & Leszczynski, J. (2004). Computer simulation of fullerenes and fullerites. In:
      Computional materials science, pp. 85--115. Elsevier B. V., Netherlands

2

Spontaneous Supersymmetry Breaking, Localization and Nicolai Mapping in Matrix Models

Fumihiko Sugino
Okayama Institute for Quantum Physics
Japan

1. Introduction

Supersymmetry (SUSY) is a symmetry between bosons and fermions. It leads to degeneracies of mass spectra between bosons and fermions. Although such degeneracies have not been observed yet, there is a possibility for SUSY being realized in nature as a spontaneously broken symmetry. From a theoretical viewpoint, SUSY provides a unified framework describing physics in high energy regime beyond the standard model (Sohnius, 1985). Spontaneous breaking of SUSY is one of the most interesting phenomena in quantum field theory. Since in general SUSY cannot be broken by radiative corrections at the perturbative level, its spontaneous breaking requires understanding of nonperturbative aspects of quantum field theory (Witten, 1981). In particular, recent developments in nonperturbative aspects of string theory heavily rely on the presence of SUSY. Thus, in order to deduce predictions to the real world from string theory, it is indispensable and definitely important to investigate a mechanism of spontaneous SUSY breaking in a nonperturbative framework of strings. Since one of the most promising approaches of nonperturbative formulations of string theory is provided by large-N matrix models (Banks et al., 1997; Dijkgraaf et al., 1997; Ishibashi et al., 1997), it will be desirable to understand how SUSY can be spontaneously broken in the large-N limit of simple matrix models as a first step. Analysis of SUSY breaking in simple matrix models would help us find a mechanism which is responsible for possible spontaneous SUSY breaking in nonperturbative string theory.

For this purpose, it is desirable to treat systems in which spontaneous SUSY breaking takes place in the path-integral formalism, because matrix models are usually defined by the path integrals, namely integrals over matrix variables. In particular, IIB matrix model defined in zero dimension can be formulated only by the path-integral formalism (Ishibashi et al., 1997). Motivated by this, we discuss in the next section the path-integral formalism for (discretized) SUSY quantum mechanics, which includes cases that SUSY is spontaneously broken. Analogously to the situation of ordinary spontaneous symmetry breaking, we introduce an external field to choose one of degenerate broken vacua to detect spontaneous SUSY breaking. The external field plays the same role as a magnetic field in the Ising model introduced to detect the spontaneous magnetization. For the supersymmetric system, we deform the boundary condition for fermions from the periodic boundary condition (PBC) to a twisted boundary condition (TBC) with twist α, which can be regarded as such an external field. If a supersymmetric system undergoes spontaneous SUSY breaking, the partition function with the PBC for all the fields, Z_{PBC}, which usually corresponds to the Witten index, is expected to vanish (Witten, 1982). Then, the expectation values of observables, which are normalized by Z_{PBC}, would be ill-defined or indefinite. By introducing the twist, the partition function is regularized and the expectation values become well-defined. It is an interesting aspect of our external field for SUSY breaking, which is not seen in spontaneous breaking of ordinary (bosonic) symmetry.

Notice that our argument can be applied to systems in less than one-dimension, for example discretized SUSY quantum mechanics with a finite number of discretized time steps. Spontaneous SUSY breaking is observed even in such simple systems with lower degrees of freedom. Also, we give some argument that an analog of the Mermin-Wagner-Coleman theorem (Coleman, 1973; Mermin & Wagner, 1966) does not hold for SUSY. Thus, cooperative phenomena are not essential to cause spontaneous SUSY breaking, which makes a difference from spontaneous breaking of the ordinary (bosonic) symmetry.

In this setup, we compute an order parameter of SUSY breaking such as the expectation value of an auxiliary field in the presence of the external field. If it remains nonvanishing after turning off the external field, it shows that SUSY is spontaneously broken because it implies that the effect of the infinitesimal external field we have introduced at the beginning remains. Here, it should be noticed that, if we are interested in the large-N behavior of SUSY matrix models, we have to take the large-N limit before turning off the external field, which is reminiscent of the thermodynamic limit of the Ising model taken before turning off the magnetic field in detecting the spontaneous Z_2 breaking.

In view of this, it is quite important to calculate the partition function in the presence of the external field in the path integral for systems which spontaneously break SUSY. Especially it would be better to calculate it in matrix models at finite N in order to observe breaking/restoration of SUSY in the large-N limit. We address this problem by utilizing two methods: localization and Nicolai mapping (Nicolai, 1979) in sections 3 and 4, respectively.

As for the localization, in section 3 we make change of integration variables in the path integral, which is always possible whether or not the SUSY is explicitly broken (the external field is on or off). It is investigated in detail how the integrand of the partition function with respect to the integral over the auxiliary field behaves as the auxiliary field approaches to zero. It plays a crucial role to understand the localization from the change of variables. For SUSY matrix models with Q-SUSY preserved, the path integral receives contributions only from the fixed points of Q-transformation, which are nothing but the critical points of superpotential, i.e. zeros of the first derivative of superpotential. However, in terms of eigenvalues of matrix variables, an interesting phenomenon arises. Localization attracts the eigenvalues to the critical points of superpotential, while the square of the Vandermonde determinant arising from the measure factor prevents the eigenvalues from collapsing. The dynamics of the eigenvalues is governed by balance of attractive force from the localization and repulsive force from the Vandermonde determinant. Without the external field, contribution to the partition function from each eigenvalue distributed around some critical point is derived for a general superpotential.

In the case that the external field is turned on, computation is still possible, but in section 4 we find that a method by the Nicolai mapping is more effective. Interestingly, the Nicolai mapping works for SUSY matrix models even in the presence of the external field which explicitly breaks SUSY. It enables us to calculate the partition function at least in the leading nontrivial order of an expansion with respect to the small external field for finite N. We can take the large-N limit of our result before turning off the external field and detect whether SUSY is spontaneously broken or not in the large-N limit. For illustration, we obtain large-N solutions for a SUSY matrix model with double-well potential.

Section 5 is devoted to summarize the results and discuss future directions.

This chapter is mainly based on the two papers (Kuroki & Sugino, 2010; 2011).

2. Preliminaries on SUSY quantum mechanics

As a preparation to discuss large-N SUSY matrix models, in this section we present some preliminary results on SUSY quantum mechanics.

Let us start with a system defined by the Euclidean (Wick-rotated) action:

$$S^{QM} = \int_0^\beta dt \left[\frac{1}{2}B^2 + iB\left(\dot{\phi} + W'(\phi)\right) + \bar{\psi}\left(\dot{\psi} + W''(\phi)\psi\right) \right], \tag{1}$$

where ϕ is a real scalar field, $\psi, \bar{\psi}$ are fermions, and B is an auxiliary field. The dot means the derivative with respect to the Euclidean time $t \in [0, \beta]$. For a while, all the fields are supposed to obey the PBC. $W(\phi)$ is a real function of ϕ called superpotential, and the prime (') represents the ϕ-derivative.

S^{QM} is invariant under one-dimensional $\mathcal{N} = 2$ SUSY transformations generated by Q and $\bar{Q}$. They act on the fields as

$$Q\phi = \psi, \quad Q\psi = 0, \quad Q\bar{\psi} = -iB, \quad QB = 0, \tag{2}$$

and

$$\bar{Q}\phi = -\bar{\psi}, \quad \bar{Q}\bar{\psi} = 0, \quad \bar{Q}\psi = -iB + 2\dot{\phi}, \quad \bar{Q}B = 2i\dot{\bar{\psi}}, \tag{3}$$

with satisfying the algebra

$$Q^2 = \bar{Q}^2 = 0, \qquad \{Q, \bar{Q}\} = 2\partial_t. \tag{4}$$

Note that S^{QM} can be written as the Q- or $Q\bar{Q}$-exact form:

$$S^{QM} = Q \int dt\, \bar{\psi} \left\{ \frac{i}{2}B - (\dot{\phi} + W'(\phi)) \right\} \tag{5}$$

$$= Q\bar{Q} \int dt \left(\frac{1}{2}\bar{\psi}\psi + W(\phi) \right). \tag{6}$$

For demonstration, let us consider the case of the derivative of the superpotential

$$W'(\phi) = g(\phi^2 - \mu^2) \qquad \text{with} \qquad g, \mu^2 \in \mathbf{R}. \tag{7}$$

For $\mu^2 < 0$, the classical minimum is given by the static configuration $\phi = 0$, with its energy $E_0 = \frac{1}{2}g^2\mu^4 > 0$ implying spontaneous SUSY breaking. Then, $B = ig\mu^2 \neq 0$ from the equation of motion, leading to $Q\bar{\psi}, \bar{Q}\psi \neq 0$, which also means the SUSY breaking.

For $\mu^2 > 0$, the classical minima $\phi = \pm\sqrt{\mu^2}$ are zero-energy configurations. It is known that the quantum tunneling (instantons) between the minima resolves the degeneracy giving positive energy to the ground state. SUSY is broken also in this case.

Next, let us consider quantum aspects of the SUSY breaking in this model. For later discussions on matrix models, it is desirable to observe SUSY breaking via the path-integral formalism, that is, by seeing the expectation value of some field. We take $\langle B \rangle$ (or $\langle B^n \rangle$

($n = 1, 2, \cdots$)) as such an order parameter. Whichever μ^2 is positive or negative, the SUSY is broken, so the ground state energy E_0 is positive. Then, for each of the energy levels E_n ($0 < E_0 < E_1 < E_2 < \cdots$), the SUSY algebra[1]

$$\{Q, \bar{Q}\} = 2E_n, \qquad Q^2 = \bar{Q}^2 = 0 \tag{8}$$

leads to the SUSY multiplet formed by bosonic and fermionic states

$$|b_n\rangle = \frac{1}{\sqrt{2E_n}} \bar{Q}|f_n\rangle, \qquad |f_n\rangle = \frac{1}{\sqrt{2E_n}} Q|b_n\rangle. \tag{9}$$

As a convention, we assume that $|b_n\rangle$ and $|f_n\rangle$ have the fermion number charges $F = 0$ and 1, respectively. Since the Q-transformation for B in (2) is expressed as $[Q, B] = 0$ in the operator formalism, we can see that

$$\langle b_n|B|b_n\rangle = \langle f_n|B|f_n\rangle \tag{10}$$

holds for each n. Then, it turns out that the unnormalized expectation value of B vanishes[2]:

$$\langle B \rangle' \equiv \int_{\text{PBC}} d(\text{fields}) \, B \, e^{-S^{\text{QM}}} = \text{Tr}\left[B(-1)^F e^{-\beta H}\right]$$
$$= \sum_{n=0}^{\infty} \left(\langle b_n|B|b_n\rangle - \langle f_n|B|f_n\rangle\right) e^{-\beta E_n} = 0. \tag{11}$$

This observation shows that, in order to judge SUSY breaking from the expectation value of B, we should choose either of the SUSY broken ground states ($|b_0\rangle$ or $|f_0\rangle$) and see the expectation value with respect to the chosen ground state. The situation is somewhat analogous to the case of spontaneous breaking of ordinary (bosonic) symmetry.

However, differently from the ordinary case, when SUSY is broken, the supersymmetric partition function vanishes:

$$Z_{\text{PBC}}^{\text{QM}} = \int_{\text{PBC}} d(\text{fields}) \, e^{-S^{\text{QM}}} = \text{Tr}\left[(-1)^F e^{-\beta H}\right] \tag{12}$$

$$= \sum_{n=0}^{\infty} \left(\langle b_n|b_n\rangle - \langle f_n|f_n\rangle\right) e^{-\beta E_n} = 0, \tag{13}$$

where the normalization $\langle b_n|b_n\rangle = \langle f_n|f_n\rangle = 1$ was used. So, the expectation values normalized by $Z_{\text{PBC}}^{\text{QM}}$ could be ill-defined (Kanamori et al., 2008a;b).

2.1 Twisted boundary condition
To detect spontaneous breaking of ordinary symmetry, some external field is introduced so that the ground state degeneracy is resolved to specify a single broken ground state. The external field is turned off after taking the thermodynamic limit, then we can judge whether spontaneous symmetry breaking takes place or not, seeing the value of the corresponding order parameter. (For example, to detect the spontaneous magnetization in the Ising model, the external field is a magnetic field, and the corresponding order parameter is the expectation value of the spin operator.)

---

[1] In the operator formalism, $\bar{Q}$, $\bar{\psi}$ are regarded as hermitian conjugate to Q, ψ, respectively.
[2] Furthermore, $\langle B^n \rangle' = 0$ ($n = 1, 2, \cdots$) can be shown.

We will do a similar thing also for the case of spontaneous SUSY breaking. For this purpose, let us change the boundary condition for the fermions to the TBC:

$$\psi(t + \beta) = e^{i\alpha}\psi(t), \qquad \bar{\psi}(t + \beta) = e^{-i\alpha}\bar{\psi}(t), \tag{14}$$

then the twist α can be regarded as an external field. Other fields remain intact. As seen shortly in section 2.1.1, the partition function with the TBC corresponds to the expression (12) with $(-1)^F$ replaced by $(-e^{-i\alpha})^F$:

$$Z_\alpha^{QM} \equiv -e^{-i\alpha}\int_{TBC} d(\text{fields})\, e^{-S^{QM}} = \text{Tr}\left[(-e^{-i\alpha})^F e^{-\beta H}\right] \tag{15}$$

$$= \sum_{n=0}^\infty \left(\langle b_n|b_n\rangle - e^{-i\alpha}\langle f_n|f_n\rangle\right) e^{-\beta E_n} = \left(1 - e^{-i\alpha}\right)\sum_{n=0}^\infty e^{-\beta E_n}. \tag{16}$$

Then, the normalized expectation value of B under the TBC becomes

$$\langle B\rangle_\alpha \equiv \frac{1}{Z_\alpha^{QM}}\text{Tr}\left[B(-e^{-i\alpha})^F e^{-\beta H}\right]$$

$$= \frac{1}{Z_\alpha^{QM}}\sum_{n=0}^\infty \left(\langle b_n|B|b_n\rangle - e^{-i\alpha}\langle f_n|B|f_n\rangle\right) e^{-\beta E_n}$$

$$= \frac{\sum_{n=0}^\infty \langle b_n|B|b_n\rangle e^{-\beta E_n}}{\sum_{n=0}^\infty e^{-\beta E_n}} = \frac{\sum_{n=0}^\infty \langle f_n|B|f_n\rangle e^{-\beta E_n}}{\sum_{n=0}^\infty e^{-\beta E_n}}. \tag{17}$$

Note that the factors $\left(1 - e^{-i\alpha}\right)$ in the numerator and the denominator cancel each other, and thus $\langle B\rangle_\alpha$ does not depend on α even for finite β. As a result, $\langle B\rangle_\alpha$ is equivalent to the expectation value taken over one of the ground states and its excitations $\{|b_n\rangle\}$ (or $\{|f_n\rangle\}$). The normalized expectation value of B under the PBC was of the indefinite form $0/0$, which is now regularized by introducing the parameter α. The expression (17) is well-defined. On the other hand, from the Q-transformation $\psi = [Q, \phi]$, we have

$$\langle b_n|\phi|b_n\rangle = \langle f_n|\phi|f_n\rangle + \frac{1}{\sqrt{2E_n}}\langle f_n|\psi|b_n\rangle. \tag{18}$$

The second term is a transition between bosonic and fermionic states via the fermionic operator ψ, which does not vanish in general. Thus, differently from $\langle B\rangle_\alpha$, the expectation value of ϕ becomes

$$\langle \phi\rangle_\alpha = \frac{1}{Z_\alpha^{QM}}\text{Tr}\left[\phi(-e^{-i\alpha})^F e^{-\beta H}\right]$$

$$= \frac{1}{Z_\alpha^{QM}}\sum_{n=0}^\infty \left(\langle b_n|\phi|b_n\rangle - e^{-i\alpha}\langle f_n|\phi|f_n\rangle\right) e^{-\beta E_n}$$

$$= \frac{\sum_{n=0}^\infty \langle f_n|\phi|f_n\rangle e^{-\beta E_n}}{\sum_{n=0}^\infty e^{-\beta E_n}} + \frac{1}{1 - e^{-i\alpha}}\frac{\sum_{n=0}^\infty \langle f_n|\psi|b_n\rangle \frac{1}{\sqrt{2E_n}} e^{-\beta E_n}}{\sum_{n=0}^\infty e^{-\beta E_n}}. \tag{19}$$

When $\langle f_n|\psi|b_n\rangle \neq 0$ for some n, the second term is α-dependent and diverges as $\alpha \to 0$. The divergence comes from the transition between $|b_n\rangle$ and $|f_n\rangle$. Since the two states are transformed to each other by the (broken) SUSY transformation, we can say that they should belong to the separate superselection sectors, in analogy to spontaneous breaking of ordinary (bosonic) symmetry. Thus, the divergence of $\langle \phi\rangle_\alpha$ as $\alpha \to 0$ implies that the superselection rule does not hold in the system.

2.1.1 Partition function with the twist α

We here show that the partition function with the TBC for the fermions (14) can be expressed in the form (15).

Let $\hat{b}, \hat{b}^\dagger$ be annihilation and creation operators of the fermions:

$$\hat{b}^2 = (\hat{b}^\dagger)^2 = 0, \qquad \{\hat{b}, \hat{b}^\dagger\} = 1, \tag{20}$$

and they are represented on the Fock space $\{|0\rangle, |1\rangle\}$ as

$$\hat{b}|0\rangle = 0, \qquad \hat{b}^\dagger|0\rangle = |1\rangle. \tag{21}$$

We assume that $|0\rangle, |1\rangle$ have the fermion numbers $F = 0, 1$, respectively. The coherent states $|\psi\rangle, \langle\bar{\psi}|$ satisfying

$$\hat{b}|\psi\rangle = \psi|\psi\rangle, \qquad \langle\bar{\psi}|\hat{b}^\dagger = \langle\bar{\psi}|\bar{\psi} \tag{22}$$

($\psi, \bar{\psi}$ are Grassmann numbers, and anticommute with $\hat{b}, \hat{b}^\dagger$.) are explicitly constructed as

$$|\psi\rangle = |0\rangle - \psi|1\rangle = e^{-\psi\hat{b}^\dagger}|0\rangle, \qquad \langle\bar{\psi}| = \langle 0| - \langle 1|\bar{\psi} = \langle 0|e^{-\hat{b}\bar{\psi}}. \tag{23}$$

Also,

$$|0\rangle = \int d\psi\,\psi|\psi\rangle, \qquad \langle 0| = \int d\bar{\psi}\,\langle\bar{\psi}|\bar{\psi}, \qquad |1\rangle = -\int d\psi\,|\psi\rangle, \qquad \langle 1| = \int d\bar{\psi}\,\langle\bar{\psi}|. \tag{24}$$

Thus, we can obtain

$$\begin{aligned}
\mathrm{Tr}\left[(-e^{-i\alpha})^F e^{-\beta H}\right] &= \langle 0|e^{-\beta H}|0\rangle - e^{-i\alpha}\langle 1|e^{-\beta H}|1\rangle \\
&= \int d\bar{\psi}d\psi\,(e^{-i\alpha} + \psi\bar{\psi})\langle\bar{\psi}|e^{-\beta H}|\psi\rangle \\
&= e^{-i\alpha}\int d\bar{\psi}d\psi\,\exp\left(e^{i\alpha}\psi\bar{\psi}\right)\langle\bar{\psi}|e^{-\beta H}|\psi\rangle.
\end{aligned} \tag{25}$$

Since the bosonic part of H is obvious, below we focus on the fermionic part $H_F = \hat{b}^\dagger W''\hat{b}$. Dividing the interval β into M short segments of length ε: $\beta = M\varepsilon$ in (25) and applying the relations

$$\langle\bar{\psi}|\psi\rangle = e^{\bar{\psi}\psi}, \qquad 1 = \int d\bar{\psi}d\psi\,|\psi\rangle e^{\psi\bar{\psi}}\langle\bar{\psi}| \tag{26}$$

to each segment, we have the following expression:

$$\mathrm{Tr}\left[(-e^{-i\alpha})^F e^{-\beta H_F}\right] = -e^{-i\alpha}\int\left(\prod_{j=1}^{M} d\psi_j d\bar{\psi}_j\right)\exp\left[-\varepsilon\sum_{j=1}^{M}\bar{\psi}_j\left(\frac{\psi_{j+1} - \psi_j}{\varepsilon} + W''\psi_j\right)\right] \tag{27}$$

with

$$\psi_{M+1} = e^{i\alpha}\psi_1, \tag{28}$$

or

$$\mathrm{Tr}\left[(-e^{-i\alpha})^F e^{-\beta H_F}\right] = -e^{-i\alpha}\int\left(\prod_{j=1}^{M} d\psi_j d\bar{\psi}_j\right)\exp\left[-\varepsilon\sum_{j=1}^{M}\left(-\frac{\bar{\psi}_j - \bar{\psi}_{j-1}}{\varepsilon} + \bar{\psi}_j W''\right)\psi_j\right] \tag{29}$$

with

$$\bar{\psi}_0 = e^{i\alpha}\bar{\psi}_M. \tag{30}$$

Since (28) and (30) correspond to (14) in the continuum limit $\varepsilon \to 0, M \to \infty$ with $\beta = M\varepsilon$ fixed, we find that the formula (15) holds.

2.2 Discretized SUSY quantum mechanics

In this subsection, we consider a discretized system of (1), namely the Euclidean time is discretized as $t = 1, \cdots, T$. The action is written as

$$S^{\text{dQM}} = Q \sum_{t=1}^{T} \bar{\psi}(t) \left\{ \frac{i}{2} B(t) - (\phi(t+1) - \phi(t) + W'(\phi(t))) \right\} \tag{31}$$

$$= \sum_{t=1}^{T} \left[\frac{1}{2} B(t)^2 + iB(t) \left\{ \phi(t+1) - \phi(t) + W'(\phi(t)) \right\} \right.$$

$$\left. + \bar{\psi}(t) \left\{ \psi(t+1) - \psi(t) + W''(\phi(t))\psi(t) \right\} \right], \tag{32}$$

where the Q-SUSY is of the same form as in (2). As is seen by the Q-exact form (31), the action is Q-invariant and the Q-SUSY is preserved upon the discretization (Catterall, 2003). On the other hand, the $\bar{Q}$-SUSY can not be preserved by the discretization in the case of $T \geq 2$.

When T is finite, the partition function or various correlators are expressed as a finite number of integrals with respect to field variables. So, at first sight, one might expect that spontaneous breaking of the SUSY could not take place, because of a small number of the degrees of freedom. In what follows, we will demonstrate that the expectation is not correct, and that the SUSY can be broken even in such a finite system.

Expressing as S_{α}^{dQM} the action (32) under the TBC

$$\phi(T+1) = \phi(1), \qquad \psi(T+1) = e^{i\alpha}\psi(1), \tag{33}$$

the partition function

$$Z_{\alpha}^{\text{dQM}} \equiv \left(\frac{-1}{2\pi} \right)^{T} \int \prod_{t=1}^{T} (dB(t)\, d\phi(t)\, d\psi(t)\, d\bar{\psi}(t))\, e^{-S_{\alpha}^{\text{dQM}}} \tag{34}$$

is computed to be

$$Z_{\alpha}^{\text{dQM}} = (-1)^{T} \left(1 - e^{i\alpha} \right) C_{T}, \tag{35}$$

$$C_{T} \equiv \int \left(\prod_{t=1}^{T} \frac{d\phi(t)}{\sqrt{2\pi}} \right) \exp \left[-\frac{1}{2} \sum_{t=1}^{T} (\phi(t+1) - \phi(t) + W'(\phi(t)))^2 \right]. \tag{36}$$

Here we used

$$\int \left(\prod_{t=1}^{T} \frac{d\phi(t)}{\sqrt{2\pi}} \right) \left[\prod_{t=1}^{T} (-1 + W''(\phi(t))) - (-1)^{T} \right]$$

$$\times \exp \left[-\frac{1}{2} \sum_{t=1}^{T} (\phi(t+1) - \phi(t) + W'(\phi(t)))^2 \right] = 0 \tag{37}$$

for the superpotential (7), which is derived from the Nicolai mapping (Nicolai, 1979). (Note the factor $\left[\prod_{t=1}^{T} (-1 + W''(\phi(t))) - (-1)^{T} \right]$ is equal to the fermion determinant under the PBC.) Also, C_{T} is positive definite.

Similarly, for the normalized expectation value

$$\langle B(t)\rangle_\alpha \equiv \frac{1}{Z_\alpha^{\text{dQM}}}\left(\frac{-1}{2\pi}\right)^T \int \prod_{t=1}^{T}\left(dB(t)\,d\phi(t)\,d\psi(t)\,d\bar\psi(t)\right)\,B(t)\,e^{-S_\alpha^{\text{dQM}}}, \qquad (38)$$

we use the Nicolai mapping to have

$$\langle B(t)\rangle_\alpha = \frac{1}{Z_\alpha^{\text{dQM}}}(-1)^T\left(1-e^{i\alpha}\right)\int \left(\prod_{t=1}^{T}\frac{d\phi(t)}{\sqrt{2\pi}}\right)(-i)\left(\phi(t+1)-\phi(t)+W'(\phi(t))\right)$$

$$\times \exp\left[-\frac{1}{2}\sum_{t=1}^{T}\left(\phi(t+1)-\phi(t)+W'(\phi(t))\right)^2\right]$$

$$= \frac{1}{C_T}\int \left(\prod_{t=1}^{T}\frac{d\phi(t)}{\sqrt{2\pi}}\right)(-i)\left(\phi(t+1)-\phi(t)+W'(\phi(t))\right)$$

$$\times \exp\left[-\frac{1}{2}\sum_{t=1}^{T}\left(\phi(t+1)-\phi(t)+W'(\phi(t))\right)^2\right]. \qquad (39)$$

The factor $(-1)^T\left(1-e^{i\alpha}\right)$ was canceled, and $\langle B(t)\rangle_\alpha$ does not depend on α, again. The result (39) is finite and well-defined. By using the Nicolai mapping, it is straightforward to generalize this result to the case of W' being a general polynomial

$$W'(\phi) = g_p\phi^p + g_{p-1}\phi^{p-1} + \cdots + g_0. \qquad (40)$$

We find that (39) holds and it is finite and well-defined for even p, and that $\lim_{\alpha\to 0}\langle B(t)\rangle_\alpha = 0$ for odd p.

2.2.1 No analog of Mermin-Wagner-Coleman theorem for SUSY

As claimed in the Mermin-Wagner-Coleman theorem (Coleman, 1973; Mermin & Wagner, 1966), continuous bosonic symmetry cannot be spontaneously broken at the quantum level in the dimensions of two or lower. In dimensions $D \leq 2$, although the symmetry might be broken at the classical level, in computing quantum corrections to a classical (nonzero) value of a corresponding order parameter, one encounters infrared (IR) divergences from loops of a massless boson. It indicates that the conclusion of the symmetry breaking from the classical value is not reliable at the quantum level any more. It is a manifestation of the Mermin-Wagner-Coleman theorem.

Here, we consider whether an analog of the Mermin-Wagner-Coleman theorem for SUSY holds or not. Naively, since loops of a massless fermion ("would-be Nambu-Goldstone fermion") would be dangerous in the dimension one or lower, we might be tempted to expect that SUSY could not be spontaneously broken at the quantum level in the dimension of one or lower. However, this expectation is not correct. Because the twist α in our setting can also be regarded as an IR cutoff for the massless fermion, the finiteness of (39) shows that $\langle B(t)\rangle_\alpha$ is free from IR divergences and well-defined at the quantum level for less than one-dimension. (For one-dimensional case, (17) has no α-dependence, thus no IR divergences.)

We can see it more explicitly in perturbative calculations. Let us consider the superpotential (7) with $\mu^2 < 0$, where the classical configuration $\phi(t) = 0$ gives $B(t) = ig\mu^2$. If the theorem held, quantum corrections should modify this classical value to zero, and there we should come across IR divergences owing to a massless fermion. Although we have obtained the finite result (39), the following perturbative analysis would clarify a role played by the massless fermion. We evaluate quantum corrections to the classical value of $B(t)$ perturbatively. Under the mode expansions

$$\phi(t) = \frac{1}{\sqrt{T}} \sum_{n=-(T-1)/2}^{(T-1)/2} \widetilde{\phi}_n \, e^{i2\pi nt/T} \qquad \text{with} \quad \widetilde{\phi}_n^* = \widetilde{\phi}_{-n},$$

$$\psi(t) = \frac{1}{\sqrt{T}} \sum_{n=-(T-1)/2}^{(T-1)/2} \widetilde{\psi}_n \, e^{i(2\pi n+\alpha)t/T},$$

$$\bar{\psi}(t) = \frac{1}{\sqrt{T}} \sum_{n=-(T-1)/2}^{(T-1)/2} \widetilde{\bar{\psi}}_n \, e^{-i(2\pi n+\alpha)t/T}, \tag{41}$$

free propagators are

$$\left\langle \widetilde{\phi}_{-n}\widetilde{\phi}_m \right\rangle_{\text{free}} = \frac{\delta_{nm}}{4\sin^2\left(\frac{\pi n}{T}\right) + M^2},$$

$$\left\langle \widetilde{\psi}_n\widetilde{\bar{\psi}}_m \right\rangle_{\text{free}} = \frac{\delta_{nm}}{e^{i(2\pi n+\alpha)/T} - 1} \tag{42}$$

with $M^2 \equiv -2g^2\mu^2$. Here we consider the case of odd T for simplicity of the mode expansion. Note that the boson is massive while the fermion is nearly massless regulated by α. Also, there are three kinds of interactions in S_α^{dQM} (after B is integrated out):

$$V_4 = \sum_{t=1}^{T} \frac{1}{2}g^2\phi(t)^4,$$

$$V_{3B} = \sum_{t=1}^{T} g\phi(t)^2 \left(\phi(t+1) - \phi(t)\right),$$

$$V_{3F} = \sum_{t=1}^{T} 2g\phi(t)\bar{\psi}(t)\psi(t). \tag{43}$$

We perturbatively compute the second term of

$$\langle B(t) \rangle_\alpha = ig\mu^2 - i\left\langle g\phi(t)^2 + \phi(t+1) - \phi(t) \right\rangle_\alpha \tag{44}$$

up to the two-loop order, and directly see that the nearly massless fermion ("would-be Nambu-Goldstone fermion") does not contribute and gives no IR singularity. It is easy to see that the tadpole $\langle \phi(t+1) - \phi(t) \rangle_\alpha$ vanishes from the momentum conservation. For $-i\langle g\phi(t)^2 \rangle_\alpha$, the one-loop contribution comes from the diagram (1B) in Figure 1, which consists only of a boson line independent of α. Also, the two-loop diagrams (2BBa), (2BBb), (2BBc) and (2BBd) do not contain fermion lines. The relevant diagrams for the IR divergence at the two-loop order are the last four (2FFa), (2FFb), (2BFa) and (2BFb), which are evaluated

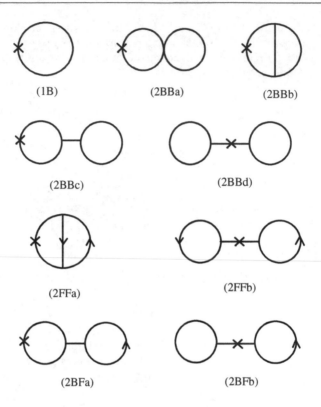

Fig. 1. One- and two-loop diagrams. The crosses represent the insertion of the operator $-ig\phi(t)^2$. The solid lines with (without) arrows mean the fermion (boson) propagators. (1B) is the one-loop diagram, and the other eight are the two-loop diagrams. The diagrams with the name "FF" ("BB") are constructed by using the interaction vertices V_{3F} twice (V_4 once or V_{3B} twice), and those with "BF" are by using each of V_{3B} and V_{3F} once.

as

$$(2FFa) = i\frac{4g^3}{T^2} \sum_{m,k=-(T-1)/2}^{(T-1)/2} \left(\frac{1}{4\sin^2\left(\frac{\pi m}{T}\right) + M^2}\right)^2 \frac{1}{e^{i(2\pi k+\alpha)/T} - 1} \frac{1}{e^{i(2\pi(m+k)+\alpha)/T} - 1},$$

$$(2FFb) = -i\frac{4g^3}{T^2}\frac{1}{M^4} \left(\sum_{m=-(T-1)/2}^{(T-1)/2} \frac{1}{e^{i(2\pi m+\alpha)/T} - 1}\right)^2,$$

$$(2BFa) = -i\frac{4g^3}{T^2}\frac{1}{M^2} \sum_{m=-(T-1)/2}^{(T-1)/2} \left(1 - \frac{M^2}{4\sin^2\left(\frac{\pi m}{T}\right) + M^2}\right) \frac{1}{4\sin^2\left(\frac{\pi m}{T}\right) + M^2}$$

$$\times \sum_{k=-(T-1)/2}^{(T-1)/2} \frac{1}{e^{i(2\pi k+\alpha)/T} - 1},$$

$$(2BFb) = -i\frac{4g^3}{T^2}\frac{1}{M^4} \sum_{m=-(T-1)/2}^{(T-1)/2} \left(1 - \frac{M^2}{4\sin^2\left(\frac{\pi m}{T}\right) + M^2}\right) \sum_{k=-(T-1)/2}^{(T-1)/2} \frac{1}{e^{i(2\pi k+\alpha)/T} - 1}.$$

$$(45)$$

Each diagram is singular as $\alpha \to 0$ due to the fermion zero-mode, however it is remarkable that the sum of them vanishes:

$$(2\text{FFa}) + (2\text{FFb}) + (2\text{BFa}) + (2\text{BFb})$$

$$= -i\frac{4g^3}{T^2}\frac{1}{M^4}\sum_{m=1}^{T-1}\left[1 - \left(\frac{M^2}{4\sin^2\left(\frac{\pi m}{T}\right) + M^2}\right)^2\right]F(m) \tag{46}$$

with

$$F(m) \equiv \sum_{k=1}^{T}\left(1 + \frac{1}{e^{i(2\pi(m+k)+\alpha)/T} - 1}\right)\frac{1}{e^{i(2\pi k+\alpha)/T} - 1}$$

$$= \sum_{k=1}^{T}\frac{1}{e^{i(2\pi k+\alpha)/T} - 1}\left[1 - \frac{e^{-i(2\pi k+\alpha)/T}}{1 - e^{i2\pi m/T}} - \frac{e^{-i(2\pi k+\alpha)/T}}{1 - e^{-i2\pi m/T}}\right]$$

$$= \sum_{k=1}^{T}e^{-i(2\pi k+\alpha)/T} = 0. \tag{47}$$

Thus, the two-loop contribution turns out to have no α-dependence, and the quantum corrections come only from the boson loops which are IR finite, that is consistent with (39). Since the classical value $ig\mu^2 = -i\frac{M^2}{2g}$ is regarded as $\mathcal{O}(g^{-1})$, and ℓ-loop contributions are of the order $\mathcal{O}(g^{2\ell-1})$, the quantum corrections can not be comparable to the classical value in the perturbation theory. Thus, the conclusion of the SUSY breaking based on the classical value continues to be correct even at the quantum level.

3. Change of variables and localization in SUSY matrix models

As argued in the previous section, in order to discuss spontaneous SUSY breaking in the path-integral formalism of (discretized) SUSY quantum mechanics, we introduce an external field to twist the boundary condition of fermions in the Euclidean time direction and observe whether an order parameter of SUSY breaking remains nonzero after turning off the external field. This motivates us to calculate the partition function in the presence of the external field. In the following, we consider a matrix-model analog of (32)

$$S^{\text{M}} = Q\sum_{t=1}^{T}N\,\text{tr}\,\bar{\psi}(t)\left\{\frac{i}{2}B(t) - (\phi(t+1) - \phi(t) + W'(\phi(t)))\right\}$$

$$= \sum_{t=1}^{T}N\,\text{tr}\left[\frac{1}{2}B(t)^2 + iB(t)\left\{\phi(t+1) - \phi(t) + W'(\phi(t))\right\}\right.$$

$$\left. + \bar{\psi}(t)\left\{\psi(t+1) - \psi(t) + QW'(\phi(t))\right\}\right], \tag{48}$$

where all variables are $N \times N$ Hermitian matrices. Under the PBC, this action is manifestly invariant under Q-transformation defined in (2). When $N = 1$, it reduces to the discretized SUSY quantum mechanics in section 2.2. We will focus on the simplest case $T = 1$ below. Under the twisted boundary condition (33) with $T = 1$, the action is

$$S_\alpha^{\text{M}} = N\,\text{tr}\left[\frac{1}{2}B^2 + iBW'(\phi) + \bar{\psi}\left(e^{i\alpha} - 1\right)\psi + \bar{\psi}QW'(\phi)\right], \tag{49}$$

and the partition function is defined by

$$Z_\alpha^M \equiv (-1)^{N^2} \int d^{N^2} B \, d^{N^2} \phi \left(d^{N^2} \psi \, d^{N^2} \bar{\psi} \right) e^{-S_\alpha^M}, \tag{50}$$

where we fix the normalization of the measure as

$$\int d^{N^2} \phi \, e^{-N \mathrm{tr} \left(\frac{1}{2} \phi^2 \right)} = \int d^{N^2} B \, e^{-N \mathrm{tr} \left(\frac{1}{2} B^2 \right)} = 1, \qquad (-1)^{N^2} \int \left(d^{N^2} \psi \, d^{N^2} \bar{\psi} \right) e^{-N \mathrm{tr} \left(\bar{\psi} \psi \right)} = 1. \tag{51}$$

Explicitly, when $W'(\phi)$ is a general polynomial (40), (49) becomes

$$S_\alpha^M = N \mathrm{tr} \left[\frac{1}{2} B^2 + i B W'(\phi) + \bar{\psi} \left(e^{i\alpha} - 1 \right) \psi + \sum_{k=1}^{p} g_k \sum_{\ell=0}^{k-1} \bar{\psi} \phi^\ell \psi \phi^{k-\ell-1} \right]. \tag{52}$$

Notice the ordering of the matrices in the last term. We see that the effect of the external field remains even after the reduction to zero dimension ($T = 1$). When $\alpha = 0$, $S_{\alpha=0}^M$ is invariant under Q and $\bar{Q}$:

$$Q\phi = \psi, \quad Q\psi = 0, \quad Q\bar{\psi} = -iB, \quad QB = 0, \tag{53}$$

and

$$\bar{Q}\phi = -\bar{\psi}, \quad \bar{Q}\bar{\psi} = 0, \quad \bar{Q}\psi = -iB, \quad \bar{Q}B = 0, \tag{54}$$

both of which become broken explicitly in S_α^M by introducing the external field α.

Now let us discuss localization of the integration in Z_α^M. Some aspects are analogous to the discretized SUSY quantum mechanics with $T \geq 2$ under the identification $N^2 = T$ from the viewpoint of systems possessing multi-degrees of freedom, while there are also interesting new phenomena specific to matrix models [3]. We make a change of variables

$$\phi = \tilde{\phi} + \bar{\epsilon}\psi, \qquad \psi = \tilde{\bar{\psi}} - i\bar{\epsilon}B, \tag{55}$$

where in the second equation, $\tilde{\bar{\psi}}$ satisfies

$$N \mathrm{tr}(B\tilde{\bar{\psi}}) = 0, \tag{56}$$

namely, $\tilde{\bar{\psi}}$ is orthogonal to B with respect to the inner product $(A_1, A_2) \equiv N \mathrm{tr}(A_1^\dagger A_2)$. Let us take a basis of $N \times N$ Hermitian matrices $\{t^a\}$ ($a = 1, \cdots, N^2$) to be orthonormal with respect to the inner product: $N \mathrm{tr}(t^a t^b) = \delta_{ab}$. More explicitly, we take

$$\bar{\epsilon} \equiv i \frac{\mathrm{tr}(B\bar{\psi})}{\mathrm{tr} B^2} = \frac{i}{\mathcal{N}_B^2} N \mathrm{tr}(B\bar{\psi}) \tag{57}$$

with $\mathcal{N}_B \equiv ||B|| = \sqrt{N \mathrm{tr}(B^2)}$ the norm of the matrix B. Notice that for general N $\bar{\psi}$ is an $N \times N$ matrix and that $\bar{\epsilon}$ does not have enough degrees of freedom to parametrize the whole space of $\bar{\psi}$. In fact, $\bar{\epsilon}$ is used to parametrize a single component of $\bar{\psi}$ parallel to B. If we write (50) as

$$Z_\alpha^M = \int d^{N^2} B \, \Xi_\alpha(B), \qquad \Xi_\alpha(B) \equiv (-1)^{N^2} \int d^{N^2} \phi \left(d^{N^2} \psi \, d^{N^2} \bar{\psi} \right) e^{-S_\alpha^M}, \tag{58}$$

---

[3] Localization in the discretized SUSY quantum mechanics is discussed in appendix A in ref. (Kuroki & Sugino, 2011).

and consider the change of the variables in $\Xi_\alpha(B)$, B may be regarded as an external variable. The measure $d^{N^2}\psi$ can be expressed by the measures associated with $\tilde{\psi}$ and $\bar{e}$ as

$$d^{N^2}\psi = \frac{i}{\mathcal{N}_B}\, d\bar{e}\, d^{N^2-1}\tilde{\psi}, \tag{59}$$

where $d^{N^2-1}\tilde{\psi}$ is explicitly given by introducing the constraint (56) as a delta-function:

$$d^{N^2-1}\tilde{\psi} \equiv (-1)^{N^2-1}d^{N^2}\tilde{\psi}\,\delta\left(\frac{1}{\mathcal{N}_B}\,N\,\mathrm{tr}(B\tilde{\psi})\right)$$

$$= (-1)^{N^2-1}\left(\prod_{a=1}^{N^2}d\tilde{\psi}^a\right)\frac{1}{\mathcal{N}_B}\sum_{a=1}^{N^2}B^a\tilde{\psi}^a. \tag{60}$$

$\tilde{\psi}^a$ and B^a are coefficients in the expansion of $\tilde{\psi}$ and B by the basis $\{t^a\}$:

$$\tilde{\psi} = \sum_{a=1}^{N^2}\tilde{\psi}^a t^a, \qquad B = \sum_{a=1}^{N^2}B^a t^a. \tag{61}$$

Notice that the measure on the RHS of (59) depends on B. When $B \neq 0$, we can safely change the variables as in (55) and in terms of them the action becomes

$$S_\alpha^M = N\,\mathrm{tr}\left[\frac{1}{2}B^2 + iBW'(\tilde{\phi}) + \tilde{\psi}\left((e^{i\alpha}-1)\psi + QW'(\tilde{\phi})\right) - (e^{i\alpha}-1)i\bar{e}B\psi\right] \tag{62}$$

with $Q\tilde{\phi} = \psi$.

3.1 $\alpha = 0$ case
Let us first consider the case of the PBC ($\alpha = 0$). $S_{\alpha=0}^M$ does not depend on $\bar{e}$ as a consequence of its SUSY invariance, because (55) reads

$$\phi = \tilde{\phi} + \epsilon Q\tilde{\phi}, \qquad \psi = \tilde{\psi} + \bar{\epsilon}Q\tilde{\psi}. \tag{63}$$

Therefore, the contribution to the partition function from $B \neq 0$

$$\tilde{Z}_{\alpha=0} = \int_{||B||\geq\varepsilon} d^{N^2}B\,\Xi_{\alpha=0}(B) \qquad (0 < \varepsilon \ll 1) \tag{64}$$

vanishes due to the integration over $\bar{e}$ according to (59). Namely, when $\alpha = 0$, the path integral of the partition function (50) is localized to $B = 0$.
For the contribution to the partition function from the vicinity of $B = 0$

$$Z_{\alpha=0}^{(0)} = \int_{||B||<\varepsilon} d^{N^2}B\,\Xi_{\alpha=0}(B), \tag{65}$$

when $W'(\phi)$ is given by (40) of degree $p \geq 2$, rescaling as

$$\tilde{\phi} = \mathcal{N}_B^{-\frac{1}{p}}\phi', \qquad \tilde{\psi} = \mathcal{N}_B^{-\frac{p-1}{p}}\psi', \tag{66}$$

we obtain

$$
Z_{\alpha=0}^{(0)} = i \left(\frac{-1}{\sqrt{2\pi}} \right)^{N^2} \left(\int_0^{\varepsilon} d\mathcal{N}_B \, \frac{1}{\mathcal{N}_B^{1+\frac{1}{p}}} \, e^{-\frac{1}{2}\mathcal{N}_B^2} \right) \int d\Omega_B \int d^{N^2}\phi' \, e^{-iN\,\mathrm{tr}(\Omega_B g_p \phi'^p)}
$$

$$
\times \int d^{N^2}\psi \int d\bar{e} \, d^{N^2-1}\bar{\psi}' \, e^{-N\,\mathrm{tr}\left[\bar{\psi}' g_p \sum_{\ell=0}^{p-1} \phi'^{\ell} \psi \phi'^{p-\ell-1} \right]} \left[1 + \mathcal{O}(\varepsilon^{1/p}) \right], \quad (67)
$$

where the measure of the B-integral was expressed in terms of polar coordinates in $\mathbf{R}^{N^2}$ as

$$
d^{N^2}B = \prod_{a=1}^{N^2} \frac{dB^a}{\sqrt{2\pi}} = \left(\frac{1}{2\pi} \right)^{\frac{N^2}{2}} \mathcal{N}_B^{N^2-1} d\mathcal{N}_B \, d\Omega_B, \quad (68)
$$

and $\Omega_B \equiv \frac{1}{\mathcal{N}_B} B$ represents a unit vector in $\mathbf{R}^{N^2}$. Since the $\bar{e}$-integral vanishes while the integration of $\mathcal{N}_B$ becomes singular at the origin, $Z_{\alpha=0}^{(0)}$ takes an indefinite form ($\infty \times 0$). When $W'(\phi)$ is linear ($p = 1$), the $\bar{\phi}$-integrals in (65) yield

$$
Z_{\alpha=0}^{(0)} = i \left(\frac{-1}{|g_1|} \right)^{N^2} \int_{||B||<\varepsilon} \left(\prod_{a=1}^{N^2} dB^a \right) \frac{1}{\mathcal{N}_B} e^{-\frac{1}{2}\mathcal{N}_B^2} \prod_{a=1}^{N^2} \delta(B^a)
$$

$$
\times \int d^{N^2}\psi \int d\bar{e} \, d^{N^2-1}\bar{\psi} \, e^{-N\,\mathrm{tr}(\bar{\psi} g_1 \psi)}, \quad (69)
$$

which is also of indefinite form – the B-integrals diverge while $\int d\bar{e}$ trivially vanishes. The indefinite form reflects that $Z_{\alpha=0}^{(0)}$ possibly takes a nonzero value if it is evaluated in a well-defined manner.

3.1.1 Unnormalized expectation values

Next, let us consider the unnormalized expectation values of $\frac{1}{N} \mathrm{tr}\, B^n$ ($n \geq 1$):

$$
\left\langle \frac{1}{N} \mathrm{tr}\, B^n \right\rangle' \equiv \int d^{N^2}B \left(\frac{1}{N} \mathrm{tr}\, B^n \right) \Xi_{\alpha=0}(B). \quad (70)
$$

Since contribution from the region $||B|| \geq \varepsilon$ is shown to be zero by the change of variables (55), we focus on the B-integration around the origin ($||B|| < \varepsilon$).
When $W'(\phi)$ is a polynomial (40) of degree $p \geq 2$, after the rescaling (66) we obtain

$$
\left\langle \frac{1}{N} \mathrm{tr}\, B^n \right\rangle' = i \left(\int_0^{\varepsilon} d\mathcal{N}_B \, \mathcal{N}_B^{n-1-\frac{1}{p}} e^{-\frac{1}{2}\mathcal{N}_B^2} \right) Y_N \left[1 + \mathcal{O}(\varepsilon^{1/p}) \right],
$$

$$
Y_N \equiv \left(\frac{-1}{\sqrt{2\pi}} \right)^{N^2} \int d\Omega_B \, \frac{1}{N} \mathrm{tr}\,(\Omega_B^n) \int d^{N^2}\phi' \, e^{-iN\,\mathrm{tr}(\Omega_B g_p \phi'^p)}
$$

$$
\times \int d^{N^2}\psi \int d\bar{e} \, d^{N^2-1}\bar{\psi}' \, e^{-N\,\mathrm{tr}\left[\bar{\psi}' g_p \sum_{\ell=0}^{p-1} \phi'^{\ell} \psi \phi'^{p-\ell-1} \right]}. \quad (71)
$$

The $\mathcal{N}_B$-integral is finite, and it is seen that Y_N definitely vanishes. Thus, the change of variables (55) is possible for any B in evaluating $\left\langle \frac{1}{N} \mathrm{tr}\, B^n \right\rangle'$ to give the result

$$
\left\langle \frac{1}{N} \mathrm{tr}\, B^n \right\rangle' = 0 \qquad (n \geq 1). \quad (72)
$$

When $W'(\phi)$ is linear, $\left\langle \frac{1}{N} \operatorname{tr} B^n \right\rangle'$ has the same expression as the RHS of (69) except the integrand multiplied by $\frac{1}{N} \operatorname{tr} B^n$. It leads to a finite result of the B-integration for $n \geq 1$, and (72) is also obtained.

Furthermore, it can be similarly shown that the unnormalized expectation values of multi-trace operators $\prod_{i=1}^{k} \frac{1}{N} \operatorname{tr} B^{n_i}$ $(n_1, \cdots, n_k \geq 1)$ vanish:

$$\left\langle \prod_{i=1}^{k} \frac{1}{N} \operatorname{tr} B^{n_i} \right\rangle' = 0. \tag{73}$$

3.1.2 Localization to $W'(\phi) = 0$, and localization versus Vandermonde

Since (73) means

$$\left\langle e^{-N \operatorname{tr}\left(\frac{u-1}{2} B^2\right)} \right\rangle' = \sum_{n=0}^{\infty} \frac{1}{n!} \left(-N^2 \frac{u-1}{2}\right)^n \left\langle \left(\frac{1}{N} \operatorname{tr} B^2\right)^n \right\rangle' = \langle 1 \rangle' = Z_{\alpha=0}^{\mathrm{M}} \tag{74}$$

for an arbitrary parameter u, we may compute $\left\langle e^{-N \operatorname{tr}\left(\frac{u-1}{2} B^2\right)} \right\rangle'$ to evaluate the partition function $Z_{\alpha=0}^{\mathrm{M}}$. It is independent of the value of u, so u can be chosen to a convenient value to make the evaluation easier.

Taking $u > 0$ and integrating B first, we obtain

$$Z_{\alpha=0}^{\mathrm{M}} = (-1)^{N^2} \int d^{N^2}\phi \left(\frac{1}{u}\right)^{\frac{N^2}{2}} e^{-N \operatorname{tr}\left[\frac{1}{2u} W'(\phi)^2\right]} \int \left(d^{N^2}\psi \, d^{N^2}\bar{\psi}\right) e^{-N \operatorname{tr}[\bar{\psi} Q W'(\phi)]}. \tag{75}$$

Then, let us consider the $u \to 0$ limit. Localization to $W'(\phi) = 0$ takes place because

$$\lim_{u \to 0} \left(\frac{1}{u}\right)^{\frac{N^2}{2}} e^{-N \operatorname{tr}\left[\frac{1}{2u} W'(\phi)^2\right]} = (2\pi)^{\frac{N^2}{2}} \prod_{a=1}^{N^2} \delta(W'(\phi)^a). \tag{76}$$

It is important to recognize that $W'(\phi)^a = 0$ for all a implies localization to a continuous space. Namely, if this condition is met, $W'(U^\dagger \phi U)^a = 0$ for $\forall U \in SU(N)$. Thus the original $SU(N)$ gauge symmetry in the matrix model makes the localization continuous in nature. This is characteristic of SUSY matrix models.

The observation above suggests that in order to localize the path integral to discrete points, we should switch to a description in terms of gauge invariant quantities. This motivates us to change the expression of ϕ to its eigenvalues and $SU(N)$ angles as

$$\phi = U \begin{pmatrix} \lambda_1 & & \\ & \ddots & \\ & & \lambda_N \end{pmatrix} U^\dagger, \qquad U \in SU(N). \tag{77}$$

This leads to an interesting situation, which is peculiar to SUSY matrix models and is not seen in the (discretized) SUSY quantum mechanics. For a polynomial $W'(\phi)$ given by (40), the partition function (75) becomes

$$Z_{\alpha=0}^{\mathrm{M}} = \left(\frac{1}{u}\right)^{\frac{N^2}{2}} \int d^{N^2}\phi \, e^{-N \operatorname{tr}\left[\frac{1}{2u} W'(\phi)^2\right]} \det\left[\sum_{k=1}^{p} g_k \sum_{\ell=0}^{k-1} \phi^\ell \otimes \phi^{k-\ell-1}\right], \tag{78}$$

after the Grassmann integrals. Note that the $N^2 \times N^2$ matrix $\sum_{k=1}^{p} g_k \sum_{\ell=0}^{k-1} \phi^\ell \otimes \phi^{k-\ell-1}$ has the eigenvalues $\sum_{k=1}^{p} g_k \sum_{\ell=0}^{k-1} \lambda_i^\ell \lambda_j^{k-\ell-1}$ $(i,j = 1, \cdots, N)$. Thus, the fermion determinant can be expressed as

$$\det\left[\sum_{k=1}^{p} g_k \sum_{\ell=0}^{k-1} \phi^\ell \otimes \phi^{k-\ell-1}\right] = \prod_{i,j=1}^{N}\left[\sum_{k=1}^{p} g_k \sum_{\ell=0}^{k-1} \lambda_i^\ell \lambda_j^{k-\ell-1}\right]$$
$$= \left(\prod_{i=1}^{N} W''(\lambda_i)\right) \prod_{i>j}\left(\frac{W'(\lambda_i) - W'(\lambda_j)}{\lambda_i - \lambda_j}\right)^2. \tag{79}$$

The measure $d^{N^2}\phi$ given in (51) can be also recast to

$$d^{N^2}\phi = \tilde{C}_N\left(\prod_{i=1}^{N} d\lambda_i\right) \triangle(\lambda)^2 \, dU, \tag{80}$$

where $\triangle(\lambda) = \prod_{i>j}(\lambda_i - \lambda_j)$ is the Vandermonde determinant, and dU is the $SU(N)$ Haar measure normalized by $\int dU = 1$. $\tilde{C}_N$ is a numerical factor depending only on N determined by

$$\frac{1}{\tilde{C}_N} = \int\left(\prod_{i=1}^{N} d\lambda_i\right) \triangle(\lambda)^2 \, e^{-N\sum_{i=1}^{N}\frac{1}{2}\lambda_i^2}. \tag{81}$$

Plugging these into (78), we obtain

$$Z_{\alpha=0}^{M} = \tilde{C}_N \int\left(\prod_{i=1}^{N} d\lambda_i\right) \left(\prod_{i=1}^{N} W''(\lambda_i)\right) \left\{\prod_{i>j}\frac{1}{u}\left(W'(\lambda_i) - W'(\lambda_j)\right)^2\right\}$$
$$\times \left(\frac{1}{u}\right)^{\frac{N}{2}} e^{-N\sum_{i=1}^{N}\frac{1}{2u}W'(\lambda_i)^2}. \tag{82}$$

In this expression, the factor in the second line forces eigenvalues to be localized at the critical points of the superpotential as $u \to 0$, while the last factor in the first line, which is proportional to the square of the Vandermonde determinant of $W'(\lambda_i)$, gives repulsive force among eigenvalues which prevents them from collapsing to the critical points. The dynamics of eigenvalues is thus determined by balance of the attractive force to the critical points originating from the localization and the repulsive force from the Vandermonde determinant. This kind of dynamics is not seen in the (discretized) SUSY quantum mechanics.

To proceed with the analysis, let us consider the situation of each eigenvalue λ_i fluctuating around the critical point $\phi_{c,i}$:

$$\lambda_i = \phi_{c,i} + \sqrt{u}\,\tilde{\lambda}_i \qquad (i = 1, \cdots, N), \tag{83}$$

where $\tilde{\lambda}_i$ is a fluctuation, and $\phi_{c,1}, \cdots, \phi_{c,N}$ are allowed to coincide with each other. Then, the partition function (82) takes the form

$$Z_{\alpha=0}^{M} = \tilde{C}_N \sum_{\phi_{c,i}} \int\left(\prod_{i=1}^{N} d\tilde{\lambda}_i\right) \prod_{i=1}^{N} W''(\phi_{c,i}) \prod_{i>j}\left(W''(\phi_{c,i})\tilde{\lambda}_i - W''(\phi_{c,j})\tilde{\lambda}_j\right)^2$$
$$\times e^{-N\sum_{i=1}^{N}\frac{1}{2}W''(\phi_{c,i})^2\tilde{\lambda}_i^2} + \mathcal{O}(\sqrt{u}). \tag{84}$$

Although only the Gaussian factors become relevant as $u \to 0$, there remain $N(N-1)$-point vertices originating from the Vandermonde determinant of $W'(\lambda_i)$ which yield a specific effect of SUSY matrix models.

In the case of $W'(\phi) = g_1 \phi$, where the corresponding scalar potential $\frac{1}{2} W'(\phi)^2$ is Gaussian, the critical point is only the origin: $\phi_{c,1} = \cdots = \phi_{c,N} = 0$. Then, (84) is reduced to

$$Z^{\mathrm{M}}_{\alpha=0} = \tilde{C}_N \int \left(\prod_{i=1}^{N} d\tilde{\lambda}_i \right) g_1^{N^2} \prod_{i>j} \left(\tilde{\lambda}_i - \tilde{\lambda}_j \right)^2 e^{-N \sum_{i=1}^{N} \frac{1}{2} g_1^2 \tilde{\lambda}_i^2}, \tag{85}$$

where no $\mathcal{O}(\sqrt{u})$ term appears since $W'(\phi)$ is linear. By using (81) we obtain

$$Z^{\mathrm{M}}_{\alpha=0} = (\mathrm{sgn}(g_1))^{N^2} = (\mathrm{sgn}(g_1))^N. \tag{86}$$

For a general superpotential, we change the integration variables as

$$\tilde{\lambda}_i = \frac{1}{W''(\phi_{c,i})} y_i, \tag{87}$$

then the integration of $\tilde{\lambda}_i$ becomes $\int_{-\infty}^{\infty} d\tilde{\lambda}_i \cdots = \frac{1}{|W''(\phi_{c,i})|} \int_{-\infty}^{\infty} dy_i \cdots$. In the limit $u \to 0$, (84) is computed to be

$$\begin{aligned}
Z^{\mathrm{M}}_{\alpha=0} &= \sum_{\phi_{c,i}} \prod_{i=1}^{N} \frac{W''(\phi_{c,i})}{|W''(\phi_{c,i})|} \left\{ \tilde{C}_N \int_{-\infty}^{\infty} \left(\prod_{i=1}^{N} dy_i \right) \Delta(y)^2 e^{-N \sum_{i=1}^{N} \frac{1}{2} y_i^2} \right\} \\
&= \sum_{\phi_{c,i}} \prod_{i=1}^{N} \mathrm{sgn}\left(W''(\phi_{c,i}) \right) \\
&= \left[\sum_{\phi_c : W'(\phi_c)=0} \mathrm{sgn}\left(W''(\phi_c) \right) \right]^N.
\end{aligned} \tag{88}$$

Note that the last factor in the first line of (88) is nothing but the partition function of the Gaussian case with $g_1 = 1$. The last line of (88) tells that the total partition function is given by the N-th power of the degree of the map $\phi \to W'(\phi)$.

Furthermore, we consider a case that the superpotential $W(\phi)$ has K nondegenerate critical points $a_1, \cdots, a_K$. Namely, $W'(a_I) = 0$ and $W''(a_I) \neq 0$ for each $I = 1, \cdots, K$. The scalar potential $\frac{1}{2} W'(\phi)^2$ has K minima at $\phi = a_1, \cdots, a_K$. When N eigenvalues are fluctuating around the minima, we focus on the situation that

the first $v_1 N$ eigenvalues λ_i ($i = 1, \cdots, v_1 N$) are around $\phi = a_1$,

the next $v_2 N$ eigenvalues $\lambda_{v_1 N + i}$ ($i = 1, \cdots, v_2 N$) are around $\phi = a_2$,

$\cdots$,

and the last $v_K N$ eigenvalues $\lambda_{v_1 N + \cdots + v_{K-1} N + i}$ ($i = 1, \cdots, v_K N$) are around $\phi = a_K$,

where $v_1, \cdots, v_K$ are filling fractions satisfying $\sum_{I=1}^{K} v_I = 1$. Let $Z_{(v_1, \cdots, v_K)}$ be a contribution to the total partition function $Z^{\mathrm{M}}_{\alpha=0}$ from the above configuration. Then,

$$Z^{\mathrm{M}}_{\alpha=0} = \sum_{v_1 N, \cdots, v_K N = 0}^{N} \frac{N!}{(v_1 N)! \cdots (v_K N)!} Z_{(v_1, \cdots, v_K)}. \tag{89}$$

(The sum is taken under the constraint $\sum_{I=1}^{K} \nu_I = 1$.) Since $Z_{(\nu_1, \cdots, \nu_K)}$ is equal to the second line of (88) with $\phi_{c,i}$ fixed as

$$
\begin{aligned}
\phi_{c,1} &= \cdots = \phi_{c,\nu_1 N} = a_1, \\
\phi_{c,\nu_1 N+1} &= \cdots = \phi_{c,\nu_1 N+\nu_2 N} = a_2, \\
&\cdots \\
\phi_{c,\nu_1 N+\cdots+\nu_{K-1}N+1} &= \cdots = \phi_{c,N} = a_K,
\end{aligned} \tag{90}
$$

we obtain

$$
Z_{(\nu_1, \cdots, \nu_K)} = \prod_{I=1}^{K} Z_{G,\nu_I}, \qquad Z_{G,\nu_I} = \left(\mathrm{sgn}\left(W''(a_I) \right) \right)^{\nu_I N}. \tag{91}
$$

Z_{G,ν_I} can be interpreted as the partition function of the Gaussian SUSY matrix model with the matrix size $\nu_I N \times \nu_I N$ describing contributions from Gaussian fluctuations around $\phi = a_I$.

3.2 $\alpha \neq 0$ case

In the presence of the external field α, let us consider $\Xi_\alpha(B)$ in (58) with the action (62) obtained after the change of variables (55). Using the explicit form of the measure (59) and (60), we obtain

$$
\Xi_\alpha(B) = (e^{i\alpha} - 1) \frac{(-1)^{N^2-1}}{\mathcal{N}_B^2} \int d^{N^2}\tilde{\phi} \left(d^{N^2}\psi \, d^{N^2}\tilde{\psi} \right) e^{-N\,\mathrm{tr}\left[\frac{1}{2}B^2 + iBW'(\tilde{\phi}) + \tilde{\psi}QW'(\tilde{\phi}) \right]}
$$
$$
\times N\,\mathrm{tr}(B\tilde{\psi})\, N\,\mathrm{tr}(B\psi)\, e^{-(e^{i\alpha}-1)\,N\,\mathrm{tr}(\tilde{\psi}\psi)}, \tag{92}
$$

which is valid for $B \neq 0$. It does not vanish in general by the effect of the twist $e^{i\alpha} - 1$. This suggests that the localization is incomplete by the twist. Although we can proceed the computation further, it is more convenient to invoke another method based on the Nicolai mapping we will present in the next section.

4. $(e^{i\alpha} - 1)$-expansion and Nicolai mapping

In the previous section, we have seen that the change of variables is useful to localize the path integral, but in the $\alpha \neq 0$ case the external field makes the localization incomplete and the explicit computation somewhat cumbersome. In this section, we instead compute Z_α^M in an expansion with respect to $(e^{i\alpha} - 1)$. For the purpose of examining the spontaneous SUSY breaking, we are interested in behavior of Z_α^M in the $\alpha \to 0$ limit. Thus it is expected that it will be often sufficient to compute Z_α^M in the leading order of the $(e^{i\alpha} - 1)$-expansion for our purpose.

4.1 Finite N

Performing the integration over fermions and the auxiliary field B in (50) with $W'(\phi)$ in (40), we have

$$
Z_\alpha^M = \int d^{N^2}\phi \, \det\left((e^{i\alpha} - 1)\mathbf{1} \otimes \mathbf{1} + \sum_{k=1}^{p} g_k \sum_{\ell=0}^{k-1} \phi^\ell \otimes \phi^{p-\ell-1} \right) e^{-N\,\mathrm{tr}\frac{1}{2}W'(\phi)^2}. \tag{93}
$$

Hereafter, let us expand this with respect to $(e^{i\alpha} - 1)$ as

$$
Z_\alpha^M = \sum_{k=0}^{N^2} (e^{i\alpha} - 1)^k Z_{\alpha,k}, \tag{94}
$$

and derive a formula in the leading order of this expansion. The change of variable ϕ as (77) recasts (93) to

$$Z_\alpha^M = \tilde{C}_N \int \left(\prod_{i=1}^N d\lambda_i\right) \triangle(\lambda)^2 \prod_{i,j=1}^N \left(e^{i\alpha} - 1 + \sum_{k=1}^p g_k \sum_{\ell=0}^{k-1} \lambda_i^\ell \lambda_j^{p-\ell-1}\right) e^{-N\sum_{i=1}^N \frac{1}{2}W'(\lambda_i)^2}, \quad (95)$$

after the $SU(N)$ angles are integrated out. Crucial observation is that we can apply the Nicolai mapping (Nicolai, 1979) for each i even in the presence of the external field

$$\Lambda_i = (e^{i\alpha} - 1)\lambda_i + W'(\lambda_i), \quad (96)$$

in terms of which the partition function is basically expressed as an unnormalized expectation value of the Gaussian matrix model

$$Z_\alpha^M = \tilde{C}_N \int \left(\prod_{i=1}^N d\Lambda_i\right) \prod_{i>j}(\Lambda_i - \Lambda_j)^2 e^{-N\sum_i \frac{1}{2}\Lambda_i^2} e^{-N\sum_i \left(-A\Lambda_i\lambda_i + \frac{1}{2}A^2\lambda_i^2\right)}, \quad (97)$$

where $A = e^{i\alpha} - 1$. However, there is an important difference from the Gaussian matrix model, which originates from the fact that the Nicolai mapping (96) is not one to one. As a consequence, λ_i has several branches as a function of Λ_i and it has a different expression according to each of the branches. Therefore, since the last factor of (97) contains $\lambda_i(\Lambda_i)$, we have to take account of the branches and divide the integration region of Λ_i accordingly. Nevertheless, we can derive a rather simple formula at least in the leading order of the expansion in terms of A owing to the Nicolai mapping (96). In the following, let us concentrate on the cases where

$$\Lambda_i \to \infty \quad \text{as} \quad \lambda_i \to \pm\infty, \quad \text{or} \quad \Lambda_i \to -\infty \quad \text{as} \quad \lambda_i \to \pm\infty, \quad (98)$$

i.e. the leading order of $W'(\phi)$ is even. In such cases, we can expect spontaneous SUSY breaking, in which the leading nontrivial expansion coefficient is relevant since the zeroth order partition function vanishes: $Z_{\alpha=0}^M = Z_{\alpha,0} = 0$. Namely, in the expansion of the last factor in (97)

$$e^{-N\sum_{i=1}^N\left(-A\Lambda_i\lambda_i + \frac{1}{2}A^2\lambda_i^2\right)} = 1 - N\sum_{i=1}^N\left(-A\Lambda_i\lambda_i + \frac{1}{2}A^2\lambda_i^2\right) + \cdots, \quad (99)$$

the first term "1" does not contribute to Z_α^M. It can be understood from the fact that it does not depend on the branches and thus the Nicolai mapping becomes trivial, i.e. The mapping degree is zero. Notice that the second term also gives a vanishing effect. For each i, we have the unnormalized expectation value of $N\left(A\Lambda_i\lambda_i - \frac{1}{2}A^2\lambda_i^2\right)$, where the Λ_j-integrals ($j \neq i$) are independent of the branches leading to the trivial Nicolai mapping. Thus, in order to get a nonvanishing result, we need a branch-dependent piece in the integrand for any Λ_i. This immediately shows that in the expansion (94), $Z_{\alpha,k} = 0$ for $k = 0, \cdots, N-1$ and that the first possibly nonvanishing contribution starts from $\mathcal{O}(A^N)$ as

$$Z_{\alpha,N} = \tilde{C}_N N^N \int \left(\prod_{i=1}^N d\Lambda_i\right) \prod_{i>j}(\Lambda_i - \Lambda_j)^2 e^{-N\sum_{i=1}^N \frac{1}{2}\Lambda_i^2} \prod_{i=1}^N (\Lambda_i\lambda_i)\Bigg|_{A=0}. \quad (100)$$

Note that the $A(= e^{i\alpha} - 1)$-dependence of the integrand comes also from λ_i as a function of Λ_i through (96). Although the integration over Λ_i above should be divided into the branches, if we change the integration variables so that we will recover the original λ_i with $A = 0$ (which we call x_i) by

$$\Lambda_i = W'(x_i), \tag{101}$$

then by construction the integration of x_i is standard and runs from $-\infty$ to ∞. Therefore, we arrive at

$$Z_{\alpha,N} = \tilde{C}_N N^N \int_{-\infty}^{\infty} \left(\prod_{i=1}^{N} dx_i \right) \prod_{i=1}^{N} (W''(x_i) W'(x_i) x_i) \prod_{i>j} (W'(x_i) - W'(x_j))^2$$

$$\times e^{-N \sum_{i=1}^{N} \frac{1}{2} W'(x_i)^2}, \tag{102}$$

which does not vanish in general. For example, taking $W'(\phi) = g(\phi^2 - \mu^2)$ we have for $N = 2$

$$Z_{\alpha,2} = 10g^2 \tilde{C}_2 I_0^2 \left[\frac{I_4}{I_0} - \frac{9}{5} \left(\frac{I_2}{I_0} \right)^2 \right], \tag{103}$$

where

$$I_n \equiv \int_{-\infty}^{\infty} d\lambda \, \lambda^n \, e^{-g^2 (\lambda^2 - \mu^2)^2} \qquad (n = 0, 2, 4, \cdots). \tag{104}$$

In fact, when $g = 1$, $\mu^2 = 1$ (double-well scalar potential case) we find

$$I_0 = 1.97373, \quad \frac{I_4}{I_0} - \frac{9}{5} \left(\frac{I_2}{I_0} \right)^2 = -0.165492 \neq 0, \tag{105}$$

hence $Z_{\alpha,2}$ actually does not vanish. In the case of the discretized SUSY quantum mechanics, we have seen in (35) that the expansion of Z_α^M with respect to $(e^{i\alpha} - 1)$ terminates at the linear order for any T. Thus, the nontrivial $\mathcal{O}(A^N)$ contribution of higher order can be regarded as a specific feature of SUSY matrix models.

We stress here that, although we have expanded the partition function in terms of $(e^{i\alpha} - 1)$ and (102) is the leading order one, it is an exact result of the partition function for any finite N and any polynomial $W'(\phi)$ of even degree in the presence of the external field. Thus, it provides a firm ground for discussion of spontaneous SUSY breaking in various settings.

4.2 Large-N

As an application of (102), let us discuss SUSY breaking/restoration in the large-N limit of our SUSY matrix models. From (102), introducing the eigenvalue density

$$\rho(x) = \frac{1}{N} \sum_{i=1}^{N} \delta(x - x_i) \tag{106}$$

rewrites the leading $\mathcal{O}(A^N)$ part of Z_α^M as

$$Z_{\alpha,N} = N^N \int \left(\prod_{i=1}^{N} dx_i \right) \exp(-N^2 F), \tag{107}$$

$$F \equiv - \int dx \, dy \, \rho(x) \rho(y) \log |W'(x) - W'(y)| + \int dx \, \rho(x) \frac{1}{2} W'(x)^2 - \frac{1}{N^2} \log \tilde{C}_N$$

$$- \frac{1}{N} \int dx \, \rho(x) \log(W''(x) W'(x) x). \tag{108}$$

In the large-N limit, $\rho(x)$ is given as a solution to the saddle point equation obtained from $\mathcal{O}(N^0)$ part of F as

$$0 = \int dy \rho(y) \frac{W''(x)}{|W'(x) - W'(y)|} - \frac{1}{2} W'(x) W''(x). \tag{109}$$

Plugging a solution $\rho_0(x)$ into F in (108), we get Z_α^M in the large-N limit in the leading order of $(e^{i\alpha} - 1)$-expansion as

$$Z_{\alpha,N} \to N^N \exp(-N^2 F_0),$$
$$F_0 = -\int dx\, dy\, \rho_0(x)\rho_0(y) \log|W'(x) - W'(y)| + \int dx\, \rho_0(x) \frac{1}{2} W'(x)^2$$
$$- \frac{1}{N^2} \log C_N, \tag{110}$$

where C_N is a factor dependent only on N which arises in replacing the integration over ϕ by the saddle point of its eigenvalue density, thus including $\tilde{C}_N$. From consideration of the Gaussian matrix model (85), C_N is calculated in appendix B in ref. (Kuroki & Sugino, 2010) as

$$C_N = \exp\left[\frac{3}{4} N^2 + \mathcal{O}(N^0)\right]. \tag{111}$$

In (110) we notice that, if we include $\mathcal{O}(1/N)$ part of F (the last term in (108)) in deriving the saddle point equation, the solution will receive an $\mathcal{O}(1/N)$ correction as $\rho(x) = \rho_0(x) + \frac{1}{N}\rho_1(x)$. However, when we substitute this into (108), $\rho_1(x)$ will contribute to F only by the order $\mathcal{O}(1/N^2)$, because $\mathcal{O}(1/N)$ corrections to F_0 under $\rho_0(x) \to \rho_0(x) + \frac{1}{N}\rho_1(x)$ vanish as a result of the saddle point equation at the leading order (109) satisfied by $\rho_0(x)$.

4.3 Example: SUSY matrix model with double-well potential

For illustration of results in the previous subsection, let us consider the SUSY matrix model with $W'(\phi) = \phi^2 - \mu^2$. The saddle point equation (109) becomes

$$\int dy\, \frac{\rho(y)}{x - y} + \int dy\, \frac{\rho(y)}{x + y} = x^3 - \mu^2 x. \tag{112}$$

Let us consider the case $\mu^2 > 0$, where the shape of the scalar potential is a double-well $\frac{1}{2}(x^2 - \mu^2)^2$.

4.3.1 Asymmetric one-cut solution

First, we find a solution corresponding to all the eigenvalues located around one of the minima $\lambda = +\sqrt{\mu^2}$. Assuming the support of $\rho(x)$ as $x \in [a, b]$ with $0 < a < b$, the equation (112) is valid for $x \in [a, b]$.

Following the method in ref. (Brezin et al., 1978), we introduce a holomorphic function

$$F(z) \equiv \int_a^b dy\, \frac{\rho(y)}{z - y}, \tag{113}$$

which satisfies the following properties:

1. $F(z)$ is analytic in $z \in \mathbf{C}$ except the cut $[a, b]$.

2. $F(z)$ is real on $z \in \mathbf{R}$ outside the cut.
3. For $z \sim \infty$,
$$F(z) = \tfrac{1}{z} + \mathcal{O}\left(\tfrac{1}{z^2}\right).$$
4. For $x \in [a, b]$,
$$F(x \pm i0) = F(-x) + x^3 - \mu^2 x \mp i\pi\rho(x).$$

Note that, if we consider the combination (Eynard & Kristjansen, 1995)

$$F_-(z) \equiv \frac{1}{2}\left(F(z) - F(-z)\right), \tag{114}$$

then the properties of $F_-(z)$ are

1. $F_-(z)$ is analytic in $z \in \mathbf{C}$ except the two cuts $[a, b]$ and $[-b, -a]$.
2. $F_-(z)$ is odd $(F_-(-z) = -F_-(z))$, and real on $z \in \mathbf{R}$ outside the cuts.
3. For $z \sim \infty$,
$$F_-(z) = \tfrac{1}{z} + \mathcal{O}\left(\tfrac{1}{z^3}\right).$$
4. For $x \in [a, b]$,
$$F_-(x \pm i0) = \tfrac{1}{2}\left(x^3 - \mu^2 x\right) \mp i\tfrac{\pi}{2}\rho(x).$$

These properties are sufficient to fix the form of $F_-(z)$ as

$$F_-(z) = \frac{1}{2}\left(z^3 - \mu^2 z\right) - \frac{1}{2}z\sqrt{(z^2 - a^2)(z^2 - b^2)} \tag{115}$$

with

$$a^2 = -2 + \mu^2, \qquad b^2 = 2 + \mu^2. \tag{116}$$

Since a^2 should be positive, the solution is valid for $\mu^2 > 2$. The eigenvalue distribution is obtained as

$$\rho_0(x) = \frac{x}{\pi}\sqrt{(x^2 - a^2)(b^2 - x^2)}. \tag{117}$$

From (117), we see that

$$\lim_{\alpha \to 0}\left(\lim_{N \to \infty}\left\langle \frac{1}{N}\operatorname{tr}\phi \right\rangle_\alpha\right) = \int_a^b dx\, x\rho_0(x) \tag{118}$$

is finite and nonsingular, differently from the situation in (19). It can be understood that the tunneling between separate broken vacua is suppressed by taking the large-N limit, and thus the superselection rule works. Note that the large-N limit in the matrix models is analogous to the infinite volume limit or the thermodynamic limit of statistical systems. In fact, this will play an essential role for restoration of SUSY in the large-N limit of the matrix model with a double-well potential.

Using (117), we compute the expectation value of $\frac{1}{N}\operatorname{tr} B$ as

$$\lim_{\alpha \to 0}\left(\lim_{N \to \infty}\left\langle \frac{1}{N}\operatorname{tr} B \right\rangle_\alpha\right) = \int_a^b dx\,(x^2 - \mu^2)\rho_0(x) = 0. \tag{119}$$

Furthermore, all the expectation values of $\frac{1}{N}\operatorname{tr} B^n$ are proven to vanish:

$$\lim_{\alpha \to 0}\left(\lim_{N \to \infty}\left\langle \frac{1}{N}\operatorname{tr} B^n \right\rangle_\alpha\right) = 0 \qquad (n = 1, 2, \cdots). \tag{120}$$

(For a proof, see appendix C in ref. (Kuroki & Sugino, 2010).) Also, the large-N free energy (110) vanishes. These evidences convince us that the SUSY is restored at infinite N.

4.3.2 Two-cut solutions

Let us consider configurations that v_+N eigenvalues are located around one minimum $\lambda = +\sqrt{\mu^2}$ of the double-well, and the remaining $v_-N(= N - v_+N)$ eigenvalues are around the other minimum $\lambda = -\sqrt{\mu^2}$.

First, we focus on the Z_2-symmetric two-cut solution with $v_+ = v_- = \frac{1}{2}$, where the eigenvalue distribution is supposed to have a Z_2-symmetric support $\Omega = [-b, -a] \cup [a, b]$, and $\rho(-x) = \rho(x)$. The equation (112) is valid for $x \in \Omega$. Due to the Z_2 symmetry, the holomorphic function $F(z) \equiv \int_\Omega dy \frac{\rho(y)}{z-y}$ has the same properties as $F_-(z)$ in section 4.3.1 except the property 4, which is now changed to

$$F(x \pm i0) = \frac{1}{2}\left(x^3 - \mu^2 x\right) \mp i\pi\rho(x) \qquad \text{for} \quad x \in \Omega. \tag{121}$$

The solution is given by

$$F(z) = \frac{1}{2}\left(z^3 - \mu^2 z\right) - \frac{1}{2}z\sqrt{(z^2 - a^2)(z^2 - b^2)}, \tag{122}$$

$$\rho_0(x) = \frac{1}{2\pi}|x|\sqrt{(x^2 - a^2)(b^2 - x^2)}, \tag{123}$$

where a, b coincide with the values of the one-cut solution (116). It is easy to see that, concerning Z_2-symmetric observables, the expectation values are the same as the expectation values evaluated under the one-cut solution. In particular, we have the same conclusion for the expectation values of $\frac{1}{N}\text{tr}\, B^n$ and the large-N free energy vanishing.

It is somewhat surprising that the end points of the cut a, b and the large-N free energy coincide with those for the one-cut solution, which is recognized as a new interesting feature of the supersymmetric models and can be never seen in the case of bosonic double-well matrix models. In bosonic double-well matrix models, the free energy of the Z_2-symmetric two-cut solution is lower than that of the one-cut solution, and the endpoints of the cuts are different between the two solutions (Cicuta et al., 1986; Nishimura et al., 2003).

Next, let us consider general Z_2-asymmetric two-cut solutions (i.e., general $v_\pm$). We can check that the following solution gives a large-N saddle point:

The eigenvalue distribution $\rho_0(x)$ has the cut $\Omega = [-b, -a] \cup [a, b]$ with a, b given by (116):

$$\rho_0(x) = \begin{cases} \frac{v_+}{\pi} x \sqrt{(x^2 - a^2)(b^2 - x^2)} & (a < x < b) \\ \frac{v_-}{\pi} |x| \sqrt{(x^2 - a^2)(b^2 - x^2)} & (-b < x < -a). \end{cases} \tag{124}$$

This is a general supersymmetric solution including the one-cut and Z_2-symmetric two-cut solutions. The expectation values of Z_2-even observables under this saddle point coincide with those under the one-cut solution, and the expectation values of $\frac{1}{N}\text{tr}\, B^n$ and the large-N free energy vanish, again. Thus, we can conclude that the SUSY matrix model with the double-well potential has an infinitely many degenerate supersymmetric saddle points parametrized by (v_+, v_-) at large N for the case $\mu^2 > 2$.

4.3.3 Symmetric one-cut solution

Here we obtain a one-cut solution with a symmetric support $[-c, c]$. As before, let us consider a complex function

$$G(z) \equiv \int_{-c}^{c} dy \frac{\rho(y)}{z - y}, \tag{125}$$

and further define

$$G_-(z) \equiv \frac{1}{2}(G(z) - G(-z)). \tag{126}$$

Then $G_-(z)$ has following properties:

1. $G_-(z)$ is odd, analytic in $z \in \mathbf{C}$ except the cut $[-c, c]$.

2. $G_-(x) \in \mathbf{R}$ for $x \in \mathbf{R}$ and $x \notin [-c, c]$.

3. $G_-(z) \to \frac{1}{z} + \mathcal{O}(\frac{1}{z^3})$ as $z \to \infty$.

4. $G_-(x \pm i0) = \frac{1}{2}(x^2 - \mu^2)x \mp i\pi\rho(x)$ for $x \in [-c, c]$.

They lead us to deduce

$$G_-(z) = \frac{1}{2}(z^2 - \mu^2)z - \frac{1}{2}\left(z^2 - \mu^2 + \frac{c^2}{2}\right)\sqrt{z^2 - c^2} \tag{127}$$

with

$$c^2 = \frac{2}{3}\left(\mu^2 + \sqrt{\mu^4 + 12}\right), \tag{128}$$

from which we find that

$$\rho_0(x) = \frac{1}{2\pi}\left(x^2 - \mu^2 + \frac{c^2}{2}\right)\sqrt{c^2 - x^2}, \quad x \in [-c, c]. \tag{129}$$

The condition $\rho_0(x) \geq 0$ tells us that this solution is valid for $\mu^2 \leq 2$, which is indeed the complement of the region of μ^2 where both the two-cut solution and the asymmetric one-cut solution exist. (129) is valid also for $\mu^2 < 0$. Given $\rho_0(x)$, it is straightforward to calculate the large-N free energy as

$$F_0 = \frac{1}{3}\mu^4 - \frac{1}{216}\mu^8 - \frac{1}{216}(\mu^6 + 30\mu^2)\sqrt{\mu^4 + 12} - \log(\mu^2 + \sqrt{\mu^4 + 12}) + \log 6, \tag{130}$$

which is positive for $\mu^2 < 2$. Also, the expectation value of $\frac{1}{N} \operatorname{tr} B$ is computed to be

$$\left\langle \frac{1}{N} \operatorname{tr} B \right\rangle = -i\left[\frac{c^4}{16}(c^2 - \mu^2) - \mu^2\right] \neq 0 \quad \text{for} \quad \mu^2 < 2. \tag{131}$$

These are strong evidence suggesting the spontaneous SUSY breaking. Also, the μ^2-derivatives of the free energy,

$$\lim_{\mu^2 \to 2-0} F_0 = \lim_{\mu^2 \to 2-0} \frac{dF_0}{d(\mu^2)} = \lim_{\mu^2 \to 2-0} \frac{d^2F_0}{d(\mu^2)^2} = 0, \quad \lim_{\mu^2 \to 2-0} \frac{d^3F_0}{d(\mu^2)^3} = -\frac{1}{2}, \tag{132}$$

show that the transition between the SUSY phase ($\mu^2 \geq 2$) and the SUSY broken phase ($\mu^2 < 2$) is of the third order.

5. Summary and discussion

In this chapter, firstly we discussed spontaneous SUSY breaking in the (discretized) quantum mechanics. The twist α, playing a role of the external field, was introduced to detect the SUSY breaking, as well as to regularize the supersymmetric partition function (essentially equivalent to the Witten index) which becomes zero when the SUSY is broken. Differently from spontaneous breaking of ordinary (bosonic) symmetry, SUSY breaking does not require cooperative phenomena and can take place even in the discretized quantum mechanics with a finite number of discretized time steps. There is such a possibility, when the supersymmetric partition function vanishes. In general, some non-analytic behavior is necessary for spontaneous symmetry breaking to take place. For SUSY breaking in the finite system, it can be understood that the non-analyticity comes from the vanishing partition function.

Secondly we discussed localization in SUSY matrix models without the external field. The formula of the partition function was obtained, which is given by the N-th power of the localization formula in the $N = 1$ case (N is the rank of matrix variables). It can be regarded as a matrix-model generalization of the ordinary localization formula. In terms of eigenvalues, localization attracts them to the critical points of superpotential, while the square of the Vandermonde determinant originating from the measure factor gives repulsive force among them. Thus, the dynamics of the eigenvalues is governed by balance of the attractive force from the localization and the repulsive force from the Vandermonde determinant. It is a new feature specific to SUSY matrix models, not seen in the (discretized) SUSY quantum mechanics. For a general superpotential which has K critical points, contribution to the partition function from $\nu_I N$ eigenvalues fluctuating around the I-th critical point ($I = 1, \cdots, K$), denoted by $Z_{(\nu_1, \cdots, \nu_K)}$, was shown to be equal to the products of the partition functions of the Gaussian SUSY matrix models $Z_{G,\nu_1} \cdots Z_{G,\nu_K}$. Here, Z_{G,ν_I} is the partition function of the Gaussian SUSY matrix model with the rank of matrix variables being $\nu_I N$, which describes Gaussian fluctuations around the I-th critical point. It is interesting to investigate whether such a factorization occurs also for various expectation values.

Thirdly, the argument of the change of variables leading to localization can be applied to $\alpha \neq 0$ case. Then, we found that α-dependent terms in the action explicitly break SUSY and makes localization incomplete. Instead of it, the Nicolai mapping, which is also applicable to the $\alpha \neq 0$ case, is more convenient for actual calculation in SUSY matrix models. In the case that the supersymmetric partition function (the partition function with $\alpha = 0$) vanishes, we obtained an exact result of a leading nontrivial contribution to the partition function with $\alpha \neq 0$ in the expansion of $(e^{i\alpha} - 1)$ for finite N. It will play a crucial role to compute various correlators when SUSY is spontaneously broken. Large-N solutions for the double-well case $W'(\phi) = \phi^2 - \mu^2$ were derived, and it was found that there is a phase transition between the SUSY phase corresponding to $\mu^2 \geq 2$ and the SUSY broken phase to $\mu^2 < 2$. It was shown to be of the third order.

For future directions, this kind of argument can be expected to be useful to investigate localization in various lattice models for supersymmetric field theories which realize some SUSYs on the lattice. Also, it will be interesting to investigate localization in models constructed in ref. (Kuroki & Sugino, 2008), which couple a supersymmetric quantum field theory to a certain large-N matrix model and cause spontaneous SUSY breaking at large N. Finally, we hope that similar analysis for super Yang-Mills matrix models (Banks et al., 1997; Dijkgraaf et al., 1997; Ishibashi et al., 1997), which have been proposed as nonperturbative definitions of superstring/M theories, will shed light on new aspects of spontaneous SUSY breaking in superstring/M theories.

6. Acknowledgements

The author would like to thank Tsunehide Kuroki for collaboration which gives a basis for main contribution to this chapter.

7. References

Banks, T.; Fischler, W.; Shenker, S. H. & Susskind, L. (1996). M theory as a matrix model: A conjecture, *Physical Review D* 55: 5112-5128.

Brezin, E.; Itzykson, C.; Parisi, G. & Zuber, J. B. (1978). Planar Diagrams, *Communications in Mathematical Physics* 59 : 35-51.

Catterall, S. (2003). Lattice supersymmetry and topological field theory, *Journal of High Energy Physics* 0305: 038.

Cicuta, G. M.; Molinari, L. & Montaldi, E. (1986). Large N Phase Transitions In Low Dimensions, *Modern Physics Letters A* 1: 125-129.

Coleman, S. R. (1973). There are no Goldstone bosons in two-dimensions, *Communications in Mathematical Physics* 31: 259-264.

Dijkgraaf, R.; Verlinde, E. P. & Verlinde, H. L. (1997). Matrix string theory, *Nuclear Physics B* 500: 43-61.

Eynard, B. & Kristjansen, C. (1995). Exact Solution of the $O(n)$ Model on a Random Lattice, *Nuclear Physics B* 455: 577-618.

Ishibashi, N.; Kawai, H.; Kitazawa, Y. & Tsuchiya, A. (1996). A large-N reduced model as superstring, *Nuclear Physics B* 498: 467-491.

Kanamori, I.; Suzuki, H. & Sugino, F. (2008). Euclidean lattice simulation for the dynamical supersymmetry breaking, *Physical Review D* 77: 091502.

Kanamori, I.; Suzuki, H. & Sugino, F. (2008). Observing dynamical supersymmetry breaking with euclidean lattice simulations, *Progress of Theoretical Physics* 119: 797-827.

Kuroki, T. & Sugino, F. (2008). Spontaneous Supersymmetry Breaking by Large-N Matrices, *Nuclear Physics B* 796: 471-499.

Kuroki, T. & Sugino, F. (2010). Spontaneous supersymmetry breaking in large-N matrix models with slowly varying potential, *Nuclear Physics B* 830: 434-473.

Kuroki, T. & Sugino, F. (2011). Spontaneous supersymmetry breaking in matrix models from the viewpoints of localization and Nicolai mapping," *Nuclear Physics B* 844: 409-449.

Mermin, N. D. & Wagner, H. (1966). Absence of ferromagnetism or antiferromagnetism in one-dimensional or two-dimensional isotropic Heisenberg models, *Physical Review Letters* 17: 1133-1136.

Nicolai, H. (1979). On A New Characterization Of Scalar Supersymmetric Theories, *Physics Letters B* 89: 341-346.

Nishimura, J.; Okubo, T. & Sugino, F. (2003). Testing the Gaussian expansion method in exactly solvable matrix models, *Journal of High Energy Physics* 0310: 057.

Sohnius, M. F. (1985). Introducing Supersymmetry, *Physics Reports* 128: 39-204.

Witten, E. (1981). Dynamical Breaking Of Supersymmetry, *Nuclear Physics B* 188: 513-554.

Witten, E. (1982). Constraints On Supersymmetry Breaking, *Nuclear Physics B* 202: 253-316.

A Fully Quantum Model of Big Bang

S. P. Maydanyuk[1], A. Del Popolo[2,3], V. S. Olkhovsky[1] and E. Recami[4,5,6]
[1]Institute for Nuclear Research, National Academy of Sciences of Ukraine
[2]Istituto di Astronomia dell' Università di Catania, Catania
[3]Dipartimento di Matematica, Università Statale di Bergamo, Bergamo
[4]INFN-Sezione di Milano, Milan
[5]Facoltà di Ingegneria, Università statale di Bergamo, Bergamo
[6]DMO/FEEC, UNICAMP, Campinas, SP, Brazil
[1]Ukraine
[2,3,4]Italy

1. Introduction

In order to understand what really happens in the formation of the Universe, many people came to the point of view that a quantum consideration of this process is necessary. After the publication of the first paper on the quantum description of Universe formation (DeWitt, 1967; Wheeler, 1968), a lot of other papers appeared in this topic (for example, see Refs. (Atkatz & Pagels, 1984; Hartle & Hawking, 1983; Linde, 1984; Rubakov, 1984; Vilenkin, 1982; 1984; 1986; Zel'dovich & Starobinsky, 1984) and some discussions in Refs. (Rubakov, 1999; Vilenkin, 1994) with references therein).

Today, among all variety of models one can select two approaches which are the prevailing ones: these are the Feynman formalism of path integrals in multidimensional spacetime, developed by the Cambridge group and other researchers, called the *"Hartle–Hawking method"* (for example, see Ref. (Hartle & Hawking, 1983)), and a method based on direct consideration of tunneling in 4-dimensional Euclidian spacetime, called the *"Vilenkin method"* (for example, see Refs. (Vilenkin, 1982; 1984; 1986; 1994)). Here, in the quantum approach we have the following picture of the Universe creation: a closed Universe with a small size is formed from "nothing" (vacuum), where by the word "nothing" one refers to a quantum state without classical space and time. A wave function is used for a probabilistic description of the creation of the Universe and such a process is connected with transition of a wave through an effective barrier. Determination of penetrability of this barrier is a key point in the estimation of duration of the formation of the Universe, and the dynamics of its expansion in the first stage. However, in the majority of these models, with the exception of some exactly solvable models, tunneling is mainly studied in details in the semiclassical approximation (see Refs. (Rubakov, 1999; Vilenkin, 1994)). An attractive side of such an approach is its simplicity in the construction of decreasing and increasing partial solutions for the wave function in the tunneling region, the outgoing wave function in the external region, and the possibility to define and to estimate in an enough simply way the penetrability of the barrier, which can be used to obtain the duration of the nucleation of the Universe. The *tunneling boundary condition*

(Vilenkin, 1994) could seem to be the most natural and clear description, where the wave function should represent an outgoing wave only in the enough large value of the scale factor a. However, is really such a wave free in the asymptotic region? In order to draw attention on the increase of the modulus of the potential with increasing value of the scale factor a and increasing magnitude of the gradient of such a potential, acting on this wave "through the barrier", then one come to a serious contradiction: *the influence of the potential on this wave increases strongly with a!* Now a new question has appeared: what should the wave represent in general in the cosmological problem? This problem connects with another and more general one in quantum physics — the real importance *to define a "free" wave inside strong fields.* To this aim we need a mathematical stable tool to study it. It is unclear whether a connection between exact solutions for the wave function at turning point and "free" wave defined in the asymptotic region is correct.

Note that the semiclassical formula of the penetrability of the barrier is constructed on the basis of wave which is defined concerning zero potential at infinity, i.e. this wave should be free outgoing in the asymptotic region. But in the cosmological problem we have opposite case, when the force acting on the wave increases up to infinity in the asymptotic region. At the same time, deformations of the shape of the potential outside the barrier cannot change the penetrability calculated in the framework of the semiclassical approach (up to the second order). An answer to such problem can be found in non-locality of definition of the penetrability in quantum mechanics, which is reduced to minimum in the semiclassical approach (i. e. this is so called "error" of the cosmological semiclassical approach).

The problem of the correct definition of the wave in cosmology is reinforced else more, if one wants to calculate the incident and reflected waves in the internal region. *Even with the known exact solution for the wave function there is uncertainty in determination of these waves!* But, namely, the standard definition of the coefficients of penetrability and reflection is based on them. In particular, we have not found papers where the coefficient of reflection is defined and estimated in this problem (which differs essentially from unity at the energy of radiation close to the height of the barrier and, therefore, such a characteristics could be interesting from a physical point of view). Note that the semiclassical approximation put serious limits to the possibility of its definition at all (Landau & Lifshitz, 1989).

Thus, in order to estimate probability of the formation of the Universe as accurately as possible, we need a fully quantum definition of the wave. Note that the non-semiclassical penetrability of the barrier in the cosmological problems has not been studied in detail and, therefore, a development of fully quantum methods for its estimation is a perspective task.

Researches in this direction exist (Acacio de Barros et al., 2007), and in these papers was estimated the penetrability on the basis of tunneling of wave packet through the barrier. However, a stationary boundary condition has uncertainty that could lead to different results in calculations of the penetrability. The stationary approach could allow to clarify this issue. It is able to give stable solutions for the wave function (and results in Ref. (Maydanyuk, 2008) have confirmed this at zero energy of radiation), using the standard definition of the coefficients of the penetrability and reflection, is more accurate to their estimation.

Aims of this Chapter are: (1) to define the wave in the quantum cosmological problem; (2) to construct the fully quantum (non-semiclassical) methods of determination of the coefficients of penetrability of the barriers and reflection from them on the basis of such a definition of the wave; (3) to estimate how much the semiclassical approach differs in the estimation of the penetrability from the fully quantum one. In order to achieve this goal, we need to construct tools for calculation of partial solutions of the wave function. In order to resolve the questions pointed out above, we shall restrict ourselves to a simple cosmological model, where the potential has a barrier and internal above-barrier region.

2. Cosmological model in the Friedmann–Robertson–Walker metric with radiation

2.1 Dynamics of Universe in the Friedmann–Robertson–Walker metric

Let us consider a simple model of the homogeneous and isotropic Universe in *Friedmann–Robertson–Walker (FRW) metric* (see Ref. (Weinberg, 1975), p. 438; also see Refs. (Brandenberger, 1999; Linde, 2005; Rubakov, 2005; Trodden & and Carroll, 2003)):

$$ds^2 = -dt^2 + a^2(t) \cdot \left(\frac{dr^2}{h(r)} + r^2(d\theta^2 + \sin^2 \theta \, d\phi^2) \right), \; h(r) = 1 - kr^2, \tag{1}$$

where t and r, θ, ϕ are time and space spherical coordinates, the signature of the metric is $(-,+,+,+)$ as in Ref. (Trodden & and Carroll, 2003) (see p. 4), $a(t)$ is an unknown function of time and k is a constant, the value of which equals $+1$, 0 or -1, with appropriate choice of units for r. Further, we shall use the following system of units: $\hbar = c = 1$. For $k = -1, 0$ the space is infinite (Universe of open type), and for $k = +1$ the space is finite (Universe of closed type). For $k = 1$ one can describe the space as a sphere with radius $a(t)$ embedded in a 4-dimensional Euclidian space. The function $a(t)$ is referred to as the *"radius of the Universe"* and is called the *cosmic scale factor*. This function contains information of the dynamics of the expansion of the Universe, and therefore its determination is an actual task.

One can find the function $a(t)$ using the Einstein equations taking into account the cosmological constant Λ in this metric (we use the signs according to the chosen signature, as in Ref. (Trodden & and Carroll, 2003) p. 8; the Greek symbols μ and ν denote any of the four coordinates t, r, θ and ϕ):

$$R_{\mu\nu} - \frac{1}{2} g_{\mu\nu} R = 8\pi \, G \, T_{\mu\nu} + \Lambda, \tag{2}$$

where $R_{\mu\nu}$ is the Ricci tensor, R is the scalar curvature, $T_{\mu\nu}$ is the energy-momentum tensor, and G is Newton's constant. From (1) we find the Ricci tensor $R_{\mu\nu}$ and the scalar curvature R:

$$R_{tt} = -3\frac{\ddot{a}}{a}, \qquad R_{rr} = \frac{a\ddot{a}}{h} + 2\frac{\dot{a}^2}{h} - \frac{h'}{hr} = \frac{2\dot{a}^2 + a\ddot{a} + 2k}{1 - kr^2},$$

$$R_{\phi\phi} = R_{\theta\theta} \sin^2 \theta, \qquad R_{\theta\theta} = a\ddot{a} \, r^2 + 2\dot{a}^2 \, r^2 - h - \frac{h'r}{2} + 1 = 2\dot{a}^2 \, r^2 + a\ddot{a} \, r^2 + 2kr^2 \tag{3}$$

$$R = g^{tt} R_{tt} + g^{rr} R_{rr} + g^{\theta\theta} R_{\theta\theta} + g^{\phi\phi} R_{\phi\phi} = \frac{6\dot{a}^2 + 6a\ddot{a} + 6k}{a^2}. \tag{4}$$

The *energy-momentum tensor* has the form (see (Trodden & and Carroll, 2003), p. 8): $T_{\mu\nu} = (\rho + p) U_\mu U_\nu + p \, g_{\mu\nu}$, where ρ and p are energy density and pressure. Here, one needs to use the normalized vector of 4-velocity $U^t = 1$, $U^r = U^\theta = U^\phi = 0$. Substituting the previously calculated components (2) of the Ricci tensor $R_{\mu\nu}$, the scalar curvature (4), the components of the energy-momentum tensor $T_{\mu\nu}$ and including the component $\rho_{\text{rad}}(a)$, describing the radiation in the initial stage (equation of state for radiation: $p(a) = \rho_{\text{rad}}(a)/3$), into the Einstein's equation (2) at $\mu = \nu = 0$), we obtain the *Friedmann equation* with the cosmological constant (see p. 8 in Ref. (Trodden & and Carroll, 2003); p. 3 in Ref. (Brandenberger, 1999)):

$$\dot{a}^2 + k - \frac{8\pi G}{3} \left\{ \frac{\rho_{\text{rad}}}{a^2(t)} + \rho_\Lambda a^2(t) \right\} = 0, \qquad \rho_\Lambda = \frac{\Lambda}{8\pi G}, \tag{5}$$

where $\dot{a}$ is derivative a at time coordinate. From here, we write a general expression for the energy density:

$$\rho(a) = \rho_\Lambda + \frac{\rho_{\text{rad}}}{a^4(t)}. \tag{6}$$

2.2 Action, lagrangian and quantization

We define the action as

$$S = \int \sqrt{-g} \left(\frac{R}{16\pi G} - \rho \right) dx^4. \tag{7}$$

Substituting the scalar curvature (4), then integrating item at $\ddot{a}$ by parts with respect to variable t, we obtain the *lagrangian*:

$$\mathcal{L}(a, \dot{a}) = \frac{3a}{8\pi G} \left(-\dot{a}^2 + k - \frac{8\pi G}{3} a^2 \rho(a) \right). \tag{8}$$

Considering the variables a and $\dot{a}$ as generalized coordinate and velocity respectively, we find a generalized momentum conjugate to a:

$$p_a = \frac{\partial \mathcal{L}(a, \dot{a})}{\partial \dot{a}} = -\frac{3}{4\pi G} a \dot{a} \tag{9}$$

and then hamiltonian:

$$h(a, p_a) = p \dot{a} - \mathcal{L}(a, \dot{a}) = -\frac{1}{a} \left\{ \frac{2\pi G}{3} p_a^2 + a^2 \frac{3k}{8\pi G} - a^4 \rho(a) \right\}. \tag{10}$$

The passage to the quantum description of the evolution of the Universe is obtained by the standard procedure of canonical quantization in the Dirac formalism for systems with constraints. In result, we obtain the *Wheeler–De Witt (WDW) equation* (see (DeWitt, 1967; Levkov et al., 2002; Wheeler, 1968)), which can be written as

$$\left\{ -\frac{\partial^2}{\partial a^2} + V(a) \right\} \varphi(a) = E_{\text{rad}} \, \varphi(a),$$

$$V(a) = \left(\frac{3}{4\pi G} \right)^2 k a^2 - \frac{3\rho_\Lambda}{2\pi G} a^4, \tag{11}$$

$$E_{\text{rad}} = \frac{3\rho_{\text{rad}}}{2\pi G},$$

where $\varphi(a)$ is wave function of Universe. This equation looks similar to the one-dimensional stationary Schrödinger equation on a semiaxis (of the variable a) at energy E_{rad} with potential $V(a)$. It is convenient to use system of units where $8\pi G \equiv M_{\text{p}}^{-2} = 1$, and to rewrite $V(a)$ in a generalized form as

$$V(a) = A a^2 - B a^4. \tag{12}$$

In particular, for the Universe of the closed type ($k = 1$) we obtain $A = 36$, $B = 12\,\Lambda$ (this potential coincides with Ref. (Acacio de Barros et al., 2007)).

2.3 Potential close to the turning points: non-zero energy case

In order to find the wave function we need to know the shape of the potential close to the turning points. Let us find the *turning points* $a_{tp,in}$ and $a_{tp,out}$ concerning the potential (12) at energy E_{rad}:

$$a_{tp,in} = \sqrt{\frac{A}{2B}} \cdot \sqrt{1 - \sqrt{1 - \frac{4BE_{rad}}{A^2}}}, \quad a_{tp,out} = \sqrt{\frac{A}{2B}} \cdot \sqrt{1 + \sqrt{1 - \frac{4BE_{rad}}{A^2}}}. \tag{13}$$

Let us expand the potential $V(a)$ (13) in powers of $q_{out} = a - a_{tp}$ (where the point $a_{tp,in}$ or $a_{tp,out}$ is used as a_{tp}. Expansion is calculated at these points), where (for small q) we restrict ourselves to the liner term:

$$V(q) = V_0 + V_1\,q, \tag{14}$$

where the coefficients V_0 and V_1 are:

$$V_0 \quad = V(a = a_{tp,in}) = V(a = a_{tp,out}) = A\,a_{tp}^2 - B\,a_{tp}^4 = E_{rad},$$

$$V_1^{(out)} = -2A \cdot \sqrt{\frac{A}{2B}\left(1 - \frac{4BE_{rad}}{A^2}\right)\left(1 + \sqrt{1 - \frac{4BE_{rad}}{A^2}}\right)}, \tag{15}$$

$$V_1^{(int)} = 2A \cdot \sqrt{\frac{A}{2B}\left(1 - \frac{4BE_{rad}}{A^2}\right)\left(1 - \sqrt{1 - \frac{4BE_{rad}}{A^2}}\right)}.$$

Now eq. (15) transforms into a new form at variable q with potential $V(q)$:

$$-\frac{d^2}{dq^2}\,\varphi(q) + V_1\,q\,\varphi(q) = 0. \tag{16}$$

3. Tunneling boundary condition in cosmology

3.1 A problem of definition of "free" wave in cosmology and correction of the boundary condition

Which boundary condition should be used to obtain a wave function that describes how the wave function leaves the barrier accurately? A little variation of the boundary condition leads to change of the fluxes concerning the barrier and, as result, it changes the coefficients of penetrability and reflection. So, a proper choice of the boundary condition is extremely important. However before, let us analyze how much the choice of the boundary condition is natural in the asymptotic region.

- In description of collisions and decay in nuclear and atomic physics potentials of interactions tend to zero asymptotically. So, in these calculations, the boundary conditions are imposed on the wave function at infinity. In cosmology we deal with another, different type of potential: its modulus increases with increasing of the scale factor a. The gradient of the potential also increases. Therefore, *here there is nothing common to the free propagation of the wave in the asymptotic region*. Thus, a direct passage of the application of the boundary condition in the asymptotic region into cosmological problems looks questionable.

- The results in Ref. (Maydanyuk, 2008), which show that when the scale factor a increases the region containing solutions for the wave function enlarges (and its two partial solutions), reinforce the seriousness of this problem. According to Ref. (Maydanyuk, 2008), the scale factor a in the external region is larger, the period of oscillations of each partial solution for the wave function is <u>smaller</u>. One has to decrease the time–step and as a consequence increase the calculation time. This increases errors in computer calculations of the wave function close the barrier (if it is previously fixed by the boundary condition in the asymptotic region). From here a natural conclusion follows on the impossibility to use practically the boundary condition at infinity for calculation of the wave (in supposition if we know it maximally accurately in the asymptotic region), if we like to pass from the semiclassical approach to the fully quantum one. Another case exists in problems of decay in nuclear and atomic physics where calculations of the wave in the asymptotic region are the most stable and accurate.

- One can add a fact that it has not been known yet whether the Universe expands at extremely large value of the scale factor a. Just the contrary, it would like to clarify this from a solution of the problem, however imposing a condition that the Universe expands in the initial stage.

So, we shall introduce the following **definition of the boundary condition** (Maydanyuk, 2010):

The boundary condition should be imposed on the wave function at such value of the scale factor a, where the potential minimally acts on the wave, determined by this wave function.

The propagation of the wave defined in such a way is close to free one for the potential and at used value of the scale factor a (we call such a wave conditionally "free"). However, when we want to give a mathematical formulation of this definition we have to answer two questions:

1. What should the free wave represent in a field of a cosmological potential of arbitrary shape? How could it be defined in a correct way close to an arbitrary selected point?

2. Where should the boundary condition be imposed?

To start with, let us consider the second question namely where we must impose the boundary condition on the wave function. One can suppose that this coordinate could be (1) a turning point (where the potential coincides with energy of radiation), or (2) a point where the gradient from the potential becomes zero, or (3) a point where the potential becomes zero. But the clear condition of free propagation of the wave is the minimal influence of the potential on this wave. So, we define these coordinate and force so (Maydanyuk, 2010):

The point in which we impose the boundary condition is the coordinate where the force acting on the wave is minimal. The force is defined as minus the gradient of the potential.

It turns out that according to such a (local) definition the force is minimal at the external turning point $a_{tp,out}$. Also, the force, acting on the wave incident on the barrier in the internal region and on the wave reflected from it, has a minimum at the internal turning point $a_{tp,in}$. Thus, we have just obtain the internal and external turning points where we should impose the boundary conditions in order to determine the waves.

3.2 Boundary condition at $a = 0$: stationary approach versus non-stationary one

A choice of the proper boundary condition imposed on the wave function is directly connected with the question: could the wave function be defined at $a = 0$, and which value should it be equal to at this point in such a case? The wave function is constructed on the basis of its two partial solutions which should be linearly independent. In particular, these two partial solutions can be real (not complex), without any decrease of accuracy in determination of the total wave function. *For any desirable boundary condition imposed on the total wave function, such methods should work.* In order to achieve the maximal linear independence between two partial solutions, we choose one solution to be increasing in the region of tunneling and another one to be decreasing in this tunneling region. For the increasing partial solution we use as starting point a_x the internal turning point $a_{tp,in}$ at $E_{rad} \neq 0$ or zero point $a_x = 0$ at $E_{rad} = 0$. For the second decreasing partial solution the starting point a_x is chosen as the external turning point $a_{tp,out}$. Such a choice of starting points turns out to give us higher accuracy in calculations of the total wave function than starting calculations of both partial solutions from zero or from only one turning point.

In order to obtain the total wave function, we need to connect two partial solutions using one boundary condition, which should be obtained from physical motivations. According to analysis in Introduction and previous section, it is natural not to define the wave function at zero (or at infinity), but to find outgoing wave at <u>finite</u> value of a in the external region, where this wave corresponds to observed Universe at present time. But, in practical calculations, we shall define such a wave at point where forces minimally act on it. This is an initial condition imposed on the outgoing wave in the external region[1].

Let us analyze a question: which value has the wave function at $a = 0$? In the paper the following ideas are used:

- *the wave function should be continuous in the whole spatial region of its definition,*
- *we have outgoing non-zero flux in the asymptotic region defined on the basis of the total wave function,*
- *we consider the case when this flux is constant.*

The non-zero outgoing flux defined at arbitrary point requires the wave function to be complex and non-zero. The condition of continuity of this flux in the whole region of definition of the wave function requires this wave function to be complex and non-zero in the entire region. If we include point $a = 0$ into the studied region, then we should obtain the non-zero and complex wave function also at such point. If we use the above ideas, then we cannot obtain zero wave function at $a = 0$. One can use notions of nuclear physics, field in which the study of such questions and their possible solutions have longer history then in quantum cosmology. As example, one can consider elastic scattering of particles on nucleus (where we have zero radial wave function at $r = 0$, and we have no divergences), and alpha decay of nucleus (where we cannot obtain zero wave function at $r = 0$). *A possible divergence of the radial wave function at zero in quantum decay problem could be explained by existence of source at a point which creates the outgoing flux in the asymptotic region (and is the source of this flux).* Now the picture becomes clearer: any quantum decay could be connected with source at zero. This is why the vanishing of the total wave function at $a = 0$, after introduction of the wall at this point (like in Ref. (Acacio de Barros et al., 2007)), is not obvious and is only one of the possibilities.

If we wanted to study physics at zero $a = 0$, we should come to two cases:

---

[1] For example, on the basis of such a boundary condition for α-decay problem we obtain the asymptotic region where the wave function is spherical outgoing wave.

- If we include the zero point into the region of consideration, we shall use to quantum mechanics with included sources. In such a case, the condition of constant flux is broken. But a more general integral formula of non-stationary dependence of the fluxes on probability can include possible sources and put them into frameworks of the standard quantum mechanics also (see eq. (19.5) in Ref. (Landau & Lifshitz, 1989), p. 80). Perhaps, black hole could be another example of quantum mechanics with sources and sinks.

- We can consider only quantum mechanics with constant fluxes and without sources. Then we should eliminate the zero point $a = 0$ from the region of our consideration. In this way, the formalism proposed in this paper works and is able to calculate the penetrability and reflection coefficients without any lost of accuracy.

This could be a stationary picture of interdependence between the wave function at zero and the outgoing flux in the asymptotic region. In order to study the non-stationary case, then we need initial conditions which should define also the evolution of the Universe. In such a case, after defining the initial state (through set of parameters) it is possible to connect zero value of wave packet at $a = 0$ (i. e. without singularity at such a point) with non-zero outgoing flux in the asymptotic region. In such direction different proposals have been made in frameworks of semiclassical models in order to describe inflation, to introduce time or to analyze dynamics of studied quantum system (for example, see (Finelli et al., 1998; Tronconi et al., 2003)).

4. Direct fully quantum method

4.1 Wave function of Universe: calculations and analysis

The wave function is known to oscillate above the barrier and increase (or decrease) under the barrier without any oscillations. So, in order to provide an effective linear independence between two partial solutions for the wave function, we look for a first partial solution increasing in the region of tunneling and a second one decreasing in this tunneling region. To start with, we define each partial solution and its derivative at a selected starting point, and then we calculate them in the region close enough to this point using the *method of beginning of the solution* presented in Subsection 4.4.1. Here, for the partial solution which increases in the barrier region, as starting point we use the internal turning point $a_{tp, in}$ at non-zero energy E_{rad} or equals to zero $a = 0$ at null energy E_{rad}, and for the second partial solution, which decreases in the barrier region, we choose the starting point to be equal to the external turning point $a_{tp, out}$. Then we calculate both partial solutions and their derivatives in the whole required range of a using the *method of continuation of the solution* presented in Subsection 4.4.2, which is improvement of the Numerov method with constant step. So, we obtain two partial solutions for the wave function and their derivatives in the whole studied region (Maydanyuk, 2010).

In order to clarify how the proposed approach gives convergent (stable) solutions, we compare our results with the paper of (Acacio de Barros et al., 2007). Let us consider the behavior of the wave function. The first partial solution for the wave function and its derivative in my calculation are presented in Fig. 1, which increase in the tunneling region and have been obtained at different values of the energy of radiation E_{rad}. From these figures one can see that the wave function satisfies the rules satisfied by the wave function inside the sub-barrier and in above-barrier regions (Olkhovsky & Recami, 1992; Olkhovsky et al., 2004; Zakhariev et al., 1990). Starting from very small a, the wave function has oscillations and its maxima increase monotonously with increasing of a. This corresponds to the behavior of the wave function in the internal region before the barrier (this becomes more obvious after essential increasing of scale, see left panel in Fig. 2). Moreover, for larger values of a, the wave

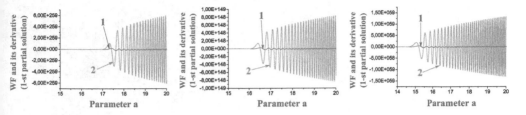

Fig. 1. The first partial solution for the wave function and its derivative at different values of the energy of radiation E_{rad}, increasing in the tunneling region. The blue plot represents the wave function; the green one, the derivative of this wave function): (a) $E_{rad} = 10$; (b) $E_{rad} = 1000$; (c) $E_{rad} = 2000$

function increases monotonously without any oscillation, that points out the transition into the tunneling region (one can see this in a logarithmic presentation of the wave function, see central panel in Fig. 2). A boundary of such a transformation in behavior of the wave function must be the point of penetration of the wave into the barrier, i. e. the internal turning point $a_{tp,in}$. Further, with increasing of a the oscillations appeared in the wave function, which could be possible inside the above barrier region only (in the right panel of Fig. 2 one can see that such a transition is extremely smooth, thing that characterizes the accuracy of the method positively). The boundary of such a transformation in the behavior of the wave function should be the external turning point $a_{tp,out}$. Like Ref. (Maydanyuk, 2008), but at arbitrary non-zero energy E_{rad} we obtain monotonous increasing of maximums of the derivative of the wave function and smooth decreasing of this wave function in the external region. One can see that the derivative is larger than the wave function. At large values of a we obtain the smooth continuous solutions up to $a = 100$ (in Ref. (Acacio de Barros et al., 2007) the maximal presented limit is $a = 30$).

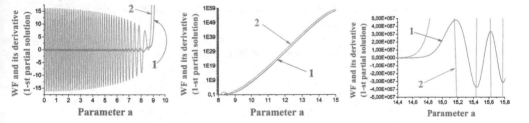

Fig. 2. The first partial solution for the wave function and its derivative at the energy of radiation $E_{rad} = 2000$. The blue line represents the wave function; the green one, the derivative of this wave function)

In Fig. 3, it is presented the second partial solution of the wave function and its derivative at different values of the energy of radiation E_{rad} According to the analysis, this solution close to the turning points, in the tunneling region, in the sub-barrier and above-barrier regions looks like the first partial solution, but with the difference that now the maxima of the wave function and their derivatives are larger essentially in the external region in a comparison with the internal region, and amplitudes in the tunneling region decrease monotonously.

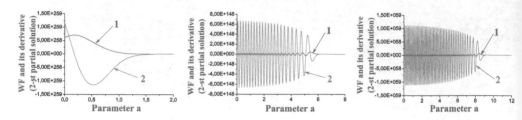

Fig. 3. The second partial solution for the wave function and its derivative at different values of the energy of radiation E_{rad}, decreasing in the tunneling region (the blue line represents the wave function; the green one represents the derivative of this wave function): (a) $E_{rad} = 10$; (b) $E_{rad} = 1000$; (c) $E_{rad} = 2000$

Comparing the previous pictures of the wave function with the results of Ref. (Acacio de Barros et al., 2007), one can see that the wave function, in this approach, is essentially more continuous, has no divergencies and its behavior is everywhere clear. From here we conclude that *the developed method for the determination of the wave function and its derivative at arbitrary energy of radiation is essentially more quick, more stable and accurate in comparison with the non-stationary quantum approach in Ref. (Acacio de Barros et al., 2007)*. Note that:

- With increasing a, the period of the oscillations, both for the wave function and its derivative, decreases uniformly in the external region and increases uniformly in the internal region (this result was partially obtained earlier in Ref. (Maydanyuk, 2008) at $E_{rad} = 0$).

- At larger distance from the barrier (i. e. for increasing values of a, in the external region, and at decreasing value of a, in the internal region) it becomes more difficult to get the convergent continuous solutions for the wave function and its derivative (this result was partially obtained earlier in Ref. (Maydanyuk, 2008) at $E_{rad} = 0$).

- *A number of oscillations of the wave function in the internal region increases with increasing of the energy of radiation E_{rad}* (this is a new result).

4.2 Definition of the wave minimally interacting with the potential

Now we shall be looking for a form of the wave function in the external region, which describes accurately the wave, whose propagation is the closest to the "free" one in the external region at the turning point $a_{tp, out}$ and is directed outside. Let us return back to eq. (16) where the variable $q = a - a_{tp, out}$ has been introduced. Changing this variable to

$$\xi = \left| V_1^{(out)} \right|^{1/3} q, \tag{17}$$

this equation is transformed into

$$\frac{d^2}{d\xi^2} \varphi(\xi) + \xi \, \varphi(\xi) = 0. \tag{18}$$

From quantum mechanics we know two linearly independent exact solutions for the function $\varphi(\xi)$ in this equation — these are the *Airy functions* Ai (ξ) and Bi (ξ) (see Ref. (Abramowitz & Stegan, 1964), p. 264–272, 291–294). Expansions of these functions into power series at small ξ, their asymptotic expansions at large $|\xi|$, their representations through Bessel functions, zeroes and their asymptotic expansions are known. We have some integrals of these functions, and also the form of the Airy functions in the semiclassical approximation (which can be applied at large $|\xi|$). In some problems of the analysis of finite solutions $\varphi(\xi)$ in the whole range of ξ it is convenient to use the integral representations of the Airy functions (see eq. (10.4.32) in Ref. (Abramowitz & Stegan, 1964), p. 265. In eq. (10.4.1) we took into account the sign and $a = 1/3$):

$$\text{Ai}\,(\pm\xi) = \frac{1}{\pi} \int\limits_0^{+\infty} \cos\left(\frac{u^3}{3} \mp \xi u\right)\,du,$$

$$\text{Bi}\,(\pm\xi) = \frac{1}{\pi} \int\limits_0^{+\infty}\left[\exp\left(-\frac{u^3}{3} \mp \xi u\right) + \sin\left(\frac{u^3}{3} \mp \xi u\right)\right]\,du. \tag{19}$$

Furthermore, we shall be interested in the solution $\varphi(\xi)$ which describes the *outgoing wave* in the range of a close to the a_{tp} point. However, it is not clear what the wave represents in general near the point a_{tp}, and which linear combination of the Ai (ξ) and Bi (ξ) functions defines it in the most accurate way.

The clearest and most natural understanding of the outgoing wave is given by the semiclassical consideration of the tunneling process. However, at the given potential the semiclassical approach allows us to define the outgoing wave in the asymptotic region only (while we can join solutions in the proximity of a_{tp} by the Airy functions). But it is not clear whether the wave, defined in the asymptotic region, remains outgoing near the a_{tp}. During the whole path of propagation outside the barrier the wave interacts with the potential, and this must inevitably lead to a deformation of its shape (like to appearance of a phase shift in the scattering of a wave by a radial potential caused by interaction in scattering theory). Does the cosmological potentials deform the wave more than the potentials used for description of nuclear collisions in scattering theory? Moreover, for the given potential there is a problem in obtaining the convergence in the calculation of the partial solutions for the wave function in the asymptotic region. According to our calculations, a small change of the range of the definition of the wave in the asymptotic region leads to a significant increase of errors, which requires one to increase the accuracy of the calculations. Therefore, we shall be looking for a way of defining the outgoing wave not in the asymptotic region, but in the closest vicinity of the point of escape, a_{tp}. In a search of solutions close to the point a_{tp}, i. e. at small enough $|\xi|$, the validity of the semiclassical method breaks down as $|\xi|$ approaches zero. Therefore, we shall not use the semiclassical approach in this paper.

Assuming the potential $V(a)$ to have an arbitrary form, we define the wave at the point a_{tp} in the following way (Maydanyuk, 2010).

Definition 1 (strict definition of the wave). *The wave is a linear combination of two partial solutions of the wave function such that the change of the modulus ρ of this wave function is approximately constant under variation of a:*

$$\left.\frac{d^2}{da^2}\,\rho(a)\right|_{a=a_{tp}} \to 0. \tag{20}$$

According to this definition, the real and imaginary parts of the total wave function have the closest behaviors under the same variation of a, and the difference between possible maximums and minimums of the modulus of the total wave function is the smallest. For some types of potentials (in particular, for a rectangular barrier) it is more convenient to define the wave less strongly.

Definition 2 (weak definition of wave):

The wave is a linear combination of two partial solutions of wave function such that the modulus ρ changes minimally under variation of a:

$$\frac{d}{da}\rho(a)\Big|_{a=a_{tp}} \to 0. \tag{21}$$

According to this definition, the change of the wave function caused by variation of a is characterized mainly by its phase (which can characterize the interaction between the wave and the potential).

Subject to this requirement, we shall look for a solution in the following form:

$$\varphi(\xi) = T \cdot \Psi^{(+)}(\xi), \tag{22}$$

where

$$\Psi^{(\pm)}(\xi) = \int\limits_0^{u_{max}} \exp \pm i\left(-\frac{u^3}{3} + f(\xi)\,u\right) du. \tag{23}$$

where T is an unknown normalization factor, $f(\xi)$ is an unknown continuous function satisfying $f(\xi) \to const$ at $\xi \to 0$, and u_{max} is the unknown upper limit of integration. In such a solution, the real part of the function $f(\xi)$ gives a contribution to the phase of the integrand function, while the imaginary part of $f(\xi)$ deforms its modulus.

Let us find the first and second derivatives of the function $\Psi(\xi)$ (a prime denotes a derivative with respect to ξ):

$$\frac{d}{d\xi}\Psi^{(\pm)}(\xi) = \pm we \int\limits_0^{u_{max}} f'u \, \exp \pm i\left(-\frac{u^3}{3} + f(\xi)u\right) du,$$

$$\frac{d^2}{d\xi^2}\Psi^{(\pm)}(\xi) = \int\limits_0^{u_{max}} \left(\pm if''u - (f')^2u^2\right)\exp \pm i\left(-\frac{u^3}{3} + f(\xi)u\right) du. \tag{24}$$

From this we obtain:

$$\frac{d^2}{d\xi^2}\Psi^{(\pm)}(\xi) + \xi\,\Psi^{(\pm)}(\xi) = \int\limits_0^{u_{max}} \left(\pm if''u - (f')^2u^2 + \xi\right)\exp \pm i\left(-\frac{u^3}{3} + f(\xi)u\right) du. \tag{25}$$

Considering the solutions at small enough values of $|\xi|$, we represent $f(\xi)$ in the form of a power series:

$$f(\xi) = \sum_{n=0}^{+\infty} f_n \xi^n, \tag{26}$$

where f_n are constant coefficients. The first and second derivatives of $f(\xi)$ are

$$f'(\xi) = \frac{d}{d\xi} f(\xi) = \sum_{n=1}^{+\infty} n f_n \, \xi^{n-1} = \sum_{n=0}^{+\infty} (n+1) f_{n+1} \, \xi^n,$$

$$f''(\xi) = \frac{d^2}{d\xi^2} f(\xi) = \sum_{n=0}^{+\infty} (n+1)(n+2) f_{n+2} \, \xi^n. \tag{27}$$

Substituting these solutions into eq. (24), we obtain

$$\frac{d^2}{d\xi^2} \Psi^{(\pm)}(\xi) + \xi \, \Psi^{(\pm)}(\xi) = \int_0^{u_{max}} \left\{ \left(\pm 2iu f_2 - u^2 f_1^2 \right) + \right.$$

$$+ \left(\pm 6iu f_3 - 4u^2 f_1 f_2 + 1 \right) \xi + + \sum_{n=2}^{+\infty} \left[\pm iu \, (n+1)(n+2) f_{n+2} - \right. \tag{28}$$

$$\left. -u^2 \sum_{m=0}^{n} (n-m+1)(m+1) f_{n-m+1} f_{m+1} \right] \xi^n \right\} \exp \pm i \left(-\frac{u^3}{3} + fu \right) du.$$

Considering this expression at small $|\xi|$, we use the following approximation:

$$\exp \pm i \left(\frac{u^3}{3} + fu \right) \rightarrow \exp \pm i \left(-\frac{u^3}{3} + f_0 u \right). \tag{29}$$

Then from eq. (18) we obtain the following condition for the unknown f_n:

$$\int_0^{u_{max}} \left(\pm 2iu f_2 - u^2 f_1^2 \right) \exp \pm i \left(-\frac{u^3}{3} + f_0 u \right) du +$$

$$+ \xi \cdot \int_0^{u_{max}} \left(\pm 6iu f_3 - 4u^2 f_1 f_2 + 1 \right) \exp \pm i \left(-\frac{u^3}{3} + f_0 u \right) du + \tag{30}$$

$$+ \sum_{n=2}^{+\infty} \xi^n \cdot \int_0^{u_{max}} \left[\pm iu \, (n+1)(n+2) f_{n+2} - u^2 \sum_{m=0}^{n} (n-m+1)(m+1) f_{n-m+1} f_{m+1} \right] \times$$

$$\times \exp \pm i \left(-\frac{u^3}{3} + f_0 u \right) du = 0.$$

Requiring that this condition is satisfied for different ξ and with different powers n, we obtain the following system:

$$\xi^0 : \qquad \int_0^{u_{max}} \left(\pm 2iu f_2 - u^2 f_1^2 \right) \exp \pm i \left(-\frac{u^3}{3} + f_0 u \right) du = 0,$$

$$\xi^1 : \qquad \int_0^{u_{max}} \left(\pm 6iu f_3 - 4u^2 f_1 f_2 + 1 \right) \exp \pm i \left(-\frac{u^3}{3} + f_0 u \right) du = 0, \tag{31}$$

$$\xi^n : \int_0^{u_{max}} \left[\pm iu \, (n+1)(n+2) f_{n+2} - u^2 \sum_{m=0}^{n} (n-m+1)(m+1) f_{n-m+1} f_{m+1} \right] \times$$

$$\times \exp \pm i \left(-\frac{u^3}{3} + f_0 u \right) du = 0.$$

Assuming that the coefficients f_0 and f_1 are known, we find the following solutions for the unknown f_2, f_3 and f_n:

$$f_2^{(\pm)} = \pm \frac{f_1^2}{2i} \cdot \frac{J_2^{(\pm)}}{J_1^{(\pm)}}, \; f_3^{(\pm)} = \pm \frac{4f_1 f_2^{(\pm)} J_2^{(\pm)} - J_0^{(\pm)}}{6i \, J_1^{(\pm)}}, \tag{32}$$

$$f_{n+2}^{(\pm)} = \frac{\sum\limits_{m=0}^{n} (n-m+1)(m+1) \, f_{n-m+1}^{(\pm)} f_{m+1}^{(\pm)}}{i(n+1)(n+2)} \cdot \frac{J_2^{(\pm)}}{J_1^{(\pm)}}, \tag{33}$$

where the following notations for the integrals have been introduced:

$$J_0^{(\pm)} = \int\limits_0^{u_{\max}} \exp \pm i \left(-\frac{u^3}{3} + f_0 u \right) du, \; J_1^{(\pm)} = \int\limits_0^{u_{\max}} u \, \exp \pm i \left(-\frac{u^3}{3} + f_0 u \right) du, \tag{34}$$

$$J_2^{(\pm)} = \int\limits_0^{u_{\max}} u^2 \, e^{\pm i \left(-\frac{u^3}{3} + f_0 u \right)} du. \tag{35}$$

Thus, we see that the solution (22) taking into account eq. (23) for the function $\varphi(\xi)$ has arbitrariness in the choice of the unknown coefficients f_0, f_1 and the upper limit of integration, $u_{\max}$. However, the solutions found, eqs. (32), define the function $f(\xi)$ so as to ensure that the equality (22) is exactly satisfied in the region of a close to the escape point a_{tp}. This proves that the function $\varphi(\xi)$ in the form (22), taking into account eq. (23) for an arbitrary choice of f_0, f_1 and $u_{\max}$ is the exact solution of the Schrödinger equation near the escape point a_{tp}. In order to write the solution $\Psi(\xi)$ in terms of the well-known Airy functions, Ai (ξ) and Bi (ξ), we choose

$$f_0 = 0, \; f_1 = 1. \tag{36}$$

For such a choice of the coefficients f_0 and f_1, the integrand function in the solution (23) (up to ξ^2) has a constant modulus and a varying phase. Therefore, one can expect that the solution (22) at the turning point a_{tp} describes the wave accurately.

4.3 Total wave function

Having obtained two linearly independent partial solutions $\varphi_1(a)$ and $\varphi_2(a)$, we can write the general solution (a prime is for the derivative with respect to a) as:

$$\varphi(a) = T \cdot \left(C_1 \, \varphi_1(a) + C_2 \, \varphi_2(a) \right), \tag{37}$$

$$C_1 = \left. \frac{\Psi \varphi_2' - \Psi' \varphi_2}{\varphi_1 \varphi_2' - \varphi_1' \varphi_2} \right|_{a=a_{tp,out}},$$

$$C_2 = \left. \frac{\Psi' \varphi_1 - \Psi \varphi_1'}{\varphi_1 \varphi_2' - \varphi_1' \varphi_2} \right|_{a=a_{tp,out}}, \tag{38}$$

where T is a normalization factor, C_1 and C_2 are complex constants found from the boundary condition introduced above: the $\varphi(a)$ function should represent an outgoing wave at turning point $a_{tp,out}$.

Fig. 4 plots the total wave function calculated in this way for the potential (12) with parameters $A = 36$, $B = 12\Lambda$ at $\Lambda = 0.01$ at different values of the energy of radiation E_{rad}. One can

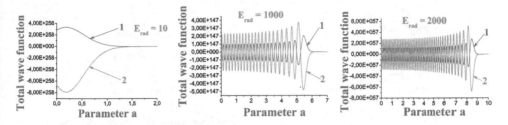

Fig. 4. The wave function at selected values of the energy of radiation E_{rad} (the blue line, represents the real part of the wave function; the green line the imaginary part of the wave function): (a) $E_{rad} = 10$; (b) $E_{rad} = 1000$; (c) $E_{rad} = 2000$

see that the number of oscillations of the wave function in the internal region increases with increasing of the energy of radiation. Another interesting property are *the larger maxima of the wave function in the internal region at smaller distances to the barrier for arbitrary energy* (result found for the first time).

In Fig. 5 it has been shown how the modulus of this wave function changes at selected values of the energy of radiation. From these figures it becomes clear why the coefficient

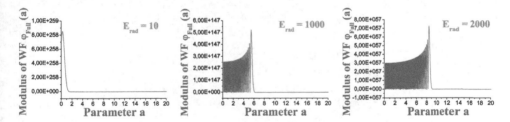

Fig. 5. The behavior of the modulus of the wave function at the selected energies of radiation E_{rad}: (a) $E_{rad} = 10$; (b) $E_{rad} = 1000$; (c) $E_{rad} = 2000$.

of penetrability of the barrier is extremely small (up to the energy $E_{rad} = 2000$). In order to estimate, how effective is the boundary condition introduced above in building up the wave on the basis of the total wave function close to the external turning point $a_{tp,out}$, it is useful to see how the modulus of this wave function changes close to this point. In Fig. 6 we plot the modulus of the found wave function close to the turning points at the energy of radiation $E_{rad} = 2000$ is shown. Here, one can see that the modulus at $a_{tp,out}$ is practically constant (see left panel in Fig. 6). It is interesting to note that the modulus of the wave function, previously defined, does not change close to the internal turning point $a_{tp,in}$, and is close to maximum (see right panel in Fig. 6).

4.4 Calculations of the wave function of Universe
4.4.1 Method of calculations of the wave function close to an arbitrary selected point a_x
Here, we look for the regular partial solution of the wave function close to an arbitrary selected point a_x. Let us write the wave function in the form:

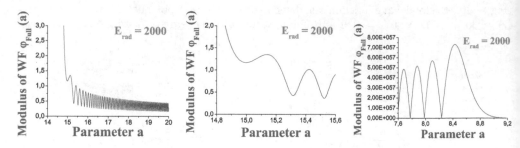

Fig. 6. The behavior of the modulus of the total wave function at the energy of radiation $E_{\text{rad}} = 2000$, close to the turning points (for $a_{\text{tp,in}} = 8.58$, $a_{\text{tp,out}} = 15.04$, see also Table 1): (a) the modulus decreases monotonously in the tunneling region, with increasing of a. It shows maxima and holes connected with the oscillations of the wave function in the external region, but the modulus is not equal to zero (thispoints out the existence of a <u>non-zero</u> flux); (b) when a increases, the modulus reaches a minimum close to the external turning point $a_{\text{tp,out}}$ (this demonstrates the practical fulfillment of the definition for the wave at such a point); (c) transition close to $a_{\text{tp,in}}$ is shown, where at increasing of a the modulus with maximums and holes is transformed rapidly into a monotonously decreasing function without maximums and holes. This is connected with transition to the region of tunneling.

$$\varphi(a) = c_2 \sum_{n=0}^{+\infty} b_n \left(a - a_x\right)^n = c_2 \sum_{n=0}^{+\infty} b_n \, \bar{a}^n, \tag{39}$$

$$\bar{a} = a - a_x$$

and rewrite the potential through the variable $\bar{a}$:

$$V(a) = C_0 + C_1 \, \bar{a} + C_2 \, \bar{a}^2 + C_3 \, \bar{a}^3 + C_4 \, \bar{a}^4, \tag{40}$$

where

$$
\begin{aligned}
C_0 &= A \, a_x^2 - B \, a_x^4, \\
C_1 &= 2a_x(A - B \, a_x^2) - 2B \, a_x^3 = 2A \, a_x - 4B \, a_x^3, \\
C_2 &= A - B \, a_x^2 - 4B \, a_x^2 - B \, a_x^2 = A - 6B \, a_x^2, \\
C_3 &= -2B \, a_x - 2B \, a_x = -4B \, a_x, \\
C_4 &= -B.
\end{aligned}
\tag{41}
$$

Substituting the wave function (39), its second derivative and the potential (40) into Schrödinger equation, we obtain recurrent relations for unknown b_n:

$$b_2 = \frac{(C_0 - E) \, b_0}{2}, \quad b_3 = \frac{(C_0 - E) \, b_1 + C_1 \, b_0}{6}, \quad b_4 = \frac{(C_0 - E) \, b_2 + C_1 \, b_1 + C_2 \, b_0}{12}, \tag{42}$$

$$b_5 = \frac{(C_0 - E) \, b_3 + C_1 \, b_2 + C_2 \, b_1 + C_3 \, b_0}{20}, \tag{43}$$

$$b_{n+2} = \frac{(C_0 - E) \, b_n + C_1 \, b_{n-1} + C_2 \, b_{n-2} + C_3 \, b_{n-3} + C_4 \, b_{n-4}}{(n+1)(n+2)} \quad \text{at } n \geq 4. \tag{44}$$

Given the values of b_0 and b_1 and using eqs. (42)–(44) one can calculate all b_n needed. At limit $E_{rad} \to 0$ and at $a_x = 0$ all found solutions for b_i transform into the corresponding solutions (40), early obtained in (Maydanyuk, 2008) at $E_{rad} = 0$. Using $c_2 = 1$, from eqs. (39) we find:

$$b_0 = \varphi(a_x), \; b_1 = \varphi'(a_x). \tag{45}$$

So, on the basis of the coefficients b_0 and b_1 one can obtain the values of the wave function and its derivative at point a_x. Imposing two different boundary conditions via b_0 and b_1, we obtain two linearly independent partial solutions $\varphi_1(a)$ and $\varphi_2(a)$ for the wave function. Using the internal turning point $a_{tp,in}$ as the starting point, we calculate the first partial solution which increases in the barrier region (we choose: $b_0 = 0.1$, $b_1 = 1$), and using the external turning point $a_{tp,out}$ as the starting point, we calculate the second partial solution which decreases in the barrier region (we choose: $b_0 = 1$, $b_1 = -0.1$). Such a choice provides effectively a linear independence between two partial solutions.

4.4.2 Method of continuation of the solution

Let us rewrite equation (18) in such a form[2]:

$$\varphi''(a) = f(a)\,\varphi(a). \tag{46}$$

Let $\{a_n\}$ be a set of equidistant points $a_n = a_0 + nh$. Denoting the values of the wave function $\varphi(a)$ at points a_n as φ_n, we have constructed an algorithm of the ninth order to determine φ_{n+1} and φ'_n when φ_n and φ_{n-1} are known:

$$
\begin{aligned}
\varphi_{n+1} &= \varphi_{n-1}\frac{g_{11}+g_{01}}{g_{01}-g_{11}} + \varphi_n \frac{g_{01}g_{10}-g_{00}g_{11}}{g_{01}-g_{11}} + O(h^9), \\
\varphi'_n &= \varphi_{n-1}\frac{2}{g_{01}-g_{11}} + \varphi_n \frac{g_{10}-g_{00}}{g_{01}-g_{11}} + O(h^9),
\end{aligned}
\tag{47}
$$

where

$$
\begin{aligned}
g_{00} &= 2 + h^2 f_n + \frac{2}{4!} h^4 \left(f''_n + f_n^2\right) + \frac{2}{6!} h^6 \left(f_n^{(4)} + 4\left(f'_n\right)^2 + 7 f_n f''_n + f_n^3\right) + \\
&\quad + \frac{2}{8!} h^8 \left(f_n^{(6)} + 16 f_n f_n^{(4)} + 26 f'_n f_n^{(3)} + 15 \left(f''_n\right)^2 + 22 f_n^2 f''_n + 28 f_n \left(f'_n\right)^2 + f_n^4\right), \\
g_{01} &= \frac{2}{4!} h^4 2 f'_n + \frac{2}{6!} h^6 \left(4 f_n^{(3)} + 6 f_n f'_n\right) + \frac{2}{8!} h^8 \left(6 f_n^{(5)} + 24 f_n f_n^{(3)} + 48 f'_n f''_n + 12 f_n^2 f'_n\right), \\
g_{10} &= \frac{2}{3!} h^3 f'_n + \frac{2}{5!} h^5 \left(f_n^{(3)} + 4 f_n f'_n\right) + \frac{2}{7!} h^7 \left(f_n^{(5)} + 11 f_n f_n^{(3)} + 15 f'_n f''_n + 9 f_n^2 f'_n\right), \\
g_{11} &= 2h + \frac{2}{3!} h^3 f_n + \frac{2}{5!} h^5 \left(3 f''_n + f_n^2\right) + \frac{2}{7!} h^7 \left(5 f_n^{(4)} + 13 f_n f''_n + 10 \left(f'_n\right)^2 + f_n^3\right).
\end{aligned}
\tag{48}
$$

A local error of these formulas at point a_n equals to:

$$\delta_n = \frac{1}{10!} h^{10} f'_n \varphi_n^{(7)}. \tag{49}$$

---

[2] Here, we used the algorithm of (Zaichenko & Kashuba, 2001)

4.5 The penetrability and reflection in the fully quantum approach

Let us analyze whether a known wave function in the whole region of its definition allows us to determine uniquely the coefficients of penetrability and reflection.

4.5.1 Problem of interference between the incident and reflected waves

Rewriting the wave function φ_{total} in the internal region through a summation of incident φ_{inc} wave and reflected φ_{ref} wave:

$$\varphi_{\text{total}} = \varphi_{\text{inc}} + \varphi_{\text{ref}}, \tag{50}$$

we consider the total flux:

$$j\left(\varphi_{\text{total}}\right) = i\left[\left(\varphi_{\text{inc}} + \varphi_{\text{ref}}\right)\nabla\left(\varphi^*_{\text{inc}} + \varphi^*_{\text{ref}}\right) - \text{h. c.}\right] = j_{\text{inc}} + j_{\text{ref}} + j_{\text{mixed}}, \tag{51}$$

where

$$j_{\text{inc}} = i\left(\varphi_{\text{inc}}\nabla\varphi^*_{\text{inc}} - \text{h. c.}\right),$$

$$j_{\text{ref}} = i\left(\varphi_{\text{ref}}\nabla\varphi^*_{\text{ref}} - \text{h. c.}\right), \tag{52}$$

$$j_{\text{mixed}} = i\left(\varphi_{\text{inc}}\nabla\varphi^*_{\text{ref}} + \varphi_{\text{ref}}\nabla\varphi^*_{\text{inc}} - \text{h. c.}\right).$$

The j_{mixed} component describes interference between the incident and reflected waves in the internal region (let us call it *mixed component of the total flux* or simply *flux of mixing*). From the constancy of the total flux j_{total} we find the flux j_{tr} for the wave transmitted through the barrier, and:

$$j_{\text{inc}} = j_{\text{tr}} - j_{\text{ref}} - j_{\text{mixed}}, \; j_{\text{tr}} = j_{\text{total}} = \text{const.} \tag{53}$$

Now one can see that *the mixed flux introduces ambiguity in the determination of the penetrability and reflection for the same known wave function.*

4.6 Determination of the penetrability, reflection and interference coefficients

In quantum mechanics the coefficients of penetrability and reflection are defined considering the potential as a whole, including asymptotic regions. However, in the radial calculation of quantum decay such a consideration depends on how the incident and reflected waves are defined inside finite internal region from the left of the barrier. The question is: does the location of such a region influence the penetrability and reflection? In order to obtain these coefficients, we shall include into definitions coordinates where the fluxes are defined (denote them as x_{left} and x_{right}):

$$T(x_{\text{left}}, x_{\text{right}}) = \frac{j_{\text{tr}}(x_{\text{right}})}{j_{\text{inc}}(x_{\text{left}})},$$

$$R(x_{\text{left}}) = \frac{j_{\text{ref}}(x_{\text{left}})}{j_{\text{inc}}(x_{\text{left}})}, \tag{54}$$

$$M(x_{\text{left}}) = \frac{j_{\text{mixed}}(x_{\text{left}})}{j_{\text{inc}}(x_{\text{left}})}.$$

So, the T and R coefficients determine the probability of transmission (or tunneling) and reflection of the wave relatively the region of the potential with arbitrary selected boundaries x_{left}, x_{right}. When x_{right} tends to the asymptotic limit, the coefficient defined before should transform into standard ones. Assuming that j_{tr} and j_{ref} are directed in opposite directions,

j_{inc} and j_{tr} — in the same directions, from eqs. (53) and (54) we obtain (Maydanyuk, 2010):

$$|T| + |R| - M = 1. \tag{55}$$

Now we see that the condition $|T| + |R| = 1$ *has sense in quantum mechanics only if there is no interference between incident and reflected waves*, and for this is enough that:

$$j_{mixed} = 0. \tag{56}$$

A new question appears: *does this condition allow to separate the total wave function into the incident and reflected components in a unique way?* It turns out that the choice of the incident and reflected waves has essential influence on the barrier penetrability, and different forms of the incident φ_{inc} and reflected φ_{ref} waves can give zero flux j_{mix}. Going from the rectangular internal well to the fully quantum treatment of the problem would become more complicated.

4.7 Wave incident on the barrier and wave reflected from it in the internal region

One can define the incident wave to be proportional to the function $\Psi^{(+)}$ and the reflected wave to be proportional to the function $\Psi^{(-)}$:

$$\varphi_{total}(a) = \varphi_{inc}(a) + \varphi_{ref}(a),$$
$$\varphi_{inc}(a) = we \cdot \Psi^{(+)}(a), \tag{57}$$
$$\varphi_{ref}(a) = R \cdot \Psi^{(-)}(a),$$

where I and R are new constants found from continuity condition of the total wave function φ_{total} and its derivative at the internal turning point $a_{tp,int}$:

$$we = \left. \frac{\varphi_{total} \Psi^{(-),\prime} - \varphi'_{total} \Psi^{(-)}}{\Psi^{(+)} \Psi^{(-),\prime} - \Psi^{(+),\prime} \Psi^{(-)}} \right|_{a = a_{tp,int}},$$
$$R = \left. \frac{\varphi'_{total} \Psi^{(+)} - \varphi_{total} \Psi^{(+),\prime}}{\Psi^{(+)} \Psi^{(-),\prime} - \Psi^{(+),\prime} \Psi^{(-)}} \right|_{a = a_{tp,int}}. \tag{58}$$

On the basis of these solutions we obtain at the internal turning point $a_{tp,int}$ the flux incident on the barrier, the flux reflected from it and the flux of mixing. The flux transmitted through the barrier was calculated at the external turning point $a_{tp,ext}$.

4.8 Penetrability and reflection: fully quantum approach versus semiclassical one

Now we shall estimate through the method described above the coefficients of penetrability and reflection for the potential barrier with parameters $A = 36$, $B = 12\,\Lambda$, $\Lambda = 0.01$ at different values of the energy of radiation E_{rad}. We shall compare the coefficient of penetrability obtained with the values given by the semiclassical method. In the semiclassical approach we shall consider two definitions of this coefficient:

$$P_{penetrability}^{WKB,(1)} = \frac{1}{\theta^2}, \quad P_{penetrability}^{WKB,(2)} = \frac{4}{\left(2\theta + 1/(2\theta)^2\right)^2}, \tag{59}$$

where

$$\theta = \exp \int_{a_{tp}^{(int)}}^{a_{tp}^{(ext)}} |V(a) - E| \, da. \tag{60}$$

One can estimate also *the duration of the formation of the Universe*, using by definition (15) in Ref. (Acacio de Barros et al., 2007):

$$\tau = 2 \, a_{tp, int} \frac{1}{P_{penetrability}}. \tag{61}$$

The results are presented in Tabl. 1. In calculations the coefficients of penetrability, reflection and mixing are defined by eqs. (54), the fluxes by eqs. (52) (calculated $P_{penetrability}^{WKB,(2)}$ coincide with $P_{penetrability}^{WKB,(1)}$ up to the first 7 digits for energies in range $0 \leq E_{rad} \leq 2500$).

From this table one can see that inside the entire range of energy, the fully quantum approach gives value for the coefficient of penetrability enough close to its value obtained by the semiclassical approach. This differs essentially from results in the non-stationary approach (Acacio de Barros et al., 2007). This difference could be explained by difference in a choice of the boundary condition, which is used in construction of the stationary solution of the wave function.

4.9 The penetrability in the FRW-model with the Chaplygin gas

In order to connect universe with dust and its accelerating stage, in Ref. (Kamenshchik et al., 2001) a new scenario with the *Chaplygin gas* was proposed. A quantum FRW-model with the Chaplygin gas has been constructed on the basis of equation of state instead of $p(a) = \rho_{rad}(a)/3$ (where $p(a)$ is pressure) by the following (see also Refs. (Bento et al., 2002; Bilic et al., 2002)):

$$p_{Ch} = -\frac{A}{\rho_{Ch}^{\alpha}}, \tag{62}$$

where A is positive constant and $0 < \alpha \leq 1$. In particular, for the standard Chaplygin gas we have $\alpha = 1$. Solution of equation of state (62) gives the following dependence of density on the scale factor:

$$\rho_{Ch}(a) = \left(A + \frac{B}{a^{3(1+\alpha)}} \right)^{1/(1+\alpha)}, \tag{63}$$

where B is a new constant of integration. Using the parameter α, this model describes transition between the stage, when Universe is filled with dust-like matter, and its accelerating expanding stage (through scenario of Chaplygin gas applied to cosmology, for details, see Refs. (Bouhmadi-Lopez & Moniz, 2005; Bouhmadi-Lopez et al., 2008; Kamenshchik et al., 2001), also historical paper (Chaplygin, 1904)).

Let us combine expression for density which includes previous forms of matter and the Chaplygin gas in addition. At limit $\alpha \to 0$ eq. (63) transforms into the ρ_{dust} component plus the ρ_{Λ} component. From such limit we find

$$A = \rho_{\Lambda}, \ B = \rho_{dust} \tag{64}$$

Energy	Penetrability $P_{\text{penetrability}}$		Time τ		Turning points	
E_{rad}	Direct method	Method WKB	Direct method	Method WKB	$a_{\text{tp, in}}$	$a_{\text{tp, out}}$
1.0	8.7126×10^{-521}	2.0888×10^{-521}	$3.8260 \times 10^{+519}$	$1.5958 \times 10^{+520}$	0.16	17.31
2.0	2.4225×10^{-520}	5.5173×10^{-521}	$1.9460 \times 10^{+519}$	$8.5448 \times 10^{+519}$	0.23	17.31
3.0	6.2857×10^{-520}	1.3972×10^{-520}	$9.1863 \times 10^{+518}$	$4.1326 \times 10^{+519}$	0.28	17.31
4.0	1.5800×10^{-519}	3.4428×10^{-520}	$4.2201 \times 10^{+518}$	$1.9367 \times 10^{+519}$	0.33	17.31
5.0	3.8444×10^{-519}	8.2935×10^{-520}	$1.9392 \times 10^{+518}$	$8.9892 \times 10^{+518}$	0.37	17.31
6.0	9.2441×10^{-519}	1.9701×10^{-519}	$8.8350 \times 10^{+517}$	$4.1455 \times 10^{+518}$	0.40	17.31
7.0	2.1678×10^{-518}	4.5987×10^{-519}	$4.0694 \times 10^{+517}$	$1.9183 \times 10^{+518}$	0.44	17.31
8.0	5.0192×10^{-518}	1.0621×10^{-518}	$1.8790 \times 10^{+517}$	$8.8797 \times 10^{+517}$	0.47	17.31
9.0	1.1604×10^{-517}	2.4316×10^{-518}	$8.6212 \times 10^{+516}$	$4.1140 \times 10^{+517}$	0.50	17.31
10.0	2.6279×10^{-517}	5.5016×10^{-518}	$4.0128 \times 10^{+516}$	$1.9168 \times 10^{+517}$	0.52	17.31
100.0	1.6165×10^{-490}	3.1959×10^{-491}	$2.0717 \times 10^{+490}$	$1.0478 \times 10^{+491}$	1.67	17.23
200.0	8.5909×10^{-465}	1.6936×10^{-465}	$5.5397 \times 10^{+464}$	$2.8100 \times 10^{+465}$	2.37	17.15
300.0	6.8543×10^{-441}	1.3419×10^{-441}	$8.5461 \times 10^{+440}$	$4.3653 \times 10^{+441}$	2.92	17.07
400.0	3.6688×10^{-418}	7.1642×10^{-419}	$1.8531 \times 10^{+418}$	$9.4900 \times 10^{+418}$	3.39	16.98
500.0	2.6805×10^{-396}	5.2521×10^{-397}	$2.8508 \times 10^{+396}$	$1.4550 \times 10^{+397}$	3.82	16.89
600.0	4.1386×10^{-375}	8.0511×10^{-376}	$2.0338 \times 10^{+375}$	$1.0454 \times 10^{+376}$	4.20	16.80
700.0	1.7314×10^{-354}	3.3810×10^{-355}	$5.2806 \times 10^{+354}$	$2.7043 \times 10^{+355}$	4.57	16.70
800.0	2.4308×10^{-334}	4.7497×10^{-335}	$4.0448 \times 10^{+334}$	$2.0701 \times 10^{+335}$	4.91	16.60
900.0	1.3213×10^{-314}	2.5761×10^{-315}	$7.9408 \times 10^{+314}$	$4.0730 \times 10^{+315}$	5.24	16.50
1000.0	3.0920×10^{-295}	6.0272×10^{-296}	$3.5999 \times 10^{+295}$	$1.8468 \times 10^{+296}$	5.56	16.40
1100.0	3.4274×10^{-276}	6.6576×10^{-277}	$3.4289 \times 10^{+276}$	$1.7652 \times 10^{+277}$	5.87	16.29
1200.0	1.9147×10^{-257}	3.7259×10^{-258}	$6.4553 \times 10^{+257}$	$3.3174 \times 10^{+258}$	6.18	16.18
1300.0	5.8026×10^{-239}	1.1253×10^{-239}	$2.2333 \times 10^{+239}$	$1.1516 \times 10^{+240}$	6.47	16.06
1400.0	9.9042×10^{-221}	1.9252×10^{-221}	$1.3683 \times 10^{+221}$	$7.0393 \times 10^{+221}$	6.77	15.93
1500.0	1.0126×10^{-202}	1.9551×10^{-203}	$1.3965 \times 10^{+203}$	$7.2333 \times 10^{+203}$	7.07	15.81
1600.0	6.2741×10^{-185}	1.2155×10^{-185}	$2.3480 \times 10^{+185}$	$1.2119 \times 10^{+186}$	7.36	15.67
1700.0	2.4923×10^{-167}	4.8143×10^{-168}	$6.1488 \times 10^{+167}$	$3.1831 \times 10^{+168}$	7.66	15.53
1800.0	6.4255×10^{-150}	1.2437×10^{-150}	$2.4783 \times 10^{+150}$	$1.2803 \times 10^{+151}$	7.96	15.38
1900.0	1.1189×10^{-132}	2.1580×10^{-133}	$1.4776 \times 10^{+133}$	$7.6619 \times 10^{+133}$	8.26	15.22
2000.0	1.3288×10^{-115}	2.5653×10^{-116}	$1.2914 \times 10^{+116}$	$6.6895 \times 10^{+116}$	8.58	15.04
2100.0	1.1105×10^{-98}	2.1357×10^{-99}	$1.6036 \times 10^{+99}$	$8.3382 \times 10^{+99}$	8.90	14.85
2200.0	6.6054×10^{-82}	1.2690×10^{-82}	$2.7988 \times 10^{+82}$	$1.4567 \times 10^{+83}$	9.24	14.64
2300.0	2.8693×10^{-65}	5.4647×10^{-66}	$6.6952 \times 10^{+65}$	$3.5154 \times 10^{+66}$	9.60	14.41
2400.0	9.1077×10^{-49}	1.7297×10^{-49}	$2.1959 \times 10^{+49}$	$1.1562 \times 10^{+50}$	10.00	14.14
2500.0	2.1702×10^{-32}	4.0896×10^{-33}	$9.6290 \times 10^{+32}$	$5.1098 \times 10^{+33}$	10.44	13.81
2600.0	3.9788×10^{-16}	7.3137×10^{-17}	$5.5322 \times 10^{+16}$	$3.0096 \times 10^{+17}$	11.00	13.37
2610.0	1.6663×10^{-14}	3.0428×10^{-15}	$1.3290 \times 10^{+15}$	$7.2780 \times 10^{+15}$	11.07	13.31
2620.0	6.9240×10^{-13}	1.2606×10^{-13}	$3.2187 \times 10^{+13}$	$1.7678 \times 10^{+14}$	11.14	13.25
2630.0	2.8842×10^{-11}	5.2116×10^{-12}	$7.7789 \times 10^{+11}$	$4.3050 \times 10^{+12}$	11.21	13.19
2640.0	1.2002×10^{-9}	2.1495×10^{-10}	$1.8825 \times 10^{+10}$	$1.0511 \times 10^{+11}$	11.29	13.12
2650.0	4.9881×10^{-8}	8.8401×10^{-9}	$4.5642 \times 10^{+8}$	$2.5754 \times 10^{+9}$	11.38	13.05
2660.0	2.0738×10^{-6}	3.6263×10^{-7}	$1.1068 \times 10^{+7}$	$6.3303 \times 10^{+7}$	11.47	12.97
2670.0	8.7110×10^{-5}	1.4836×10^{-5}	$2.6596 \times 10^{+5}$	$1.5615 \times 10^{+6}$	11.58	12.87
2680.0	3.6953×10^{-3}	6.0519×10^{-4}	$6.3369 \times 10^{+3}$	$3.8693 \times 10^{+4}$	11.70	12.76
2690.0	1.5521×10^{-1}	2.4634×10^{-2}	$1.5293 \times 10^{+2}$	$9.3602 \times 10^{+2}$	11.86	12.61

Table 1. The penetrability $P_{\text{penetrability}}$ of the barrier and the duration τ of the formation of the Universe defined by eq. (61) in the fully quantum and semiclassical approaches

and obtain the following generalized density:

$$\rho\left(a\right) = \left(\rho_\Lambda + \frac{\rho_{\text{dust}}}{a^{3(1+\alpha)}}\right)^{1/(1+\alpha)} + \frac{\rho_{\text{rad}}}{a^4(t)}. \tag{65}$$

Now we have:

$$\dot{a}^2 + k - \frac{8\pi G}{3}\left\{a^2\left(\rho_\Lambda + \frac{\rho_{\text{dust}}}{a^{3(1+\alpha)}}\right)^{1/(1+\alpha)} + \frac{\rho_{\text{rad}}}{a^2(t)}\right\} = 0. \tag{66}$$

After quantization we obtain the Wheeler-De Witt equation

$$\left\{-\frac{\partial^2}{\partial a^2} + V_{\text{Ch}}\left(a\right)\right\}\varphi(a) = E_{\text{rad}}\,\varphi(a),\; E_{\text{rad}} = \frac{3\,\rho_{\text{rad}}}{2\pi G}, \tag{67}$$

where

$$V_{\text{Ch}}\left(a\right) = \left(\frac{3}{4\pi G}\right)^2 k\,a^2 - \frac{3}{2\pi G}\,a^4\left(\rho_\Lambda + \frac{\rho_{\text{dust}}}{a^{3(1+\alpha)}}\right)^{1/(1+\alpha)}. \tag{68}$$

For the Universe of closed type (at $k = 1$) at $8\pi G \equiv M_p^{-2} = 1$ we have (see eqs. (6)–(7) in Ref. (Bouhmadi-Lopez & Moniz, 2005)):

$$V_{\text{Ch}}\left(a\right) = 36\,a^2 - 12\,a^4\left(\Lambda + \frac{\rho_{\text{dust}}}{a^{3(1+\alpha)}}\right)^{1/(1+\alpha)},\; E_{\text{rad}} = 12\,\rho_{\text{rad}}. \tag{69}$$

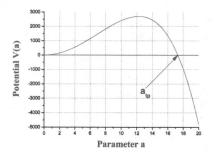 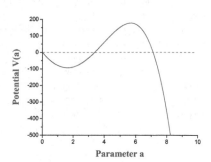

Fig. 7. Cosmological potentials with and without Chaplygin gas: Left panel is for potential $V(a) = 36\,a^2 - 12\,\Lambda\,a^4$ with parameter $\Lambda = 0.01$ (turning point $a_{tp} = 17.320508$ at zero energy $E_{\text{rad}} = 0$), Right panel is for potential (69) with parameters $\Lambda = 0.01$, $\rho_{\text{dust}} = 30$, $\alpha = 0.5$ (minimum of the hole is -93.579 and its coordinate is 1.6262, maximum of the barrier is 177.99 and its coordinate is 5.6866).

Let us expand the potential (69) close to arbitrary selected point $\bar{a}$ by powers of $q = a - \bar{a}$ and restrict ourselves to linear terms:

$$V_{\text{Ch}}\left(q\right) = V_0 + V_1 q. \tag{70}$$

For coefficients V_0 and V_1 we find:

$$V_0 = V_{Ch}(a = \bar{a}),$$

$$V_1 = \left.\frac{dV_{Ch}(a)}{da}\right|_{a=\bar{a}} = 72\,a + 12\,a^3 \left\{-4\Lambda - \frac{\rho_{dust}}{a^3\,(1+\alpha)}\right\} \cdot \left(\Lambda + \frac{\rho_{dust}}{a^3\,(1+\alpha)}\right)^{-\alpha/(1+\alpha)} \tag{71}$$

and eq. (67) has the form:

$$-\frac{d^2}{dq^2}\,\varphi(q) + (V_0 - E_{rad} + V_1\,q)\,\varphi(q) = 0. \tag{72}$$

After the change of variable

$$\zeta = |V_1|^{1/3}\,q, \quad \frac{d^2}{dq^2} = \left(\frac{d\zeta}{dq}\right)^2 \frac{d^2}{d\zeta^2} = |V_1|^{2/3}\,\frac{d^2}{d\zeta^2} \tag{73}$$

eq. (72) becomes:

$$\frac{d^2}{d\zeta^2}\,\varphi(\zeta) + \left\{\frac{E_{rad} - V_0}{|V_1|^{2/3}} - \frac{V_1}{|V_1|}\,\zeta\right\}\,\varphi(\zeta) = 0. \tag{74}$$

After the new change

$$\xi = \frac{E_{rad} - V_0}{|V_1|^{2/3}} - \frac{V_1}{|V_1|}\,\zeta \tag{75}$$

we have

$$\frac{d^2}{d\xi^2}\,\varphi(\xi) + \xi\,\varphi(\xi) = 0. \tag{76}$$

From eqs. (73) and (75) we have:

$$\xi = \frac{E_{rad} - V_0}{|V_1|^{2/3}} - \frac{V_1}{|V_1|^{2/3}}\,q. \tag{77}$$

Using such corrections after inclusion of the density component of the Chaplygin gas, we have calculated the wave function and on its basis the coefficients of penetrability, reflection and mixing by the formalism presented above. Now following the method of Sec. 3.1, we have defined the incident and reflected waves relatively to a new boundary which is located in the minimum of the hole in the internal region. Results are presented in Tabl. 3. One can see that penetrability changes up to 100 times, in such a coordinate, in dependence on the location of the boundary or in the internal turning point (for the same barrier shape and energy E_{rad})! This confirms that the coordinate where incident and reflected waves are defined has essential influence on estimation of the coefficients of penetrability and reflection. This result shows that the method proposed in the present paper has physical sense. In the next Tabl. 4, we demonstrate the fulfillment of the property (55) inside the entire energy range, which is calculated on the basis of the coefficients of penetrability, reflection and mixing obtained before.

5. Multiple internal reflections fully quantum method

5.1 Passage to non-stationary WDW equation: motivations

Tunneling is a pure quantum phenomenon characterized by the fact that a particle crosses through a classically-forbidden region of the barrier. By such a reason, the process of incidence of the particle on the barrier and its further tunneling and reflection are connected by unite cause-effect relation. So, the dynamical consideration of the tunneling process through cosmological barriers is a natural one (Aharonov, 2002; Esposito, 2003; Jakiel et al., 1999; Olkhovsky & Recami, 1992; Olkhovsky et al., 1995; 2004; 2005; Olkhovsky & Recami, 2008; Olkhovsky, 2011; Recami, 2004). The rejection of the dynamical consideration of tunneling from quantum cosmology limits the possible connection between initial stage, when the wave is incident on the barrier, and next propagation of this wave. This leads to uncertainties in determination of penetrability and rates. According to quantum mechanics, a particle is a quantum object having properties both particle and wave. In the classically forbidden regions the wave properties of the studied object are evident. So, the wave description of tunneling is natural.

So, we define a non-stationary generalization of WDW equation as

$$\left(\frac{\partial^2}{\partial a^2} - V_{\text{eff}}(a) \right) \Psi(a, \tau) = -i \frac{\partial}{\partial \tau} \Psi(a, \tau), \tag{78}$$

where τ is a new variable describing dynamics of evolution of the wave function being analog of time. According to quantum mechanics, the penetrability and reflection are stationary characteristics, and such characteristics, obtained in the following, are independent on the parameter τ. Note that all these characteristics are solutions of stationary WDW equation, while non-stationary consideration of multiple packets moving along barrier gives clear understanding of the process.

In order to give a basis to readers to estimate ability of the approach developed in this paper, let us consider results in (Monerat et al., 2007) (see eq. (19)). Here was studied the non-stationary WDW equation

$$\left(\frac{1}{12} \frac{\partial^2}{\partial a^2} - V_{\text{eff}}(a) \right) \Psi(a, \tau) = -i \frac{\partial}{\partial \tau} \Psi(a, \tau) \tag{79}$$

with the potential for the closed FRW model with the included generalized Chaplygin gas.

$$V_{\text{eff}}(a) = 3 a^2 - \frac{a^4}{\pi} \sqrt{\bar{A} + \frac{\bar{B}}{a^6}} \tag{80}$$

After change of variable $a_{\text{new}} = a_{\text{old}} \sqrt{12}$ the non-stationary eq. (79) transforms into our eq. (78) since the V_{eff} potential is independent on the τ variable (such a choice allows a correspondence between energy levels, convenient in comparative analysis). The potential (79) after such a transformation is shown in figs. 8. We shall analyze the behavior of the wave function.

5.2 Tunneling of the packet through a barrier composed from arbitrary number of rectangular steps

Now let us come to another more difficult problem, namely that a packet penetrating through the radial barrier of arbitrary shape in a cosmological problem. In order to apply the idea of multiple internal refections for study the packet tunneling through the real barrier, we have to generalize the formalism of the multiple internal reflections presented above (Maydanyuk, 2011). We shall assume that the total potential has successfully been approximated by finite

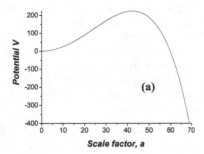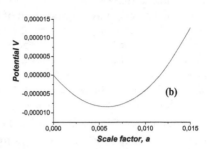

Fig. 8. Behavior of the potential (80) after change $a_{\text{new}} = a_{\text{old}} \sqrt{12}$ at $\bar{A} = 0.001$ and $\bar{B} = 0.001$ (choice of parameters see in fig. 1, tables I and II in (Monerat et al., 2007)): (a) shape of the barrier ($V_{\text{max}} = 223.52$ at $a = 42.322$); (b) there is a little internal well close to zero ($V_{\text{min}} = -8.44$ at $a = 0.00581$)

number N of rectangular steps:

$$
V(a) = \begin{cases}
V_1, \text{ at } a_{\text{min}} < a \leq a_1 & \text{(region 1),} \\
V_2, \text{ at } a_1 < a \leq a_2 & \text{(region 2),} \\
\ldots \ldots & \ldots \\
V_N, \text{ at } a_{N-1} < a \leq a_{\text{max}} & \text{(region } N\text{),}
\end{cases}
\tag{81}
$$

where V_i are constants ($i = 1 \ldots N$). Let us assume that the packet starts to propagate outside inside the region with some arbitrary number M (for simplicity, we denote its left boundary a_{M-1} as a_{start}) from the left of the barrier. We are interested in solutions for energies above that of the barrier while the solution for tunneling could be obtained after by change $i\xi_i \rightarrow k_i$. A general solution of the wave function (up to its normalization) has the following form:

$$
\varphi(a) = \begin{cases}
\alpha_1 e^{ik_1 a} + \beta_1 e^{-ik_1 a}, \\
\quad \text{at } a_{\text{min}} \leq a \leq a_1 \quad \text{(region 1),} \\
\ldots \\
\alpha_{M-1} e^{ik_{M-1} a} + \beta_{M-1} e^{-ik_{M-1} a}, \\
\quad \text{at } a_{M-2} \leq a \leq a_{M-1} \quad \text{(region } M-1\text{),} \\
e^{ik_M a} + A_R e^{-ik_M a}, \\
\quad \text{at } a_{M-1} < a \leq a_M \quad \text{(region } M\text{),} \\
\alpha_{M+1} e^{ik_{M+1} a} + \beta_{M+1} e^{-ik_{M+1} a}, \\
\quad \text{at } a_M \leq a \leq a_{M+1} \quad \text{(region } M+1\text{),} \\
\ldots \\
\alpha_{n-1} e^{ik_{N-1} a} + \beta_{N-1} e^{-ik_{N-1} a}, \\
\quad \text{at } a_{N-2} \leq a \leq a_{N-1} \quad \text{(region } N-1\text{),} \\
A_T e^{ik_N a}, \text{ at } a_{N-1} \leq a \leq a_{\text{max}} \quad \text{(region } N\text{),}
\end{cases}
\tag{82}
$$

where α_j and β_j are unknown amplitudes, A_T and A_R are unknown amplitudes of transmission and reflection, $k_i = \frac{1}{\hbar}\sqrt{2m(E - V_i)}$ are complex wave numbers. We have fixed the normalization so that the modulus of the starting wave $e^{ik_M a}$ equals to one. We look for a solution of such a problem by the approach of the multiple internal reflections.

Let us consider the initial stage when the packet starts to propagate to the right in the region with number M. According to the method of the multiple internal reflections, propagation of the packet through the barrier is considered by steps of its propagation relatively to each boundary (see (Cardone et al., 2006; Maydanyuk et al., 2002a; Maydanyuk, 2003; Maydanyuk & Belchikov, 2011), for details). Each next step in such a consideration of propagation of the packet will be similar to the first $2N - 1$ steps. From analysis of these steps recurrent relations are found for calculation of all unknown amplitudes $A_T^{(n)}$, $A_R^{(n)}$, $\alpha_j^{(n)}$ and $\beta_j^{(n)}$ for arbitrary step n (for region with number j), summation of these amplitudes are calculated. We shall look for the unknown amplitudes, requiring the wave function and its derivative to be continuous at each boundary. We shall consider the coefficients $T_1^{\pm}$, $T_2^{\pm}$ ... and $R_1^{\pm}$, $R_2^{\pm}$ ... as additional factors to amplitudes $e^{\pm ika}$. Here, the bottom index denotes the number of the region, upper (top) signs "+" and "−" denote directions of the wave to the right or to the left, correspondingly. To begin with, we calculate $T_1^{\pm}$, $T_2^{\pm}$ ... $T_{N-1}^{\pm}$ and $R_1^{\pm}$, $R_2^{\pm}$ ... $R_{N-1}^{\pm}$:

$$
T_j^+ = \frac{2k_j}{k_j + k_{j+1}} e^{i(k_j - k_{j+1})a_j}, \quad T_j^- = \frac{2k_{j+1}}{k_j + k_{j+1}} e^{i(k_j - k_{j+1})a_j},
$$

$$
R_j^+ = \frac{k_j - k_{j+1}}{k_j + k_{j+1}} e^{2ik_j a_j}, \quad R_j^- = \frac{k_{j+1} - k_j}{k_j + k_{j+1}} e^{-2ik_{j+1} a_j}.
$$

(83)

Analyzing all possible "paths" of the propagations of all possible packets inside the barrier and internal well, we obtain (Maydanyuk, 2011):

$$
\sum_{n=1}^{+\infty} A_{\text{inc}}^{(n)} = 1 + \tilde{R}_M^+ \tilde{R}_{M-1}^- + \tilde{R}_M^+ \tilde{R}_{M-1}^- \cdot \tilde{R}_M^+ \tilde{R}_{M-1}^- + \ldots =
$$

$$
= 1 + \sum_{m=1}^{+\infty} (\tilde{R}_M^+ \tilde{R}_{M-1}^-)^m = \frac{1}{1 - \tilde{R}_M^+ \tilde{R}_{M-1}^-},
$$

$$
\sum_{n=1}^{+\infty} A_T^{(n)} = \left(\sum_{n=1}^{+\infty} A_{\text{inc}}^{(n)} \right) \cdot \left\{ \tilde{T}_{N-2}^+ T_{N-1}^+ + \right.
$$

$$
\left. + \tilde{T}_{N-2}^+ \cdot R_{N-1}^+ \tilde{R}_{N-2}^- \cdot T_{N-1}^+ + \ldots \right\} =
$$

$$
= \left(\sum_{n=1}^{+\infty} A_{\text{inc}}^{(n)} \right) \cdot \tilde{T}_{N-1}^+,
$$

(84)

$$
\sum_{n=1}^{+\infty} A_R^{(n)} = \tilde{R}_M^+ + \tilde{R}_M^+ \cdot \tilde{R}_{M-1}^- \tilde{R}_M^+ +
$$

$$
+ \tilde{R}_M^+ \cdot \tilde{R}_{M-1}^- \tilde{R}_M^+ \cdot \tilde{R}_{M-1}^- \tilde{R}_M^+ + \ldots =
$$

$$
= \tilde{R}_M^+ \cdot \left(1 + \sum_{m=1}^{+\infty} (\tilde{R}_{M-1}^- \tilde{R}_M^+)^m \right) =
$$

$$
= \frac{\tilde{R}_M^+}{1 - \tilde{R}_{M-1}^- \tilde{R}_M^+} = \left(\sum_{n=1}^{+\infty} A_{\text{inc}}^{(n)} \right) \cdot \tilde{R}_M^+,
$$

where

$$
\tilde{R}_{j-1}^+ = R_{j-1}^+ + T_{j-1}^+ \tilde{R}_j^+ T_{j-1}^- \left(1 + \sum_{m=1}^{+\infty} (\tilde{R}_j^+ R_{j-1}^-)^m\right) =
$$

$$
= R_{j-1}^+ + \frac{T_{j-1}^+ \tilde{R}_j^+ T_{j-1}^-}{1 - \tilde{R}_j^+ R_{j-1}^-},
$$

$$
\tilde{R}_{j+1}^- = R_{j+1}^- + T_{j+1}^- \tilde{R}_j^- T_{j+1}^+ \left(1 + \sum_{m=1}^{+\infty} (R_{j+1}^+ \tilde{R}_j^-)^m\right) = \tag{85}
$$

$$
= R_{j+1}^- + \frac{T_{j+1}^- \tilde{R}_j^- T_{j+1}^+}{1 - R_{j+1}^+ \tilde{R}_j^-},
$$

$$
\tilde{T}_{j+1}^+ = \tilde{T}_j^+ T_{j+1}^+ \left(1 + \sum_{m=1}^{+\infty} (R_{j+1}^+ \tilde{R}_j^-)^m\right) = \frac{\tilde{T}_j^+ T_{j+1}^+}{1 - R_{j+1}^+ \tilde{R}_j^-}.
$$

Choosing as starting points, the following:

$$
\tilde{R}_{N-1}^+ = R_{N-1}^+,
$$

$$
\tilde{R}_M^- = R_M^-, \tag{86}
$$

$$
\tilde{T}_M^+ = T_M^+,
$$

we calculate the coefficients $\tilde{R}_{N-2}^+ \dots \tilde{R}_M^+$, $\tilde{R}_{M+1}^- \dots \tilde{R}_{N-1}^-$ and $\tilde{T}_{M+1}^+ \dots \tilde{T}_{N-1}^+$.
We shall consider propagation of all packets in the region with number M, to the left. Such packets are formed in result of all possible reflections from the right part of potential, starting from the boundary a_M. In the previous section to describe their reflection from the left boundary R_0 to the right one, we used coefficient R_0^-. Now since we want to pass from simple boundary a_{M-1} to the left part of the potential well starting from this point up to $a_{\min}$, we generalize the coefficient R_{M-1}^- to $\tilde{R}_{M-1}^-$. The middle formula in (85) is applicable when we use eqs. (83) for definition of $T_i^{\pm}$ and $R_i^{\pm}$. Finally, we determine coefficients α_j and β_j:

$$
\sum_{n=1}^{+\infty} \alpha_j^{(n)} = \tilde{T}_{j-1}^+ \left(1 + \sum_{m=1}^{+\infty} (R_j^+ \tilde{R}_{j-1}^-)^m\right) =
$$

$$
= \frac{\tilde{T}_{j-1}^+}{1 - R_j^+ \tilde{R}_{j-1}^-} = \frac{\tilde{T}_j^+}{T_j^+},
$$

$$
\sum_{n=1}^{+\infty} \beta_j^{(n)} = \tilde{T}_{j-1}^+ \left(1 + \sum_{m=1}^{+\infty} (\tilde{R}_j^+ \tilde{R}_{j-1}^-)^m\right) R_j^+ = \tag{87}
$$

$$
= \frac{\tilde{T}_{j-1}^+ R_j^+}{1 - \tilde{R}_j^+ \tilde{R}_{j-1}^-} = \frac{\tilde{T}_j^+ R_j^+}{T_j^+},
$$

the amplitudes of transmission and reflection:

$$
A_T = \sum_{n=1}^{+\infty} A_T^{(n)}, \qquad A_R = \sum_{n=1}^{+\infty} A_R^{(n)},
$$

$$
\alpha_j = \sum_{n=1}^{+\infty} \alpha_j^{(n)} = \frac{\tilde{T}_j^+}{T_j^+}, \ \beta_j = \sum_{n=1}^{+\infty} \beta_j^{(n)} = \alpha_j \cdot R_j^+ \tag{88}
$$

and coefficients T and R describing penetration of the packet from the internal region outside and its reflection from the barrier

$$T_{MIR} \equiv \frac{k_N}{k_M} |A_T|^2 = |A_{\text{inc}}|^2 \cdot T_{\text{bar}}, \; T_{\text{bar}} = \frac{k_N}{k_M} |\tilde{T}_{N-1}^+|^2,$$
$$R_{MIR} \equiv |A_R|^2 = |A_{\text{inc}}|^2 \cdot R_{\text{bar}}, \qquad R_{\text{bar}} = |\tilde{R}_M^+|^2. \tag{89}$$

Choosing $a_{\min} = 0$, we assume full propagation of the packet through such a boundary (with no possible reflection) and we have $R_0^- = -1$ (it could be interesting to analyze results with varying R_0^-). We use the test:

$$\frac{k_N}{k_M} |A_T|^2 + |A_R|^2 = 1 \quad \text{or} \quad T_{MIR} + R_{MIR} = 1. \tag{90}$$

Now if energy of the packet is located below then height of one step with number m, then the following change

$$k_m \to i \, \xi_m \tag{91}$$

should be used to describe the transition of this packet through such a barrier with its tunneling. In the case of a barrier consisting from two rectangular steps of arbitrary heights and widths we have already obtained coincidence between amplitudes calculated by method of MIR and the corresponding amplitudes found by standard approach of quantum mechanics up to first 15 digits. Even increasing the number of steps up to some thousands has the right accuracy to fulfill the property (90).

In particular, we reconstruct completely the pictures of the probability and reflection presented in figs. 9 (a) and (b), figs. 10 (a) and (b), figs. 11 (b), but using such a standard technique. So, the *result concerning the oscillating dependence of the penetrability on the position of the starting point* a_{start} *in such figures is independent on the fully quantum method chosen for calculations.*

This is an important test which confirms reliability of the method MIR. So, we have obtained full coincidence between all amplitudes, calculated by method MIR and by standard approach of quantum mechanics. This is why we generalize the method MIR for description of tunneling of the packet through potential, consisting from arbitrary number of rectangular barriers and wells of arbitrary sizes (Maydanyuk, 2011).

5.3 Results

We have applied the above method to analyze the behavior of the packet tunneling through the barrier (80) (we used $a_{\text{new}} \to \sqrt{12}\, a_{\text{old}}$). The first interesting result is *a visible change of the penetrability on the displacement of the starting point* $a_{\min} \leq a \leq a_1$, *where we put the packet.* Using the possibility of decreasing the width of intervals up to an enough small value (and choosing, for convenience, the width of each interval to be the same), we choose $a_{\min}$ as *starting point* (and denote it as a_{start}), from where the packet begins to propagate outside. We have analyzed how the position of such a point influences the penetrability. In fig. 9 (a) one can see that the penetrability strongly changes in dependence of a_{start} for arbitrary values of energy of radiation E_{rad}: it has oscillating behavior (Maydanyuk, 2011). Difference between its minimums and maximums is minimal at a_{start} in the center of the well (i. e. its change tends to zero in the center of the well), this difference increases with increasing value of a_{start} and achieves the maximum close to the turning point. With this result, we may conclude that exists a *dependence of penetrability on the starting point* a_{start} *of the packet.* The coefficients of

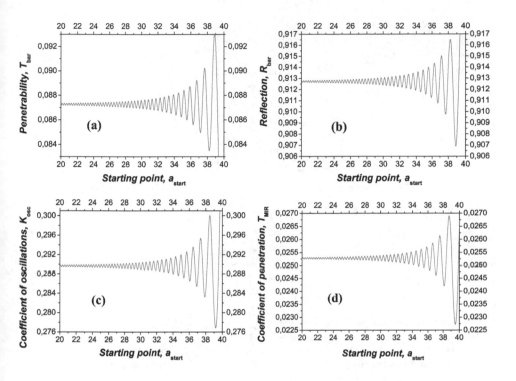

Fig. 9. Dependencies of the coefficients of the penetrability T_{bar} (a), reflection R_{bar} (b), coefficient of oscillations K_{osc} (c) and coefficient of penetration T_{MIR} (d) in terms of the position of the starting point a_{start} for the energy $E = 220$ ($A = 0.001$, $B = 0.001$, $a_{max} = 70$. The total number of intervals is 2000, for all presented cases the achieved accuracy is $|T_{bar} + R_{bar} - 1| < 10^{-15}$). These figures clearly demonstrate oscillating (i.e. not constant) behavior of all considered coefficients on a_{start}.

reflection, oscillations and penetration on the position of the starting point a_{start} are presented in next figs. 9 (b), (c), (d) and have similar behavior.

Usually, in cosmological quantum models the penetrability is determined by the barrier shape. In the non-stationary approach one can find papers where the role of the initial condition is analyzed in calculations of rates, penetrability etc.[3] But, the stationary limit does not give us any choice on which to work. We conclude: (a) the penetrability should be connected with the initial condition (not only in non-stationary consideration, but also in the stationary one). (b) Even in the stationary consideration, the penetrability of the barrier should be determined in dependence on the initial condition.

The first question is how much these results are reliable. In particular, how stable will such results be if we shift the external boundary outside? The results of such calculations are presented in fig. 10, where it is shown how the penetrability changes with a_{max} (for clearness sake, we have fixed the starting point $a_{start} = 10$, (Maydanyuk, 2011)). One can see that

---

[3] Such papers are very rare and questions about dynamics have not been studied deeply.

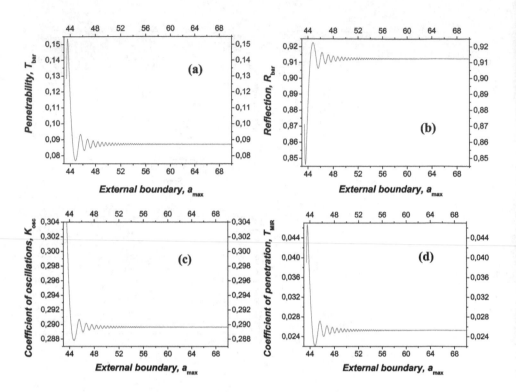

Fig. 10. Dependencies of the coefficients of penetrability (a), reflection (b), oscillations (c) and penetration (d) on the position of the external region, a_{max} for the energy $E = 223$ ($A = 0.001, B = 0.001$). For all presented values we have achieved accuracy $|T_{bar} + R_{bar} - 1| < 1 \cdot 10^{-15}$ (the maximum number of intervals is 2000).

all calculations are well convergent, that confirms efficiency of the method of the multiple internal reflections. On the basis of such results we choose $a_{max} = 70$ for further calculations. However, one can see that inclusion of the external region can change the coefficients of penetrability and penetration up to 2 times for the chosen energy level.

The second question is how strong this affects the calculations of the penetrability. If it was small than, the semiclassical approaches would have enough good approximation. From figs. 9 it follows that the penetrability is not strongly changed in dependence on shift of the starting point. However, such small variations are connected with relatively small height of the barrier and depth of the well, while they would be not small for another choice of parameters (the coefficient of oscillation and penetration turn out to change at some definite energies of radiation, see below). So, this effect is supposed to be larger at increasing height of the barrier and depth of the well, and also for near-barrier energies (i. e. for energies comparable with the barrier height, and above-barrier energies of radiation).

We have analyzed how these characteristics change in dependence on the energy of radiation. We did not expect the results that we got (see figs. 11). The coefficient of penetration has oscillations with peaks clearly shown (Maydanyuk, 2011). These peaks are separated by

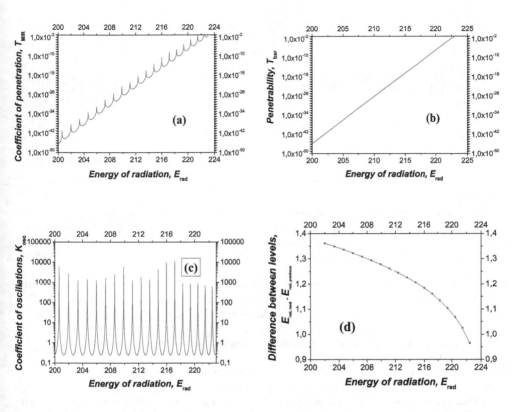

Fig. 11. Dependencies of the coefficient of the coefficient of penetration T_{MIR} (a), the coefficient of the penetrability T_{bar} (b), coefficient of oscillations K_{osc} (c) and difference $E_{res,next} - E_{res,previous}$ between two closest energy peaks (d) on the E_{rad} energy (we have choose: $A = 0.001$ and $B = 0.001$, $a_{start} = 10$, $a_{max} = 70$, number of intervals inside the scale axis a 1000, number of intervals of energy 100000). Inside the energy region $E_{rad} = 200 - 223$ we observe 19 resonant peaks in the dependencies of coefficients T_{MIR} and K_{osc} while the penetrability increases monotonously with increasing the E_{rad} energy.

similar distances and could be considered as resonances in energy scale. So, by using the fully quantum approach we observed for the first time clear pictures of resonances which could be connected with some early unknown quasi-stationary states. At increasing energy of radiation the penetrability changes monotonously and determines a general tendency of change of the coefficient of penetration, while the coefficient of oscillations introduces the peaks. Now the reason of the presence of resonances has become clearer: oscillations of the packet inside the internal well produce them, while the possibility of the packet to penetrate through the barrier (described by the penetrability of the barrier) has no influence on them. In general, we observe 134 resonant levels inside energy range $E_{rad} = 0$–200, and else 19 levels inside $E_{rad} = 200$–223.

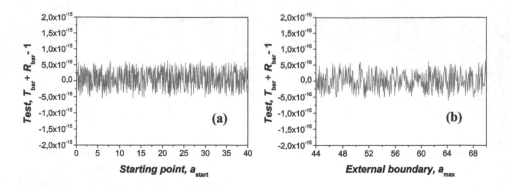

Fig. 12. Accuracy of the obtained penetrability T_{bar} and reflection R_{bar} for the energy $E = 220$ used in previous figs. 9 and 10. As a test, we calculate $T_{bar} + R_{bar} - 1$ in dependence on the position of the starting point a_{start} (a) and the external boundary a_{max} (b) ($A = 0.001, B = 0.001$, total number of intervals is 2000).

In the last fig. 12 one can see that we have achieved $|T_{bar} + R_{bar} - 1| < 10^{-15}$ inside whole region of changes of a_{start} and a_{max} (such data were used in the previous figs. 9 and 10). This is the accuracy of the method of the multiple internal reflections in obtaining T_{bar} and R_{bar}.

5.4 The fully quantum penetrability versus semiclassical one in cosmology: a quick comparison

Does the penetrability, determined according to the semiclassical theory by a shape of the barrier between two turning points, give exhaustive answers and the best estimations of rates of evolution of universe? If we look at figs. 9 (a), we shall see that this is not the case. The penetrability is depended on the position (coordinate) of maximum of the packet which begins to propagate outside at time moment $t = 0$. So, the penetrability should be a function of some parameters of the packet at beginning. For the first time, it has been demonstrated the difference between the fully quantum approach and the semiclassical one. However, let us perform a general analysis (Maydanyuk, 2011).

(1) If we wanted to check the semiclassical approach, we should miss some of the parameters. One can use test of $T + R = 1$ (where T and R are the penetrability through the barrier and reflection from it). But, note that the semiclassical approximation neglects the reflected waves in quantum mechanics (see (Landau & Lifshitz, 1989), eq. (46.10), p. 205, p. 221–222). Therefore, we cannot use the test above for checking T in the semiclassical theory.

(2) If we would like to determine the reflection coefficient, then we should find a more accurate semiclassical approximation (in order to take into account both decreasing and increasing components of the wave function in the tunneling region). In such a case, we shall face another problem, namely the presence of a non-zero interference between the incident and reflected waves. Now the relation $T + R = 1$ cannot be used as test, and one needs to take the third component M of interference into account (see (Maydanyuk, 2010)). If we improperly separate the exactly known full wave function in the incident and reflected waves[4], the interference component should increase without limit. In such a case, the penetrability and reflection

---

[4] However, the semiclassical approaches have no apparatus for such an analysis.

can freely exceed unit and increase without limit. What is now the general meaning of the penetrability?

(3) We shall give only some examples from quantum mechanics. (i) If we consider two-dimensional penetration of the packet through the simplest rectangle barrier (with finite size), we shall see that the penetrability is directly dependent on direction of tunneling of the packet. So, the penetrability is not a single value but a function. (ii) If we consider one-dimensional tunneling of the packet through the simplest rectangular barrier, we shall obtain "interference picture" of its amplitude in the transmitted region, which is dependent on time and space coordinates and is an exact analytical solution. Of course, the stationary part of such a result exactly coincides with well known stationary solutions (Maydanyuk, 2003).

(4) A tunneling boundary condition (Vilenkin, 1994) seems to be natural and clear, where the wave function should represent an outgoing wave at large scale factor a. However, is such a wave free? In contrast to problems of quantum atomic and nuclear physics, in cosmology we deal with potentials, which modules increase with increasing the scale factor a (their gradients increase, which have sense of force acting on the wave). Therefore, in quantum cosmology we should define the boundary condition on the basis of the waves propagating inside strong fields (see (Maydanyuk, 2010)).

These points destroy the semiclassical basis of the cosmological models. Now the statement concerning reliability of the semiclassical approach become a question of " faith" (note that this is widespread (Maydanyuk, 2010; 2011)). The semiclassical approach could be compared with *"black box"*, where deeper and more detailed information about the dynamics of the universe is hidden.

6. A brief review on the problems of the Universe origin

In the science history and in the science philosophy of XX-XXI cc. (especially in the field of the natural sciences, beginning from physics) there has been a lot of interesting things, which had not obtained a sufficiently complete elucidation and analysis yet. Firstly, under the influence of scientific and technological progress a great attention has been paid to the justification of such direction in the science philosophy as the scientific realism (i.e. the correspondence of the science to the reality), which has successively acquired three forms: the naive realism, the usual realism and the critical science realism. Secondly, some new important problems of physics (especially the problem of the essentially probabilistic description of the reality of the microscopic world, the problem of the essential influence of the observer on the reality, the collapse of the wave function) had been revealed in the development of quantum mechanics, *the continuously complicated interpretation of the Universe origin and the expansion after the Big Bang,* and also no succeeded attempt in explaining the origin of the biological life in terms of physics and other natural sciences, all being with a variety of interpretation versions, connected with the world-views of the researchers.

As to "great" and "grand" problems of natural sciences: There is an extensive introduction in the large number of open problems in many fields of physics, published by the Russian physicist V. Ginzburg in (Ginzburg, 1999) which is rather interesting to study. Inside this large list of open problems of modern physics (and in a certain degree of modern natural sciences), represented by V. Ginzburg repeatedly in Russian editions, some of them are marked him "great" or "grand" problems. Between namely these problems we would like to underline three of them.

a) *The problem of interpretation and comprehension of quantum mechanics (even of the non-relativistic quantum theory) remains still topical.* The majority of critics of quantum mechanics are unsatisfied with the probabilistic nature of its predictions. One can add here also the questions and paradoxes of the theory of quantum measurements theory, especially like the wave-function reduction. The appearance of quantum mechanics, and, in particular, the discussion of N. Bohr with A. Einstein (lasting many years), had seriously undermined the traditional forms of the naive realism in the philosophy of the scientific realism and had strongly influenced (and are continuating to influence) not only on physics but also on other kinds of knowledge in the sense of the dependence of the reality on the observer and, moreover, on our understanding of the human knowledge at all. More lately the new interpretation of quantum mechanics is appeared: in it the hypothesis of many universes, which are the exactly same as ours, permits to avoid the wave-function reduction.

b) *The relationship between physics and biology and, specifically, the problem of reductionism.* The main problem, according to V. Ginzburg, is connected with the explanation of the origin of the biologic life and the origin of the human abstract thinking (but the second one is connected not with biology but with the origin of the human spiritual life which is far beyond natural sciences). V. Ginzburg assumes that for a possible explanation of the origin of the biologic life one can naturally imagine a certain jump which is similar to some kind of phase transition (or, may be, certain synergetic process). But there are other points of view too.

c) *The cosmological problem* (in other words, *the problem of the Universe origin*). According to V. Ginzburg, it is also a grand problem, or strictly speaking, a great complex of cosmic problems many of which is far from the solution.

We did also analyzed in (Olkhovsky, 2010) these three problems in the context of other aspects, first of all regarding the increasing discussions between the supporters of two different meta-theoretical, meta-philosophical doctrines: either the beginning of the Universe formation from vacuum ("nothing") is either a result of the irrational randomness after passing from other space-time dimensions or from other universe, caused by some unknown process, or a result of the creation of the expanding Universe (together with the laws of its functioning) by the supreme intelligent design from *nigilo*.

6.1 Schematic description of the problems connected with the Universe origin and expansion

Earlier, after Enlightenment till approximately 1920, scientists in the natural sciences did usually consider the Universe as eternally existing and eternally moving. Now the most convincing arguments against the model of the eternally existing Universe are:

(a) the second law of thermodynamics which does inevitably bring to heat death of Universe,

(b) the observed cosmic microwave background.

The most surprising conclusion of the revealed non-stationary state of the Universe is the existence of the "beginning", under which the majority of physicists understand the beginning of the Universe expansion.

The cosmologic problem as the problem of the origin and evolution of the Universe has initiated to be analyzed by A. Einstein (after 1917) and now it is connected with papers of many other physicists. The first several authors had been G. Lemaitre (who proposed what became known as the Big Bang theory of the origin of the Universe, although he called it his "hypothesis of the primeval atom"), A. Friedman and G. Gamow.

And what namely had been in the "beginning"? Gamow had assumed in 1921 that the expansion had initiated from the super-condensed hot state as a result of the Big Bang, to which he and others had ascribed the time moment $t = 0$, i.e. the beginning of the Universe history. The initial state in this model is in fact postulated. The nature of the initial super-condensed hot Universe state is not known. Such initial point (or super-small region), in which the temperature, pressure, energy density etc had reached the anomalous huge (almost infinite) values, can be considered as a particular point, where the "physical" processes cannot be described by physical equations and in fact are excluded from the model analysis. Under these conditions the theory of grand unification (or superunification) of all four known interactions (strong, weak, electromagnetic and gravitational) is assumpted to act. But no satisfactory superunification has yet been constructed. The superstring theory claims the role of such superunification, but this goal has not yet been achieved (Ginzburg, 1999).

Strictly speaking, namely in the region of this point (from $t = 0$ till $t_0 = 10^{-44}$ sec., where t_0 is the Planck time) is arising the general problem of the world origin and also the choice dilemma: the beginning of the Universe formation from vacuum ("nothing") is either a result of the irrational randomness after passing from other space-time dimensions or from other universe, caused by some unknown process, or a result of the creation of the expanding Universe (together with the laws of its functioning) by the supreme intelligent design from nigilo.

The framework for the standard cosmologic model relies on Einstein's general relativity and on simplifying assumptions (such as homogeneity and isotropy of space). There are even non-standard alternative models. Now there are many supporters of Big Bang models. The number of papers and books on standard versions of the cosmologic Big Bang models is too enormous for citing in this short paper (it is possible to indicate, only for instance, (Hartle & Hawking, 1983; Kragh, 1996; Peacock, 1999; Vilenkin, 1994) for the initial reading in cosmology of the Universe and in the different quasi-classical and quantum approaches in cosmology for description of the creation and the initial expansion of the Universe). However, there is no well-supported model describing the Universe history prior to 10^{-15} sec. or so. Apparently a new unified theory of quantum gravitation is needed to break this barrier but the theory of quantum gravitation is only schematically constructed in the quasilinear approximation. Understanding this earliest era in the history of the Universe is currently one of the most important unsolved problems in physics. Further, over the time interval 10^{-35} sec., which is much larger than the Planck time and so can still be considered classically, the Universe was expanding (inflating) much more rapidly than in the known Friedman models. After the inflation, the Universe had been as though developing in accord with the Friedman's scenario (Ginzburg, 1999). It may be possible to deduce what happened before inflation through observational tests yet to be discovered, and a crucial role at the inflation stage could be played the so-called Λ-term added to the Einstein equations of the General Relativity.

A lot of observations testify that there is exists non-luminous matter in the Universe which manifests itself owing to its gravitational interaction and is present everywhere — both in the galaxies and in the intergalactic space. And what is the nature of dark mass? According to the very popular hypothesis, the role of dark matter is played by the hypothetical WIMPs (Weakly Interacting Massive Particles) with masses higher than protons (Ginzburg, 1999). There are also exist some other candidates for the role of dark matter (for instance, pseudoscalar particles — axions) (Ellis, 1998). Cosmic strings can be also mentioned (Ginzburg, 1999).

The possibility of the existence of the above-mentioned Λ-term in equations of the General Relativity is now frequently referred to as "dark energy" or quintessence. For $\Lambda > 0$ it "works" as "antigravity" (against the normal gravitational attraction) and testifies to the acceleration of the Universe expansion in our epoch (Armendariz-Picon et al., 2000; Ginzburg, 1999).

Moreover, it is worth to underline that many physicists consider that the second law of thermodynamics is universal for all closed systems, including also our Universe as a whole (which is closed in naturalistic one-world view). Therefore the heat death is inevitable (see, for instance, (Ginzburg, 1999) and especially (Adams & Laughlin, 1997)).

There are also versions of the non-standard versions of the cosmologic Big Bang models (Albrecht & Magueijo, 1999; Moffat, 1993; Petit, 1988; Petit & Viton, 1989; Setterfield & Norman, 1987; Troitskii, 1987). We shall shortly refer to these models, noting that at least one of them (by B. Setterfield and T. Norman (Setterfield & Norman, 1987)) clearly speaks on the young Universe: They indicate that after the Big Bang the light speed had been gradually decreased approximately $10^6 - 10^7$ times and it was deduced that the velocities of the electromagnetic and radioactive decays had been gradually decreased near 10^7 times too. In (Setterfield & Norman, 1987; Troitskii, 1987) it was deduced that after the inflation the Universe had not been really expanding.

6.2 On the anthropic principle

From 1973 (and particularly after eighties) the term *"anthropic principle"*, introduced by B. Carter, has become to acquire in the science and out of the science a certain popularity (Barrow & Tipler, 1986; Carter, 1974). Carter and other authors had been noted that physical constants must have values in the very narrow interval in order the existence of the biologic life can become possible, and that the measured values of these constants are really found in this interval. In other words, the Universe seems to be exactly such as it is necessary for the origin of the life. If physical constants would be even slightly other, then the life could be impossible. After meeting such testimonies, a number of scientists had formulated several interpretations of anthropic principle each of which brings the researchers to the worldview choice in its peculiar way. We shall consider here two of them. According to the *weak anthropic principle (WAP)*, the observed values of physical and cosmological constants caused by the necessary demand that the regions, where the organic life would be developed, ought to be possible. And in the context of WAP there is the possibility of choice between two alternatives:

1. Either someone does irrationally believe that there are possible an infinity of universes, in the past, in the present and in the future, and we exist and are sure in the existence of our Universe namely because the unique combination of its parameters and properties could permit our origin and existence.

2. Or someone does (also irrationally) believe that our unique Universe is created by Intelligent Design of a Creator (God) and the human being is also created by Creator in order to govern the Universe.

According to the *strong anthropic principle (SAP)*, the Universe has to have such properties which permit earlier or later the development of life. This form of the anthropic principle does not only state that the universe properties are limited by the narrow set of values, compatible with the development of the human life, but does also state that this limitation is necessary for such purpose. So, one can interpret such tuning of the universe parameters as the testimony of the supreme intelligent design of a certain creative basis. There is also a rather unexpected interpretation of SAP, connected with the eastern philosophy, but it is not widely known.

7. Conclusions and perspectives

In this Chapter the closed Friedmann–Robertson–Walker model with quantization in the presence of a positive cosmological constant and radiation was studied. We have solved it numerically and have determined the tunneling probability for the birth of an asymptotically de-Sitter, inflationary Universe as a function of the radiation energy. Note the following.

1. A fully quantum definition of the wave which propagates inside strong field and which interact minimally with them, has been formulated for the first time, and approach for its determination has been constructed.

2. A new stationary approach for the determination of the incident, reflected and transmitted waves relatively to the barrier has been constructed. The tunneling boundary condition has been corrected.

3. A quantum stationary method of determination of coefficients of penetrability and reflection relatively to the barrier with analysis of uniqueness of solution has been developed, where for the first time non-zero interference between the incident and reflected waves has been taken into account and for its estimation the coefficient of mixing has been introduced.

4. In this chapter a development of the method of multiple internal reflections is presented (see Refs. (Cardone et al., 2006; Maydanyuk et al., 2002a;b; Maydanyuk, 2003; Olkhovsky & Maydanyuk, 2000), also Refs. (Anderson, 1989; Fermor, 1966; McVoy et al., 1967)). When the barrier is composed from arbitrary number n of rectangular potential steps, the exact analytical solutions for amplitudes of the wave function, the penetrability T_{bar} through the barrier and the reflection R_{bar} from it are found. At $n \to \infty$ these solutions can be considered as exact limits for potential with the barrier and well of arbitrary shapes.

In such a quantum approach the penetrability of the barrier for the studied quantum cosmological model with parameters $A = 36$, $B = 12\Lambda$ ($\Lambda = 0.01$) has been estimated with a comparison with results of other known methods. Note the following.

1. The modulus of the coefficient of mixing is less 10^{-19}. This points out that *there is no interference between the found incident and reflected waves close to the internal turning point.*

2. On the basis of the calculated coefficients we reconstruct a property (55) inside the whole studied range of energy of radiation (see Fig. 12).

3. The probability of penetration of the packet from the internal well outside with its tunneling through the barrier of arbitrary shape is determined. We call such coefficient as *coefficient of penetration. This coefficient is separated on the penetrability and a new coefficient, which characterizes oscillating behavior of the packet inside the internal well and is called* coefficient of oscillation. *The formula found, seems to be the fully quantum analogue of the semiclassical formula of Γ width of decay in quasistationary state proposed in Ref. (Gurvitz & Kälbermann, 1987).* Here, the coefficient of oscillations is the fully quantum analogue for the semiclassical F factor of formation and the coefficient of penetration is analogue for the semiclassical Γ width.

4. The penetrability of the barrier visibly changes in dependence of the position of the starting point R_{start} inside the internal well, where the packet begins to propagate (see figs. 9). We note the following peculiarities: the penetrability has oscillating behavior, difference between its minimums and maximums is minimal at R_{start} in the center of the well, with increasing R_{start} this difference increases achieving to maximum near the turning

point. The coefficients of reflection, oscillations and penetration have similar behavior. We achieve coincidence (up to the first 15 digits) between the amplitudes of the wave function obtained by such a method, and the corresponding amplitudes obtained by the standard approach of quantum mechanics (see Appendix B in (Maydanyuk & Belchikov, 2011) where solutions for amplitudes were calculated in general quantum decay problem). This confirms that this result does not depend on a choice of the fully quantum method applied for calculations. Such a peculiarity is shown in the fully quantum considerations and it is hidden after imposing the semiclassical restrictions.

5. The coefficient of penetration has oscillating dependence on the energy of radiation. Here, peaks are clearly shown. They are localized at similar distances (see figs. 11). So, for the first time we have obtained in the fully quantum approach a clear and stable picture of resonances, which indicate the presence of some early unknown quasistationary states. If the energy of radiation increases, the penetrability is monotonously changed. It describes a general tendency of behavior of the coefficient of penetration, while the coefficient of oscillations gives peaks. Now the reason of existence of resonances becomes clear: oscillations of the packet inside the internal well give rise to them. In particular, we establish 134 such resonant levels inside range E_{rad} = 0–223 for the barrier (8) with parameters $A = 0.001$ and $B = 0.001$.

6. A dependence of the penetrability on the starting point has maxima and minima. This allows to predict some definite initial values of the scale factor, when the universe begins to expand. Such initial data is direct result of quantization of the cosmological model.

7. The modulus of the wave function in the internal and external regions has minima and maxima which were clearly established in (Maydanyuk, 2008; 2010). This indicates, in terms of values of the scale factor, where the probable "appearance" of the universe is maximal or minimal. So, the radius of the universe during its expansion changes not continuously, but consequently passes through definite discrete values connected with these maxima. It follows that space-time of universe on the first stage after quantization seems to be rather discrete than continuous. According to results (Maydanyuk, 2008; 2010; 2011), difference between maxima and minima is slowly smoothed with increasing of the scale factor a. In this way, we obtain the continuous structure of the space-time at latter times. The discontinuity of space-time is direct result of quantization of cosmological model. This new phenomenon is the most strongly shown on the first stage of expansion and disappears after imposition of the semiclassical approximations.

8. Acknowledgments

One of us [ER] acknowledges partial support by INFN, and a fellowship by CAPES.

9. References

Spravochnik po spetsialnim funktsiyam s formulami, grafikami we matematicheskimi tablitsami, Pod redaktsiei Abramowitza M. we Stegan I. A. (Nauka, Moskva, 1979), 832 p.. — [in Russian; eng. variant: *Handbook of mathematical functions with formulas, graphs and mathematical tables*, Edited by Abramowitz M. and Stegan I. A., National bureau of standards, Applied math. series – 55, 1964].

Acacio de Barros, J., Correa Silva, E. V., Monerat, G. A., Oliveira-Neto, G., Ferreira Filho, L. G. & Romildo Jr., P. (2007). Tunneling probability for the birth of an asymptotically de Sitter universe, *Phys. Rev.* Vol. D75: 104004. URL: gr-qc/0612031.

Adams, F. C. & Laughlin, G. (1997). A Dying Universe: the Long-Term Fate and Evolution of Astrophysical Objects, *Rev. Mod. Phys.* Vol. 69: 337–372.

Aharonov, Y., Erez, N. & Resnik, B. (2002). Superoscillations and tunneling times, *Phys. Rev.* Vol. A65: 052124.

Albrecht, A. & Magueijo, J. (1999). A time varying speed of light as a solution to cosmological puzzles, *Phys. Rev.* Vol. D59: 043516. URL: astro-ph/9811018.

Anderson, A. (1989). Multiple scattering approach to one-dimensional potential problems, *Am. Journ. Phys.* Vol. 57 (No. 3): 230–235.

Armendariz-Picon, C., Mukhanov, V. & Steinhardt, P. J. (2000). Dynamical solution to the problem of a small cosmological constant and late-time cosmic acceleration, *Phys. Rev. Lett.* Vol. 85: 4438.

Atkatz, D. & Pagels, H. (1982). Origin of the Universe as a quantum tunneling effect, *Phys. Rev.* Vol. D 25 (No. 8): 2065–2073.

Barrow, J. D. & Tipler, F. J. (1986). *The Anthropic Cosmological Principle*, Clarendon Press, Oxford, 1986.

Bento, M. C., Bertolami, O. & Sen, A. A. (2002). *Phys. Rev.* Vol. D 66: 043507. URL: gr-qc/0202064.

Bilic, N., Tupper, G. B. & Viollier, R. D. (2002). *Phys. Lett.* Vol. B535: 17. URL: astro-ph/0111325.

Bouhmadi-Lopez, M. & Moniz, P. V. (2005). FRW quantum cosmology with a generalized Chaplygin gas, *Phys. Rev.* Vol. D71 (No. 6): 063521 [16 pages]. URL: gr-qc/0404111.

Bouhmadi-López, M., Gonzáles-Diaz, P. F. & Martin-Moruno, P. (2008). On the generalised Chaplygin gas: worse than a big rip or quieter than a sudden singularity? *Int. Journ. Mod. Phys.* Vol. D17 (No. 4): 2269–2290. URL: arXiv:0707.2390.

Brandenberger, R. H. (2000). Inflationary cosmology: progress and problems, *Lectures at the International School on Cosmology* (Kish Island, Iran, Jan. 22 — Feb. 4 1999; Proceedings: Kluwer, Dordrecht): p. 48. URL: hep-ph/9910410.

Cardone, F., Maidanyuk, S. P., Mignani, R. & Olkhovsky, V. S. (2006). Multiple internal reflections during particle and photon tunneling, *Found. Phys. Lett.* Vol. 19 (No. 5): 441–457.

Carter, B. (1974). Large number coincidences and the anthropic principle in cosmology, *IAU Symposium 63: Confrontation of cosmological theories with observational data*, Dordrecht, Reidel.

Casadio, R., Finelli, F., Luzzi, M. & Venturi, G. (2005). Improved WKB analysis of cosmological perturbations, *Phys. Rev.* Vol. D71 (No. 4): 043517 [12 pages]. URL: gr-qc/0410092.

Casadio, R., Finelli, F., Luzzi, M. & Venturi, G. (2005). Improved WKB analysis of slow-roll inflation, *Phys. Rev.* Vol. D72 (No. 10): 103516 [10 pages]. URL: gr-qc/0510103.

Chaplygin, S. (1904). *Sci. Mom. Moskow Univ. Math. Phys.* Vol. 21: 1.

DeWitt, B. S. (1967). Quantum theory of gravity. I. The canonical theory, *Phys. Rev.* Vol. 160 (No. 5): 1113–1148.

Ellis, J. (1998). Particle components of dark matter, *Proc. Natl. Acad. Sci. USA* Vol. 95: 53.

Esposito, S. (2003). Multibarrier tunneling, *Phys. Rev.* Vol. E67 (No. 1): 016609. URL: quant-ph/0209018.

Fermor, J. H., (1966). Quantum-mechanical tunneling, *Am. Journ. Phys.* Vol. 34: 1168–1170.

Finelli, F., Vacca, G. P. & Venturi, G. (1998). Chaotic inflation from a scalar field in nonclassical states, *Phys. Rev.* Vol. D58: 103514 [14 pages].

Ginzburg, V. L. (1999). What problems of physics and astrophysics seem now to be especially important and interesting (30 years later, already on the verge of XXI century), *Physics - Uspekhi* Vol. 42: 353–272.

Gurvitz, S. A. & Kälbermann, G. (1987). Decay width and the shift of a quasistationary state, *Phys. Rev. Lett.* Vol. 59: 262–265.

Hartle, J. B. & Hawking, S. W. (1983). Wave function of the Universe, *Phys. Rev.* Vol. D28 (No. 12): 2960–2975.

Jakiel, J., Olkhovsky, V. S. & Recami, E. (1999). Validity of the Hartman effect for all kinds of mean tunnelling times, in *Mysteries, Puzzles and Paradoxes in Quantum Mechanics*, edited by Bonifacio R. (Am. Inst. Phys.; Woodbury, N. Y.), pp. 299–302.

Kamenshchik, A. Y., Moschella, U. & Pasquier, V. (2001). *Phys. Lett.* Vol. B511: 265. URL: gr-qc/0103004.

Kragh, H. (1996). *Cosmology and Controversy*, Princeton University Press (NJ, ISBN 0-691-02623-8).

Landau, L. D. & Lifshitz, E. M. *Quantum mechanics, course of Theoretical Physics*, Vol. 3 (Nauka, Mockva, 1989), p. 768 — [in Russian; eng. variant: Oxford, Uk, Pergamon, 1982].

Levkov, D., Rebbi, C. & Rubakov, V. A. (2002). Tunneling in quantum cosmology: numerical study of particle creation, *Phys. Rev.* Vol. D66 (No. 8): 083516. URL: gr-qc/0206028.

Linde, A. (1984). Quantum creation of the inflationary Universe, *Lett. Nuov. Cim.* Vol. 39 (No. 2): 401–405.

Linde, A. (1990). Particle physics and inflationary cosmology (Harwood, Chur, Switzerland): 362 pp.; Contemporary Concepts in Physics, v. 5. URL: hep-th/0503203.

Luzzi, M. (2007). Semiclassical Approximations to Cosmological Perturbations, *Ph. D. thesis* (Advisor: Prof. Giovanni Venturi, University of Bologna): 148 pages. URL: arXiv:0705.3764.

Maydanyuk, S. P., Olkhovsky, V. S. & Zaichenko A. K., (2002). The method of multiple internal reflections in description of tunneling evolution of nonrelativistic particles and photons, *Journ. Phys. Stud.* Vol. 6 (No. 1): 1–16. URL: nucl-th/0407108.

Maydanyuk, S. P., Olkhovsky V. S. & Belchikov S. V., (2002). The method of multiple internal reflections in description of nuclear decay, *Probl. At. Sci. Tech. (Voprosi atomnoi nauki we tehniki, RFNC-VNIIEF, Sarov, Russia)* Vol. 1: 16–19. URL: nucl-th/0409037.

Maydanyuk, S. P. (2003). Time analysis of tunneling processes in nuclear collisions and decays, *Ph. D. dissertation* (Supervisor: Prof. V. S. Olkhovsky, Kiev), p. 147 [in Ukrainian].

Maydanyuk, S. P. (2008). Wave function of the Universe in the early stage of its evolution, *Europ. Phys. Journ.* Vol. C57 (No. 4): 769–784. URL: arxiv.org:0707.0585.

Maydanyuk, S. P. (2010). A fully quantum method of determination of penetrability and reflection coefficients in quantum FRW model with radiation, *Int. Journ. Mod. Phys.* Vol. D19 (No. 4): 392–435. URL: arXiv:0812.5081.

Maydanyuk, S. P. (2011). Resonant structure of space-time of early universe, *Europ. Phys. Journ. Plus* Vol. 126 (No. 8): 76. URL: arXiv:1005.5447.

Maydanyuk, S. P. & Belchikov, S. V. (2011). Problem of nuclear decay by proton emission in fully quantum consideration: calculations of penetrability and role of boundary conditions, *Journ. Mod. Phys.* Vol. 2: 572–585.

McVoy, K. W., Heller, L. & Bolsterli, M. (1967). Optical analysis of potential well resonances, *Rev. Mod. Phys.* Vol. 39 (No. 1): 245–258.

Moffat, J. (1993). Superluminary universe: A possible solution to the initial value problem in cosmology, *Int. J. Mod. Phys.* Vol. D2 (No. 3): 351–366. URL: gr-qc/9211020.

Monerat, G. A., Oliveira-Neto, G., Correa Silva, E. V. et al. (2007). Dynamic of the early universe and the initial condition for inflation in a model with radiation and a Chaplygin gas, *Phys. Rev.* Vol. D76: 024017 [11 pages].

Olkhovsky, V. S. & Recami, E. (1992). Recent developments in the time analysis of tunnelling processes, *Physics Reports* Vol. 214 (No. 214): 339–357.

Olkhovsky, V. S., Recami, E., Raciti, F. & Zaichenko, A. K. (1995). Tunnelling times, the dwell time, and the "Hartman effect", *J. de Phys. (France)* I Vol. 5: 1351–1365.

Olkhovsky, V. S. & Maydanyuk, S. P. (2000). Method of multiple internal reflections in description of tunneling evolution through barriers, *Ukr. Phys. Journ.* Vol. 45 (No. 10): 1262–1269. URL: nucl-th/0406035.

Olkhovsky, V. S., Recami, E. & Salesi, G. (2002). Tunneling through two successive barriers and the Hartman effect, *Europhysics Letters* Vol. 57: 879–884. URL: quant-ph/0002022.

Olkhovsky, V. S., Recami, E. & Jakiel, J. (2004). Unified time analysis of photon and nonrelativistic particle tunnelling, *Phys. Rep.* Vol. 398: 133–178.

Olkhovsky, V. S., Recami, E. & Zaichenko, A. K. (2005). Resonant and non-resonant tunneling through a double barrier, *Europhysics Letters* Vol. 70: 712–718. URL: quant-th/0410128.

Olkhovsky, V. S. & Recami, E. (2008). New developments in the study of time as a quantum observable, *Int. J. Mod. Phys.* Vol. B22: 1877–1897.

Olkhovsky, V. S. (2010). A retrospective view on the history of the natural sciences in XX-XXI, *Natural Science* Vol. 2 (No. 3): 228–245.

Olkhovsky, V. S. (2011). On time as a quantum observable canonically conjugate to energy, *Physics – Uspekhi* Vol. 54 (No. 11): 829–835.

Peacock, J. (1999). *Cosmological Physics*, Cambridge University Press (ISBN 0521422701).

Petit, J. P. (1988). An interpretation of cosmological model with variable light velocity, *Mod. Phys. Lett.* Vol. A3 (No. 16): 1527–1532.

Petit, J. P. & Viton, M. (1989). Gauge cosmological model with variable light velocity. Comparizon with QSO observational data, *Mod. Phys. Lett.* Vol. A4: 2201–2210.

Recami, E. (2004). Superluminal tunneling through successive barriers. Does QM predict infinite group-velocities? *Journal of Modern Optics* Vol. 51: 913–923.

Rubakov, V. A. (1984). Quantum mechanics in the tunneling universe, *Phys. Lett.* Vol. B 148 (No. 4–5): 280–286.

Rubakov, V. A. (1999). Quantum cosmology, Proceedings: *Structure formation in the Universe* (Edited by R. G. Crittenden and N. G. Turok, Kluwer): 63–74. URL: gr-qc/9910025.

Rubakov, V. A. (2005). Introduction to cosmology (RTN Winter School of Strings, Supergravity and Gauge Theories, January 31 – February 4, SISSA, Trieste, Italy): 58 p. URL: pos.sissa.it

Setterfield, B. & Norman, T. (1987). *The atomic constants, light and time*, Paper for SRI International.

Trodden, M. & Carroll, S. M. (2003) *TASI Lectures: Introduction ot cosmology*, Lectures at the Theoretical Advanced Study Institutes in elementary particle physics (TASI-2003, *Recent Trends in String Theory*, University of Colorado at Boulder, 1–27 Jun 2003; Edited by J. M. Maldacena; Hackensack, World Scientific, 2005; 548 pp.), 82 pp. URL: astro-ph/0401547.

Troitskii, V. S. (1987). *Astrofizika i kosmicheskaya nauka* Vol. 139: 389–411 [in Russian].

Tronconi, A., Vacca, G. P. & Venturi, G. (2003). Inflaton and time in the matter-gravity system, *Phys. Rev.* Vol. D67: 063517 [8 pages].

Vilenkin, A. (1982). Creation of universes from nothing, *Phys. Lett.* Vol. B117 (No. 1–2): 25–28.

Vilenkin, A. (1984). Quantum creation of universes, *Phys. Rev.* Vol. D 30 (No. 2): 509–511.

Vilenkin, A. (1986). Boundary conditions in quantum cosmology, *Phys. Rev.* Vol. D 33 (No. 12): 3560–3569.

Vilenkin, A. (1994). Approaches to quantum cosmology, *Phys. Rev.* Vol. D 50 (No. 12): 2581–2594. URL: gr-qc/9403010.

Weinberg, S. *Gravitatsiya we kosmologiya: printsipi we prilozheniya obschei teorii otnositel'nosti* (Mir, Moskva, 1975), 696 p.. — [in Russian; eng. variant: Weinberg, S. *Gravitation and cosmology: principles and applications of the General theory of relativity*, MIT, John Wiley and Sons, New York - London - Sydney - Toronto, 1972].

Wheeler, J. A. (1968). *Batelle Rencontres*, Benjamin, New York.

Zaichenko, A. K. & Kashuba, I. E. (2001). Evaluation of the parabolic cylinder function in the context of nuclear physics, Kyiv, 14 p. (Preprint/ National Academy of Sciences of Ukraine. Institute for Nuclear Research; KINR-01-3).

Zakhariev, B. N., Kostov, N. A. & Plehanov, E. B. (1990). Exactly solbable one- and manychannel models (Quantum intuition lessons), *Physics of elementary particles and atomic nuclei* Vol. 21 (Iss. 4): 914–962. — [in Russian].

Zel'dovich, Ya. B., & Starobinsky, A. A. (1984). Quantum creation of a universe in a nontrivial topology, *Sov. Astron. Lett.* Vol. 10 (No. 3): 135.

4

Correspondences of Scale Relativity Theory with Quantum Mechanics

Călin Gh. Buzea[1], Maricel Agop[2] and Carmen Nejneru[3]
[1]National Institute of Research and Development for Technical Physics
[2]Department of Physics, Technical "Gh. Asachi" University
[3]Materials and Engineering Science, Technical "Gh. Asachi" University
Romania

1. Introduction

We perform a critical analysis of some quantum mechanical models such as the hydrodynamic model (Madelung's model), de Broglie's theory of double solution etc., specifying both mathematical and physical inconsistencies that occur in their construction.

These inconsistencies are eliminated by means of the fractal approximation of motion (physical objects moving on continuous and non-differentiable curves, i.e. fractal curves) developed in the framework of Scale Relativity (SR) (Nottalle, L., 1993; Chaline, J. et al, 2009; Chaline, J. et al, 2000; Nottale, L., 2004; Nottale, L. & Schneider J., 1984; Nottale, L., 1989; Nottale, L., 1996). The following original results are obtained: i) separation of the physical motion of objects in wave and particle components depending on the scale of resolution (differentiable as waves and non-differentiable as particles) - see paragraphs 5-7; ii) solidar motion of the wave and particle (wave-particle duality) - see paragraph 8, the mechanisms of duality (in phase wave-particle coherence, paragraphs 9 and 10 and wave-particle incoherence, see paragraph 11); iii) the particle as a clock, its incorporation into the wave and the implications of such a process - see paragraphs 12 and 13; iv) Lorentz-type mechanisms of wave-particle duality - see paragraph 14.

The original results of this work are published in references (Harabagiu A. et al , 2010; Agop, M. et al, 2008; Harabagiu, A. & Agop, M., 2005;Harabagiu, A. et al, 2009; Agop, M. et al, 2008). Explicitly, Eulerian's approximation of motions on fractal curves is presented in (Agop, M. et al, 2008), the hydrodynamic model in a second order approximation of motion in (Harabagiu, A. & Agop, M., 2005), wave-particle duality for „coherent" fractal fluids with the explanation of the potential gap in (Harabagiu, A. et al, 2009), the physical self-consistence of wave-particle duality in various approximations of motion and for various fractal curves in (Agop, M. et al, 2008). A unitary treatment of both the problems listed above and their various mathematical and physical extensions are developed in (Harabagiu A. et al , 2010).

2. Hydrodynamic model of quantum mechanics (Madelung's model)

Quantum mechanics is substantiated by the Schrödinger wave equation (Țițeica, S., 1984; Felsager, B., 1981; Peres, A., 1993; Sakurai J.J. & San Fu Taun, 1994)

$$i\hbar\frac{\partial\Psi}{\partial t}=U\Psi-\frac{\hbar^2}{2m_0}\Delta\Psi \tag{1}$$

where $\hbar$ is the reduced Planck's constant, m_0 the rest mass of the test particle, U the external scalar field and Ψ the wave-function associated to the physical system. This differential equation is linear and complex.

Starting from this equation, Madelung (Halbwacs, F., 1960; Madelung R., 1927) constructed the following model. One separates real and imaginary parts by choosing Ψ of the form:

$$\Psi(\mathbf{r},t)=R(\mathbf{r},t)e^{iS(\mathbf{r},t)} \tag{2}$$

which induces the velocity field:

$$\mathbf{v}=\frac{\hbar}{m_0}\nabla S \tag{3}$$

and the density of the probability field:

$$\rho(\mathbf{r},t)=R^2(\mathbf{r},t) \tag{4}$$

Using these fields one gets the hydrodynamic version of quantum mechanics (Madelung's model)

$$\frac{\partial}{\partial t}(m_0\rho\mathbf{v})+\nabla(m_0\rho\mathbf{vv})=-\rho\nabla(U+Q) \tag{5}$$

$$\frac{\partial\rho}{\partial t}+\nabla\cdot(\rho\mathbf{v})=0 \tag{6}$$

where

$$Q=-\frac{\hbar^2}{2m_0}\frac{\Delta\sqrt{\rho}}{\sqrt{\rho}} \tag{7}$$

is called the quantum potential. Equation (5) corresponds to the momentum conservation law and equation (6) to the conservation law of the probability's density field (quantum hydrodynamics equations).

We have the following: i) any micro-particle is in constant interaction with an environment called „subquantic medium" through the quantum potential Q, ii) the „subquantic medium" is identified with a nonrelativistic quantum fluid described by the equations of quantum hydrodynamics. In other words, the propagation of the Ψ field from wave mechanics is replaced by a fictitious fluid flow having the density ρ and the speed $\mathbf{v}$, the fluid being in a field of forces $\nabla(U+Q)$. Moreover, the following model of particle states (Bohm D. & Hiley B.J., 1993; Dörr D. et al,1992; Holland P.R., 1993; Albert D.Z., 1994; Berndl K. et al, 1993; Berndl K. et al, 1994; Bell J.S., 1987; Dörr D. et al, 1993): Madelung type fluid in „interaction" with its own „shell" (there is no space limitation of the fluid, though of the particle).

3. DeBroglie's theory of double solution. The need for introducing the model of Bohm and Vigier

One of the key observations that de Broglie left in the development of quantum mechanics, is the difference between the relativistic transformation of the frequency of a wave and that of a clock's frequency (de Broglie L., 1956; de Broglie L., 1957; de Broglie L., 1959; de Broglie L., 1963; de Broglie L., 1964; de Broglie L., 1980). It is well known that, if v_0 is the frequency of a clock in its own framework, the frequency confered by an observer who sees it passing with the speed $v = \beta c$ is

$$v_c = v_0\sqrt{1 - \beta^2} \ .$$

This is what is called the phenomenon of "slowing down of horologes". This phenomenon takes place due to the relative motion of horologes. On the contrary, if a wave within a certain reference system is a stationary one, with frequency v_0 and is noticed in a reference system animated with speed $v = \beta c$, as compared with the first one, it will appear as a progressive wave that propagates in the sense of the relative motion, with frequency

$$v = \frac{v_0}{\sqrt{1 - \beta^2}}$$

and with the phase speed

$$V = \frac{c}{\beta} = \frac{c^2}{v} \ .$$

If the corpuscle, according to relation $W = hv$, is given an internal frequency

$$v_0 = \frac{m_o c^2}{h}$$

and if we admit that within the appropriate system of the corpuscle the associated wave is a stationary one, with frequency v_0, all the fundamental relations of undulatory mechanics and in particular $\lambda = \frac{h}{p}$, in which p is the impulse of the corpuscle, are immediately obtained from the previous relations.

Since de Broglie considers that the corpuscle is constantly located in the wave, he notices the following consequence: the motion of the corpuscle has such a nature that it ensures the permanent concordance between the phase of the surrounding wave and the internal phase of the corpuscle considered as a small horologe. This relation can be immediately verified in the simple case of a corpuscle in uniform motion, accompanied by a monochromatic plain wave. Thus, when the wave has the general form

$$\Psi = A(x,y,z,t)e^{\frac{2\pi i}{h}\Phi(x,y,z,t)}$$

in which A and Φ are real, the phase concordance between the corpuscle and its wave requires that the speed of the corpuscle in each point of its trajectory be given by the relation

$$\mathbf{v} = -\frac{1}{m_0}\nabla\Phi$$

Nevertheless it was not enough to superpose the corpuscle with the wave, imposing it to be guided by the propagation of the wave: the corpuscle had to be represented as being incorporated in the wave, i.e. as being a part of the structure of the wave. De Broglie was thus directed to what he himself called the theory of "double solution". This theory admits that the real wave is not a homogeneous one, that it has a very small area of high concentration of the field that represents the corpuscle and that, besides this very small area, the wave appreciably coincides with the homogeneous wave as formulated by the usual undulatory mechanics.

The phenomenon of guiding the particle by the surrounding undulatory field results from the fact that the equations of the field are not linear ones and that this lack of linearity, that almost exclusively shows itself in the corpuscular area, solidarizes the motion of the particle with the propagation of the surrounding wave (de Broglie L., 1963; de Broglie L., 1964; de Broglie L., 1980).

Nevertheless there is a consequence of "guidance" upon which we should insist. Even if a particle is not submitted to any external field, if the wave that surrounds it is not an appreciably plain and monochromatic one (therefore if this wave has to be represented through a superposition of monochromatic plain waves) the motion that the guidance formula imposes is not rectilinear and uniform. The corpuscle is subjected by the surrounding wave, to a force that curves its trajectory: this "quantum force" equals the gradient with the changed sign of the quantum potential Q given by (7). Therefore, the uniform motion of the wave has to be superposed with a "Brownian" motion having random character that is specific to the corpuscle.

Under the influence of Q, the corpuscle, instead of uniformly following one of the trajectories that are defined by the guidance law, constantly jumps from one of these trajectories to another, thus passing in a very short period of time, a considerably big number of sections within these trajectories and, while the wave remains isolated in a finite area of the space, this zigzag trajectory hurries to explore completely all this region. In this manner, one can justify that the probability of the particle to be present in a volume element $d\tau$ of the physical space is equal to $|\Psi|^2 d\tau$. This is what Bohm and Vigier did in their statement: therefore they showed that the probability of repartition in $|\Psi|^2$ must take place very quickly. The success of this demonstration must be correlated with the characteristics if "Markov's chains."(Bohm, D., 1952; Bohm D. & Hiley B.J., 1993; Bohm D., 1952; Bohm D., 1953).

4. Comments

In his attempt to built the theory of the double solution, de Broglie admits certain assertions (de Broglie L., 1956; de Broglie L., 1957; de Broglie L., 1959; de Broglie L., 1963; de Broglie L., 1964; de Broglie L., 1980):

i.   the frequency of the corpuscle that is assimilated to a small horologe must be identified with the frequency of the associated progressive wave;

ii. the coherence of the inner phase of the corpuscle-horologe with the phase of the associated wave;

iii. the corpuscle must be "incorporated" into the progressive associated wave through the "singularity" state. Thus, the motion of the corpuscle "solidarizes" with the propagation of the associated progressive wave. Nevertheless, once we admit these statements, de Broglie's theory does not answer a series of problems, such as, for example:

1. What are the mechanisms through which either the undulatory feature or the corpuscular one impose, either both of them in the stationary case as well as in the non stationary one?;
2. The limits in the wave-corpuscle system of the corpuscular component as well as the undulatory one and their correspondence;
3. How is the "solidarity" between the motion of the corpuscle and the one of the associated progressive wave naturally induced?
iv. What are the consequences of this "solidarity"? And we could continue ... . Moreover, Madelung's theory (Halbwacs, F., 1960; Madelung R., 1927) brings new problems. How can we built a pattern of a corpuscle (framework + Madelung liquid) endlessly extended in space?

Here are some of the "drawbacks" of the patterns in paragraphs 2 and 3 which we shall analyze and remove by means of introducing the fractal approximation of the motion.

5. The motion equation of the physical object in the fractal approximation of motion. The Eulerian separation of motion on resolution scales

The fractal approximation of motion refers to the movement of physical objects (wave + corpuscle) on continuous and non differentiable curves (fractal curves). This approximation is based on the scale Relativity theory (RS) (Nottalle, L., 1993; Chaline, J. et al, 2009; Chaline, J. et al, 2000; Nottalle, L., 2004, Nottalle, L. & Schneider J., 1984; Nottalle, L., 1989; Nottalle, L., 1996). Thus, the fractal differential operator can be introduced

$$\frac{\hat{d}}{dt} = \frac{\partial}{\partial t} + \hat{\mathbf{V}} \cdot \nabla - i\frac{\lambda^2}{2\tau}\left(\frac{dt}{\tau}\right)^{(2/D_F)-1}$$ (8)

where $\hat{\mathbf{V}}$ is the complex speed field

$$\hat{V} = V - iU$$ (9)

λ is the scale length, dt is the temporary resolution scale, τ is the specific time to fractal-non fractal transition, and D_F is the arbitrary and constant fractal dimension. Regarding the fractal dimension, we can use any of Hausdorff-Bezicovici, Minkowski-Bouligand or Kolmogoroff dimensions, etc. (Budei, L., 2000; Barnsley, M., 1988; Le Mehante A., 1990; Heck, A. & Perdang, J.M., 1991; Feder, J. & Aharony, A., 1990; Berge, P. et al, 1984; Gouyet J.F., 1992; El Naschie, M.S. et al, 1995; Weibel, P. et al, 2005; Nelson, E., 1985; Nottalle, L., 1993; Chaline, J. et al, 2009; Chaline, J. et al, 2000; Nottalle, L., 2004; Agop, M. et al, 2009). The only restriction refers to the maintaining of the same type of fractal dimension during the dynamic analysis. The real part of the speed field V is differentiable and independent as compared with the resolution scale, while the imaginary scale U is non differentiable (fractal) and depends on the resolution scale.

Now we can apply the principle of scale covariance by substituting the standard time derivate (d/dt) with the complex operator $\hat{d}/dt$. Accordingly, the equation of fractal space-time geodesics (the motion equation in second order approximation, where second order derivates are used) in a covariant form:

$$\frac{d\hat{\mathbf{V}}}{dt} = \frac{\partial\hat{\mathbf{V}}}{\partial t} + \hat{\mathbf{V}} \cdot \nabla\hat{\mathbf{V}} - i\frac{\lambda^2}{2\tau}\left(\frac{dt}{\tau}\right)^{(2/D_F)-1}\nabla^2\hat{\mathbf{V}} \equiv 0$$ (10)

This means that the sum of the local acceleration $\partial\hat{\mathbf{V}}/\partial t$, convection $\hat{\mathbf{V}}\cdot\nabla\hat{\mathbf{V}}$ and "dissipation" $\nabla^2\hat{\mathbf{V}}$ reciprocally compensate in any point of the arbitrarily fractal chosen trajectory of a physical object.

Formally, (10) is a Navier-Stokes type equation, with an imaginary viscosity coefficient,

$$\eta = i\frac{\lambda^2}{2\tau}\left(\frac{dt}{\tau}\right)^{(2/D_F)-1} \tag{11}$$

This coefficient depends on two temporary scales, as well as on a length scale. The existence of a pure imaginary structured coefficient specifies the fact that "the environment" has rheological features (viscoelastic and hysteretic ones (Chioroiu, V. et al, 2005; Ferry, D. K. & Goodnick, S. M., 2001; Imry, Y., 2002)).

For

$$\frac{\lambda^2}{2\tau}\cdot\left(\frac{dt}{\tau}\right)^{(2/D_F)-1} \to 0 \tag{12}$$

equation (10) reduces to Euclidian form (Harabagiu A. et al , 2010; Agop, M. et al, 2008):

$$\frac{\partial\hat{\mathbf{V}}}{\partial t}+\hat{\mathbf{V}}\cdot\nabla\hat{\mathbf{V}} \equiv 0 \tag{13}$$

and, hence, separating the real part from the imaginary one

$$\frac{\partial\mathbf{V}}{\partial t}+\mathbf{V}\cdot\nabla\mathbf{V}-\mathbf{U}\cdot\nabla\mathbf{U}=0$$
$$\frac{\partial\mathbf{U}}{\partial t}+\mathbf{U}\cdot\nabla\mathbf{V}+\mathbf{V}\cdot\nabla\mathbf{U}=0 \tag{14a,b}$$

Equation (14a) corresponds to the law of the impulse conservation at differentiable scale (the undulatory component), while (14b) corresponds to the same law, but at a non differentiable scale (corpuscular component). As we will later show, in the case of irotational movements (14) it will be assimilated to the law of mass conservation.

6. Rotational motions and flow regimes of a fractal fluid

For rotational motions, $\nabla\times\hat{\mathbf{V}} \neq 0$ relation (10) with (9) through separating the real part from the imaginary one, i.e. through separating the motions at a differential scale (undulatory characteristic) and non differential one (corpuscular characteristic), results (Harabagiu A. et al , 2010)

$$\frac{\partial\mathbf{V}}{\partial t}+\mathbf{V}\cdot\nabla\mathbf{V}-\mathbf{U}\cdot\nabla\mathbf{U}-\frac{\lambda^2}{2\tau}\left(\frac{dt}{\tau}\right)^{(2/D_F)-1}\Delta\mathbf{U}=0$$
$$\frac{\partial\mathbf{U}}{\partial t}+\mathbf{U}\cdot\nabla\mathbf{V}+\mathbf{V}\cdot\nabla\mathbf{U}+\frac{\lambda^2}{2\tau}\left(\frac{dt}{\tau}\right)^{(2/D_F)-1}\Delta\mathbf{V}=0 \tag{15a,b}$$

According to the operator relations

$$\mathbf{V} \cdot \nabla \mathbf{V} = \nabla\left(\frac{\mathbf{V}^2}{2}\right) - \mathbf{V} \times (\nabla \times \mathbf{V})$$

$$\mathbf{U} \cdot \nabla \mathbf{U} = \nabla\left(\frac{\mathbf{U}^2}{2}\right) - \mathbf{U} \times (\nabla \times \mathbf{U}) \qquad \text{(16a-c)}$$

$$\mathbf{U} \cdot \nabla \mathbf{V} + \mathbf{V} \cdot \nabla \mathbf{U} = \nabla(\mathbf{U} \cdot \mathbf{V}) - \mathbf{V} \times (\nabla \times \mathbf{U}) - \mathbf{U} \times (\nabla \times \mathbf{V})$$

equations (15) take equivalent forms

$$\frac{\partial \mathbf{V}}{\partial t} + \nabla\left(\frac{\mathbf{V}^2}{2} - \frac{\mathbf{U}^2}{2}\right) - \mathbf{V} \times (\nabla \times \mathbf{V}) - \mathbf{U} \times (\nabla \times \mathbf{U}) - \frac{\lambda^2}{2\tau}\left(\frac{dt}{\tau}\right)^{(2/D_F)-1} \Delta \mathbf{U} = 0$$

$$\frac{\partial \mathbf{U}}{\partial t} + \nabla(\mathbf{V} \cdot \mathbf{U}) - \mathbf{V} \times (\nabla \times \mathbf{U}) - \mathbf{U}(\nabla \times \mathbf{V}) + \frac{\lambda^2}{2\tau}\left(\frac{dt}{\tau}\right)^{(2/D_F)-1} \Delta \mathbf{V} = 0 \qquad \text{(17a,b)}$$

We can now characterize the flow regimes of the fractal fluid at different scales, using some classes of Reynolds numbers. At a differential scale we have

$$R(differential-nondifferential) = R(D-N) = \frac{|V \cdot \nabla V|}{D|\Delta U|} \approx \frac{V^2 l^2}{DUl} \qquad \text{(18)}$$

$$R(nondifferential-nondifferential) = R(N-N) = \frac{|U \cdot \nabla U|}{D|\Delta U|} \approx \frac{Ul}{D} \qquad \text{(19)}$$

with

$$D = \frac{\lambda^2}{2\tau}\left(\frac{dt}{\tau}\right)^{(2/D_F)-1} \qquad \text{(20)}$$

and at nondifferential scale

$$R \text{ (differential-non differential-differential transition)} = R(TDN-D) = \frac{|U \cdot \nabla V|}{D|\Delta V|} \approx \frac{UL}{D} \qquad \text{(21)}$$

$$R \text{ (non differential-differential-differential transition)} = R(TND-D) = \frac{|V \cdot \nabla U|}{D|\Delta V|} \approx \frac{UL^2}{Dl} \qquad \text{(22)}$$

In previous relations V, L, D, are the specific parameters, while U, l, D are the parameters of the non differential scale. The parameters V, U are specific speeds, L, l specific lengths and D is a viscosity coefficient. Moreover, the common "element" for R(D-N), R(N-N), R(TDN-D) and R(TND-D) is the "viscosity" which, through (20) is imposed by the resolution scale.

Equations (15) are simplified in the case of the stationary motion for small Reynolds numbers. Thus, equation (15) for small R (D-N) becomes

$$-\mathbf{U} \cdot \nabla \mathbf{U} - \frac{\lambda^2}{2\tau}\left(\frac{dt}{\tau}\right)^{(2/D_F)-1} \Delta \mathbf{U} = 0 \qquad \text{(23)}$$

and for small R(N-N)

$$-\mathbf{V}\cdot\nabla\mathbf{V} - \frac{\lambda^2}{2\tau}\left(\frac{dt}{\tau}\right)^{(2/D_F)-1}\Delta U = 0 \tag{24}$$

Equation (15b) for small R(TDN-D) takes the form

$$\mathbf{V}\cdot\nabla U + \frac{\lambda^2}{2\tau}\left(\frac{dt}{\tau}\right)^{(2/D_F)-1}\Delta\mathbf{V} = 0 \tag{25}$$

and for small R(TND-D)

$$U\cdot\nabla\mathbf{V} + \frac{\lambda^2}{2\tau}\left(\frac{dt}{\tau}\right)^{(2/D_F)-1}\Delta\mathbf{V} = 0 \tag{26}$$

7. Irotational motions of a fractal fluid. The incorporation of the associate wave corpuscle through the solidarity of movements and generation of Schrodinger equation

For irotational motions

$$\nabla \times \hat{\mathbf{V}} = 0 \tag{27}$$

which implies

$$\nabla \times \mathbf{V} = 0, \nabla \times U = 0 \tag{28 a,b}$$

equation (10) (condition of solidarity of movements) becomes (Harabagiu A. et al , 2010)

$$\frac{\partial\mathbf{V}}{\partial t} + \nabla\left(\frac{\mathbf{V}^2}{2}\right) - i\frac{\lambda^2}{2\tau}\left(\frac{dt}{\tau}\right)^{(2/D_F)-1}\Delta\mathbf{V} = 0 \tag{29}$$

Since through (27) the complex speed field is expressed by means of a scalar function gradient Φ,

$$\hat{\mathbf{V}} = \nabla\Phi \tag{30}$$

equation (29) taking into account the operator identities

$$\frac{\partial}{\partial t}\nabla = \nabla\frac{\partial}{\partial t}, \quad \nabla\Delta = \Delta\nabla \tag{31}$$

takes the form

$$\nabla\left[\frac{\partial\Phi}{\partial t} + \frac{1}{2}(\nabla\Phi)^2 - i\frac{\lambda^2}{2\tau}\left(\frac{dt}{\tau}\right)^{(2/D_F)-1}\Delta\Phi\right] = 0 \tag{32}$$

or furthermore, through integration

$$\frac{\partial \Phi}{\partial t} + \frac{1}{2}(\nabla \Phi)^2 - i\frac{\lambda^2}{2\tau}\left(\frac{dt}{\tau}\right)^{(2/D_F)-1} \Delta \Phi = F(t) \tag{33}$$

where F(t) is an arbitrary function depending only on time.
In particular, for Φ having the form

$$\Phi = -2i\frac{\lambda^2}{2\tau}\left(\frac{dt}{\tau}\right)^{(2/D_F)-1} \ln \Psi \tag{34}$$

where Ψ is a new complex scalar function, equation (46), with the operator identity

$$\frac{\Delta \Psi}{\Psi} = \Delta \ln \Psi + (\nabla \ln \Psi)^2 \tag{35}$$

takes the form :

$$\frac{\lambda^4}{4\tau^2}\left(\frac{dt}{\tau}\right)^{(4/D_F)-2} \Delta \Psi + i\frac{\lambda^2}{2\tau}\left(\frac{dt}{\tau}\right)^{(2/D_F)-1} \frac{\partial \Psi}{\partial t} + \frac{F(t)}{2}\Psi = 0 \tag{36}$$

The Schrodinger "geodesics" can be obtained as a particular case of equation (36), based on the following hypothesis (conditions of solidarity of the motion, incorporating the associated wave corpuscle):

i.   the motions of the micro-particles take place on fractal curves with the fractal dimension D_F=2, i.e. the Peano curves (Nottalle, L., 1993; Nottalle, L., 2004);

ii. $d_\pm\xi^i$ are the Markov-Wiener type stochastic variables (Nottalle, L., 1993; Nottalle, L., 2004) that satisfy the rule

$$\langle d_\pm\xi^i d_\pm\xi^l \rangle = \pm\delta^{il}\frac{\lambda^2}{\tau}dt \tag{37}$$

iii. space scale λ and temporary one τ are specific for the Compton scale

$$\lambda = \frac{\hbar}{m_0 c}, \quad \tau = \frac{\hbar}{m_0 c^2} \tag{38}$$

with m_0 the rest mass of the microparticle, c the speed of light in vacuum and $\hbar$ the reduced Planck constant. The parameters (38) should not be understood as "structures" of the standard space-time, but as standards of scale space-time; iv) function F(t) from (36) is null. Under these circumstances, (36) is reduced to the standard form of Schrodinger's equation (Țițeica, S., 1984; Peres, A., 1993)

$$\frac{\hbar^2}{2m_0}\Delta \Psi + i\hbar\frac{\partial \Psi}{\partial t} = 0 \tag{39}$$

In such a context, the scale potential of the complex speeds plays the role of the wave function.

8. Extended hydrodynamic model of scale relativity and incorporation of associated wave corpuscle through fractal potential. The correspondence with Madelung model

Substituting the complex speed (9) with the restriction (27) and separating the real part with the imaginary one, we obtain the set of differential equations (Harabagiu A. et al , 2010)

$$m_0 \frac{\partial \mathbf{V}}{\partial t} + m_0 \nabla \left(\frac{\mathbf{V}^2}{2} \right) = -\nabla(Q)$$

$$\frac{\partial \mathbf{U}}{\partial t} + \nabla(\mathbf{V} \cdot \mathbf{U}) + \frac{\lambda^2}{2\tau} \left(\frac{dt}{\tau} \right)^{(2/D_F)-1} \Delta \mathbf{V} = 0$$

(40a,b)

where Q is the fractal potential, expressed as follows

$$Q = -\frac{m_0 \mathbf{U}^2}{2} - \frac{m_0}{2} \frac{\lambda^2}{2\tau} \left(\frac{dt}{\tau} \right)^{(2/D_F)-1} \nabla \cdot \mathbf{U}$$

(41)

For

$$\Psi = \sqrt{\rho} e^{iS}$$

(42)

with $\sqrt{\rho}$ an amplitude and S a phase, then (34) under the form

$$\Phi = -i \frac{\lambda^2}{\tau} \left(\frac{dt}{\tau} \right)^{(2/D_F)-1} \ln\left(\sqrt{\rho} e^{iS} \right)$$

implies the complex speed fields of components

$$\mathbf{V} = \frac{\lambda^2}{2\tau} \left(\frac{dt}{\tau} \right)^{(2/D_F)-1} \nabla S, \quad \mathbf{U} = \frac{\lambda^2}{2\tau} \left(\frac{dt}{\tau} \right)^{(2/D_F)-1} \nabla \ln \rho$$

(43a,b)

From the perspective of equations (43), the equation (40) keeps its form, and the fractal potential is given by the simple expression

$$Q = -m_0 \frac{\lambda^2}{\tau} \left(\frac{dt}{\tau} \right)^{(2/D_F)-1} \frac{\Delta\sqrt{\rho}}{\sqrt{\rho}}$$

(44)

Again through equations (43), equation (40b) takes the form:

$$\nabla \left(\frac{\partial \ln \rho}{\partial t} + \mathbf{V} \cdot \nabla \ln \rho + \nabla \cdot \mathbf{V} \right) = 0$$

or, still, through integration with $\rho \neq 0$

$$\frac{\partial \rho}{\partial t} + \nabla \cdot (\rho \mathbf{V}) = T(t)$$

(45)

with T(t), an exclusively time dependent function
Equation (40) corresponds to the impulse conservation law at differential scale (the classical one), while the impulse conservation law at non differential scale is expressed through (45) with $T(t) \equiv 0$, as a probability density conservation law
Therefore, equations

$$m_0 \left(\frac{\partial \mathbf{V}}{\partial t} + \mathbf{V} \cdot \nabla \left(\frac{\mathbf{V}^2}{2} \right) \right) = -\nabla(Q)$$

$$\frac{\partial \rho}{\partial t} + \nabla \cdot (\rho \mathbf{V}) = 0$$

(46a,b)

with Q given by (41) or (44) forms the set of equations of scale relativity extended hydrodynamics in fractal dimension D_F. We mention that in references (Nottalle, L., 1993; Chaline, J. et al, 2009; Chaline, J. et al, 2000; Nottalle, L., 2004) the model has been extended only for D_F=2. The fractal potential (41) or (44) is induced by the non differentiability of space-time.
In an external scalar field U, the system of equations (46) modifies as follows

$$m_0 \left[\frac{\partial \mathbf{V}}{\partial t} + \nabla \left(\frac{\mathbf{V}^2}{2} \right) \right] = -\nabla(Q+U)$$

$$\frac{\partial \rho}{\partial t} + \nabla(\rho \mathbf{V}) = 0$$

(47a,b)

Now the quantum mechanics in hydrodynamic formula (Madelung's model (Halbwacs, F., 1960)) is obtained as a particular case of relations (47), using the following hypothesis: i) the motion of the micro-particles takes place on Peano curves with D_F=2; ii) $d_\pm \xi^i$ are the Markov-Wiener variables (Nottalle, L., 1993; Chaline, J. et al, 2009; Chaline, J. et al, 2000; Nottalle, L., 2004); iii) the time space scale is a Compton one. Then, (38) have the expressions

$$\mathbf{V} = \frac{\hbar}{m_0} \nabla S, \quad \mathbf{U} = \frac{\hbar}{2m_0} \nabla \ln \rho$$

(48)

and (41),

$$Q = -\frac{m_0 \mathbf{U}^2}{2} - \frac{\hbar}{2} \nabla \cdot \mathbf{U}$$

(49)

9. "Mechanisms" of duality through coherence in corpuscle-wave phase

In the stationary case, the system of equations (46) becomes (Harabagiu A. et al , 2010)

$$\nabla \left(\frac{\mathbf{V}^2}{2} + Q \right) = 0$$

$$\nabla(\rho \mathbf{V}) = 0$$

(50a,b)

or, still, through integration

$$\frac{\mathbf{V}^2}{2} + Q = E = const.$$

$$\rho \mathbf{V} = const.$$

(51a,b)

Let us choose the null power density in (51b). Then there is no impulse transport at differential scale between corpuscle and wave. Moreover, for $\rho \neq 0$

$$\mathbf{V} = 0$$

(52)

which implies through relation (43)

$$S = const.$$

(53)

In other words, the fluid becomes coherent (the fluid particles have the same phase). Such a state is specific for quantum fluids (Ciuti C. & Camsotto I., 2005; Benoit Deveand, 2007), such as superconductors, superfluids, etc. (Felsager, B., 1981; Poole, C. P. et al, 1995). Under such circumstances, the phase of the corpuscle considered as a small horologe equals the phase of the associated wave (coherence in corpuscle-wave phase).

At non-differential scale, equation (51) , with restriction (52) takes the form

$$Q = -\frac{2m_0 D^2 \Delta \sqrt{\rho}}{\sqrt{\rho}} = -\frac{m_0 \mathbf{U}^2}{2} - m_0 D \nabla \mathbf{U} = E = const$$

$$D = \frac{\lambda^2}{\tau} \left(\frac{dt}{\tau} \right)^{(2/D_F)-1}$$

(54 a,b)

or, still, by applying the gradient operator

$$\mathbf{A} = \nabla(\sqrt{\rho})$$

(55)

$$\Delta \mathbf{A} + \frac{E}{2m_0 D^2} \mathbf{A} = 0$$

(56)

We distinguish the following situations
i.   For E>0 and with substitution

$$\frac{1}{\Lambda^2} = \frac{E}{2m_0 D^2}$$

(57)

equation (56) becomes

$$\Delta \mathbf{A} + \frac{1}{\Lambda^2} \mathbf{A} = 0$$

(58)

Therefore:
1.   the space oscillations of field $\mathbf{A}$ and, therefore the space associated with the motion of coherent fluid particles is endowed with regular non homogeneities (of lattice type). In other words, the field $\mathbf{A}$ crystallizes ("periodicizes") the space. The one dimensional space "crystal" has the constant of the network

$$\Lambda = \frac{\lambda^2}{2\tau}\left(\frac{dt}{\tau}\right)^{(2/D_F)-1}\left(\frac{2m_0}{E}\right)^{1/2} \tag{59}$$

that depends both on the "viscosity" – $i\eta$ given by (11) and on the energy of the particle;

2. the one dimensional geodesics of the "crystallized" space given by the expression

$$\rho(x) = A^2 \sin^2(kx + \delta) \tag{60}$$

implies both fractal speed

$$U_x = D\frac{d\ln\rho}{dx} = 2Dkctg(kx + \delta) \tag{61}$$

and fractal potential

$$Q_x = -\frac{m_0 U_x^2}{2} - m_0 D\frac{dU_x}{dx} = -2m_0 D^2 k^2 ctg^2(kx + \delta)$$
$$+2m_0 D^2 k^2 \frac{1}{\sin^2(kx + \delta)} = 2m_0 D^2 k^2 \tag{62}$$

with A and δ and the integration constants

$$k = \frac{1}{\Lambda} \tag{63}$$

3. for the movements of microparticles on Peano curves ($D_F=2$) at Compton scale

$$D = 2m_0\hbar ,$$

therefore, through (62) under the form

$$Q_x = 2m_0 D^2 k^2 = \frac{p_x^2}{2m_0}, \quad p_x = 2m_0 Dk \tag{64a,b}$$

de Broglie "quantum" impulse is found

$$p_x = \frac{\hbar}{\Lambda} \tag{65}$$

4. the dominant of the undulatory characteristic is achieved by the "self diffraction" mechanism of the fractal field, ρ, on the one dimensional space "crystal" of constant Λ induced by the same field. Indeed, relation (61) with notations

$$\Phi = kx + \delta , \quad k = \frac{1}{\Lambda} \tag{66a,b}$$

in approximation $\Phi \ll 1$, i.e. for $tg\Phi \approx \sin\Phi$ and using Nottale's relation (Nottalle, L., 1993; Chaline, J. et al, 2009; Chaline, J. et al, 2000; Nottale, L., 2004; Nottale, L. &

Schneider J., 1984; Nottale, L., 1989; Nottale, L., 1996) $2D / U_x \approx n\lambda$ it takes the common form (Bragg's relation)

$$\Lambda \sin \Phi \approx n\lambda; \tag{67}$$

This result is in concordance with the recently expressed opinion in (Mandelis A. et al, 2001; Grössing G., 2008; Mandelis A., 2000);

5. there is impulse transfer on the fractal field between the corpuscle and the wave;
6. according to Taylor's criterion (Popescu, S., 2004) self-organization (crystallization and self diffraction of the space) appears when the energy of the system is minimal. This can be immediately verified using relation (51a);

ii. For E=0, equations (51a) and (56) have the same form

$$\Delta\sqrt{\rho}=0 \quad \Delta\mathbf{A} = 0 \tag{68}$$

It follows that:

1. the geodesics are expressed through harmonic functions and the particle finds itself in a critical state, i.e. the one that corresponds to the wave-corpuscle transition;
2. in the one –dimensional case, the geodesics have the form

$$\rho(x) = kx + \delta \tag{69}$$

which induces the fractal speed field

$$U_x = \frac{D}{kx + \delta} \tag{70}$$

namely the null value of the fractal potential

$$Q_x = -\frac{m_0}{2} \frac{D^2}{(kx+\delta)^2} + \frac{m_0}{2} \frac{D^2}{(kx+\delta)^2} = 0 \tag{71}$$

3. although the energy is null, there is impulse transfer between corpuscle and wave on the fractal component of the speed field

iii. For E<0 and with notations

$$\frac{1}{\bar{\Lambda}^2} = \frac{\bar{E}}{2m_0 D^2}, \quad E = -\bar{E} \tag{72}$$

equation (56) takes the form

$$\Delta\mathbf{A} - \frac{1}{\bar{\Lambda}^2}\mathbf{A} = 0 \tag{73}$$

The following aspects result:

1. field A is expelled from the structure, its penetration depth being

$$\bar{\Lambda} = \frac{\lambda^2}{2\tau}\left(\frac{dt}{\tau}\right)^{(2/D_F)-1}\left(\frac{2m_0}{\bar{E}}\right)^{1/2} \tag{74}$$

2. the one-dimensional geodesics of the space are described through function

$$\rho(x) = \overline{A}^2 sh^2(\overline{k}x + \overline{\delta})$$ (75)

and lead to the fractal speed

$$U_x = 2D\overline{k}cth(\overline{k}x + \overline{\delta})$$ (76)

the fractal potential respectively

$$Q_x = -2m_0 D^2 \overline{k}^2 cth^2(\overline{k}x + \overline{\delta}) + 2m_0 D^2 \frac{\overline{k}^2}{sh^2(\overline{k}x + \overline{\delta})} = -2m_0 D^2 \overline{k}^2$$ (77)

where $\overline{A}$, $\overline{\delta}$ are two integration constants and

$$\overline{k} = \frac{1}{\overline{\Lambda}}$$ (78)

3. the dominant of the corpuscular characteristic is accomplished by means of "self-expulsion" mechanism of the fractal field from its own structure that it generates (that is the corpuscle), the penetration depth being $\overline{\Lambda}$. The identification

$$Q_x = -2m_0 D^2 \overline{k}^2 = \frac{\overline{p}^2}{2m_0}$$ (79)

implies the purely imaginary impulse

$$\overline{p} = -2im_0 D\overline{k}$$ (80)

that suggests ultra rapid virtual states (ultra rapid motions in the wave field, resulting in the "singularity"of the field, i.e. the corpuscle). As a matter of fact, if we consider de Broglie's original theory (motions on Peano curves with $D_F=2$, at Compton's scale), singularity (the corpuscle) moves "suddenly" and chaotically in the wave field, the wave-corpuscle coupling being accomplished through the fractal potential. The corpuscle "tunnels" the potential barrier imposed by the field of the associate progressive wave, generating particle-antiparticle type pairs (ghost type fields (Bittner E.R., 2000)). Nevertheless this model cannot specify the type of the physical process by means of which we reach such a situation: it is only the second quantification that can do this (Ciuti C. & Camsotto I., 2005; Benoit Deveand Ed., 2007; Mandelis A. et al, 2001; Grössing G., 2008; Mandelis A., 2000; Bittner E.R., 2000);

4. there is an impulse transfer between the corpuscle and the wave on the fractal component of the speed field, so that all the attributes of the differential speed could be transferred on the fractal speed.

   All the above results indicate that wave-particle duality is an intrinsic property of space and not of the particle.

10. Wave-corpuscle duality through flowing stationary regimes of a coherent fractal fluid in phase. The potential well

According to the previous paragraph, let us study the particle in a potential well with infinite width and walls. Then the speed complex field has the form (Harabagiu A. et al, 2010; Agop, M. et al, 2008; Harabagiu, A. & Agop, M., 2005; Harabagiu, A. et al, 2009)

$$\hat{V}_x = V_x - iU_x = 0 - 2iD\left(\frac{n\pi}{a}\right)ctg\left(\frac{n\pi}{a}\right)x \tag{81}$$

and generates the fractal potential (the energy of the structure) under the form of the noticeable

$$Q_n = 2m_0 D^2\left(\frac{n\pi}{a}\right)^2 = E_n \tag{82}$$

The last relation (82) allows the implementation of Reynold's criterion

$$R(n) = \frac{V_c a}{D} = 2n\pi, \quad V_c = \left(\frac{2E_n}{m_0}\right)^{\frac{1}{2}} \tag{83a,b}$$

For movements on Peano curves ($D_F=2$) at Compton scale $(2m_0 D = \hbar)$ (83) with substitutions

$$m_0 V_c = \Delta P_x, \quad a = \Delta x \tag{84a,b}$$

and n=1 reduces to Heinsenberg's relation of uncertainty under equal form

$$\Delta p_x \Delta x = \frac{h}{2} \tag{85}$$

while for $n < +\infty$ it implies a Ruelle-Takens' type criterion of evolution towards chaos (Ruelle D. & Takens, F., 1971; Ruelle, D., 1975). Therefore, the wave-corpuscle duality is accomplished through the flowing regimes of a fractal fluid that is coherent in phase. Thus, the laminar flow (small n) induces a dominant ondulatory characteristic, while the turbulent flow (big n) induces a dominant corpuscular characteristic.

11. Wave-corpuscle duality through non-stationary regimes of an incoherent fractal fluid

In the one dimensional case the equations of hydrodynamics (46) take the form

$$m_0\left(\frac{\partial V}{\partial t} + V \cdot \frac{\partial V}{\partial x}\right) = -\frac{\partial}{\partial x}\left[-2m_0 D^2 \frac{1}{\rho^{1/2}} \frac{\partial^2}{\partial x^2}\left(\rho^{1/2}\right)\right]; \frac{\partial \rho}{\partial t} + \frac{\partial}{\partial x}(\rho V) = 0 \tag{86a,b}$$

Imposing the initial conditions

$$V(x,t=0) = c = const$$

$$\rho(x,t=0) = \frac{1}{\pi^{1/2}\alpha}e^{-\left(\frac{x}{\alpha}\right)^2} = \rho_0 \tag{87a,b}$$

and on the frontier

$$V(x = ct, t) = c$$
$$\rho(x = -\infty, t) = \rho(x = +\infty, t) = 0 \qquad \text{(88a,b)}$$

the solutions of the system (86), using the method in (Munceleanu, C.V. et al, 2010), have the expressions

$$\rho(x,t) = \frac{1}{\pi^{1/2} \left[\alpha^2 + \left(\dfrac{2D}{\alpha}t\right)^2\right]^{1/2}} \exp\left[\frac{(x-ct)^2}{\alpha^2 + \left(\dfrac{2D}{\alpha}t\right)^2}\right]$$

$$V = \frac{c\alpha^2 + \left(\dfrac{2D}{\alpha}\right)^2 tx}{\alpha^2 + \left(\dfrac{2D}{\alpha}t\right)^2} \qquad \text{(89a,b)}$$

The complex speed field is obtained

$$\hat{V} = V - iU = \frac{c\alpha^2 + \left(\dfrac{2D}{\alpha}\right)^2 tx}{\alpha^2 + \left(\dfrac{2D}{\alpha}t\right)^2} + 2iD\frac{x-ct}{\alpha^2 + \left(\dfrac{2D}{\alpha}t\right)^2} \qquad \text{(90)}$$

and the field of fractal forces

$$F = 4m_0 D^2 \frac{(x-ct)}{\left[\alpha^2 + \left(\dfrac{2D}{\alpha}t\right)^2\right]^2} \qquad \text{(91)}$$

Therefore:
i.   both differential scale speed V and non-differential one U are not homogeneous in x and t. Under the action of fractal force F, the corpuscle is assimilated to the wave, is a part of its structure, so that it joins the movement of the corpuscle with the propagation of the associated progressive wave;
ii. the timing of the movements at the two scales, V=U implies the space-time homographic dependence

$$x = \frac{c\alpha^2}{2D} \frac{1 + \dfrac{2D}{\alpha^2}t}{1 - \dfrac{2D}{\alpha^2}t} \qquad \text{(92)}$$

in the field of forces

$$F = \frac{2m_0 Dc}{\left(1 - \frac{2D}{\alpha^2}t\right)\left[\alpha^2 + \left(\frac{2D}{\alpha}t\right)^2\right]} \tag{93}$$

Considering that the type (92) changes are implied in gravitational interaction (Ernst, F.J., 1968; Ernst, F.J., 1971), it follows that the solidarity of the corpuscle movement with the movement of the associated progressive wave is accomplished by means of the appropriate gravitational field of the physical object;

iii. the uniform movement V=c is obtained for null fractal force F=0 and fractal speed U=0, using condition x=ct. The fractal forces in the semi space. $-\infty \leq x \leq \bar{x}$ and $\bar{x} \leq x \leq +\infty$ are reciprocally compensated.

$$F\Big|_{-\infty}^{\bar{x}} = F\Big|_{\bar{x}}^{+\infty}$$

This means that the corpuscle in "free" motion simultaneously polarizes the "environment" of the wave behind $x \leq ct$ and in front of $x \geq ct$, in such a manner that the resulting force has a symmetrical distribution as compared with the plane that contains the position of the noticeable object $\bar{x} = ct$ at any time moment t. Under such circumstances, the physical object uniformly moves (the corpuscle is located in the field of the associated wave).

12. The corpuscle as a horologe and its incorporation in the associated wave. Consequences

According to de Broglie's theory, the corpuscle must be associated to a horologe having the frequency equal to that of the associated progressive wave. Mathematically we can describe such an oscillator through the differential equation

$$\ddot{q} + \omega^2 q = 0 \tag{94}$$

where ω defines the natural frequency of the oscillator as it is dictated by the environment (the wave), and the point above the symbol referes to the differential as compared with time. The most general solution of equation (94) generally depends not on two arbitrary constants, as it is usually considered, but on three: the initial relevant coordinate, the initial speed and the phase of the harmonic oscillatory within the ensemble that structurally represents the environment (the isolated oscillator is an abstraction !). Such a solution gives the relevant co-ordinate

$$q(t) = h e^{i(\omega t + \Phi)} + \bar{h} e^{-i(\omega t + \Phi)} \tag{95}$$

where $\bar{h}$ refers to the complex conjugate of h and Φ is an initial phase specific to the individual movement of the oscillator. Such a notation allows us to solve a problem that we could name "the oscillators with the same frequency", such as Planck's resonators' ensemble-the basis of the quantum theory arguments in their old shape. That is, given an ensemble of oscillators having the same frequency in a space region, which is the relation between them?

The mathematical answer to this problem can be obtained if we note that what we want here is to find a mean to pass from a triplet of numbers –the initial conditions- of an oscillator towards the same triplet of another oscillator with the same frequency. This process (passing) implies a simple transitive continuous group with three parameters that can be built using a certain definition of the frequency. We start from the idea that the ratio of two fundamental solutions of equation (94) is a solution of Schwarts' non linear equation (Agop, M. & Mazilu, N., 1989; Agop, M. & Mazilu, N., 2010; Mihăileanu, N., 1972)

$$\frac{d}{dt}\left(\frac{\ddot{\tau}_0}{\dot{\tau}_0}\right) - \frac{1}{2}\left(\frac{\ddot{\tau}_0}{\dot{\tau}_0}\right)^2 = 2\omega^2, \quad \tau_0(t) \equiv e^{-2i\omega t} \tag{96}$$

This equation proves to be a veritable definition of frequency as a general characteristic of an ensemble of oscillators that can be scanned through a continuous group of three parameters. Indeed equation (96) is invariant to the change of the dependent variable

$$\tau(t) = \frac{a\tau_0(t) + b}{c\tau_0(t) + d} \tag{97}$$

which can be verified through direct calculation. Thus, $\tau(t)$ characterizes another oscillator with the same frequency which allows us to say that, starting from a standard oscillator we can scan the whole ensemble of oscillators of the same frequency when we let loose the three ratios a: b: c: d in equation (97). We can make a more precise correspondence between a homographic change and an oscillator, by means of associating to each oscillator a personal $\tau(t)$ through equation

$$\tau_1(t) = \frac{h + \bar{h}k\tau_0(t)}{1 + k\tau_0(t)} \qquad k \equiv e^{-2i\Phi} \tag{98}$$

Let us notice that τ_0, τ_1 can be freely used one instead the other, which leads to the next group of changes for the initial conditions

$$h' \rightarrow \frac{ah + b}{ch + d} \quad \bar{h}' \rightarrow \frac{a\bar{h} + b}{c\bar{h} + d} \quad k' \rightarrow k \cdot \frac{c\bar{h} + d}{ch + d} \qquad a, b, c, d \in R \tag{99a-d}$$

This is a simple transitive group: one and only one change of the group (the Barbilian group (Agop, M. & Mazilu, N., 1989; Agop, M. & Mazilu, N., 2010; Barbilian, D., 1935; Barbilian, D., 1935; Barbilian, D., 1938; Barbilian, D., 1971)) corresponds to a given set of values (a/c, b/c, d/c).

This group admits the 1-differential forms, absolutely invariant through the group (Agop, M. & Mazilu, N., 1989)

$$\omega_0 = i\left(\frac{dk}{k} - \frac{dh + d\bar{h}}{h - \bar{h}}\right), \quad \omega_1 = \bar{\omega}_2 = \frac{dh}{k(h - \bar{h})} \tag{100}$$

and the 2- differential form

$$\frac{ds^2}{\alpha^2} = \omega_0^2 - 4\omega_1\omega_2 = -\left(\frac{dk}{k} - \frac{dh - d\bar{h}}{h - \bar{h}}\right)^2 + 4\frac{dh\,d\bar{h}}{\left(h - \bar{h}\right)^2}, \quad \alpha = const. \tag{101}$$

respectively.

If we restrict the definition of a parallelism of directions in Levi-Civita manner (Agop, M. & Mazilu, N., 1989)

$$d\varphi = -\frac{du}{v} \tag{102}$$

with

$$h=u+iv \,, \overline{h} = u-iv, \, k=e^{-i\varphi} \tag{103}$$

Barbilian's group invariates the metrics of Lobacevski's plane (Agop, M. & Mazilu, N., 1989),

$$\frac{ds^2}{\alpha^2} = -\frac{du^2+dv^2}{v^2} \tag{104}$$

Metrics (104) coincides with the differential invariant that is built with the complex scalar field of the speed,

$$\frac{ds^2}{\alpha^2} \equiv d\phi d\overline{\phi} = (2Dds - iDd\ln\rho)(2Dds + iDd\ln\rho) = 4D^2(ds)^2 + D^2\left(\frac{d\rho}{\rho}\right)^2 \tag{105}$$

which admits the identities

$$\alpha = D, \; 2ds \equiv d\Phi = -\frac{du}{v}, \; d\ln\rho \equiv d\ln v \tag{106a-c}$$

Now, through a Matzner-Misner type principle one can obtain Ernst's principle of generating the symmetrical axial metrics (Ernst, F.J., 1968; Ernst, F.J., 1971)

$$\delta\int\frac{\nabla h\nabla\overline{h}}{(h-\overline{h})^2}\gamma^{1/2}d^3x = 0 \tag{106d}$$

where $\gamma = \det\gamma_{\alpha\beta}$ with $\gamma_{\alpha\beta}$ the metrics of the "environment".

Therefore, the incorporation of the corpuscle in the wave, considering that it functions as a horologe with the same frequency as that of the associated progressive wave, implies gravitation through Einstein's vacuum equations (equivalent to Ernst's principle (106d)). On the contrary, when the frequencies do not coincide, there is an induction of Stoler's group from the theory of coherent states (the parameter of the change is the very ratio of frequencies when creation and annihilation operators refer to a harmonic oscillator (Agop, M. & Mazilu, N., 1989)).

Let us note that the homographic changes (99) generalize the result (92). Moreover, if $a,b,c,d \, \epsilon\mathbb{Z}$ then the Ernst type equations describe supergravitation N=1 (Green, M.B. et al, 1998).

13. Informational energy through the fractal potential of complex scalar speed field. The generation of forces

The informational energy of a distribution is defined through the known relation (Mazilu N. & Agop M., 1994),

$$E = -\int \rho \ln \rho dx \tag{107}$$

where $\rho(x)$ is the density of distributions, and we note by x, on the whole, the random variables of the problem, dx being the elementary measure of their field.

This functional represents a measure of the uncertainty degree, when defining the probabilities, i.e. it is positive, it increases when uncertainty also increasases taken in the sense of expanding distribution and it is additive for sources that are independent as compared to uncertainity. If we admit the maximum of informational energy in the inference against probabilities, having at our disposal only a partial piece of information this is equivalent to frankly admitting the fact that we cannot know more. Through this, the distributions that we obtain must be at least displaced, as compared to the real ones, because there is no restrictive hypothesis regarding the lacking information. In other words, such a distribution can be accomplished in the highest number of possible modalities. The partial piece of information we have at our disposal, is given, in most cases, in the form of a f(x) function or of more functions.

$$\overline{f} = \int \rho(x) f(x) dx \tag{108}$$

Relation (108), together with the standard relation of distribution density

$$\int \rho(x) dx = 1 \tag{109}$$

are now constraints the variation of the functional (107) has to subject to, in order to offer the distribution density corresponding to the maximum of informational energy. In this concrete case, Lagrange's non determined multipliers method directly leads to the well known exponential distribution

$$\rho(x) = \exp(-x - \mu f(x)) \tag{110}$$

Let us notice that through the fractal component of the complex scalar of speed field

$$\Phi = D \ln \rho \tag{111}$$

expression (107), ignoring the scale factor D, is identical with the average mean of (111)

$$E = -\frac{\overline{\Phi}}{D} = -\int \rho \ln \rho dx \tag{112}$$

In the particular case of a radial symmetry, imposing the constraints

$$\overline{r} = \int \rho(r) r dr \tag{113}$$

$$\int \rho(r) dr \equiv 1 \tag{114}$$

the distribution density $\rho(r)$ through the maximum of informational energy implies the expression

$$\rho(r) = \exp(-\lambda - \mu r), \quad \lambda, \mu = const. \tag{115}$$

or in notations

$$\exp(-\lambda) \equiv \rho_0, \quad \mu = 2/a \tag{116}$$

$$\rho(r) = \rho_0 e^{-\frac{2r}{a}} \tag{117}$$

Then the fractal speed

$$u = D\frac{d}{dr}(\ln \rho) = -\frac{2D}{a} = const \tag{118}$$

through the fractal potential

$$Q = -\frac{m_0 u^2}{2} - m_0 D^2\left[\frac{d^2}{dr^2}(\ln \rho) + \frac{2}{r}\frac{d}{dr}(\ln \rho)\right] = -\frac{2m_0 D^2}{a}\left(\frac{1}{a} - \frac{2}{r}\right) \tag{119}$$

implies the fractal field of central forces

$$F(r) = -\frac{dQ}{dr} = -\frac{4m_0 D^2}{ar^2} \tag{120}$$

Consequently, the fractal "medium" by maximization of the informational energy becomes a source of central forces (gravitational or electric type).

14. Lorenz type mechanism of wave-corpuscle duality in non stationary systems

Impulse conservation law

Let us rewrite the system of equations (15) for an external scalar field U under the form

$$\frac{\partial \mathbf{V}}{\partial t} + \mathbf{V} \cdot \nabla \mathbf{V} - \mathbf{U} \cdot \nabla U - D\Delta U = -\nabla U$$

$$\frac{\partial \mathbf{U}}{\partial t} + \mathbf{V} \cdot \nabla \mathbf{U} + \mathbf{U} \cdot \nabla \mathbf{V} + D\Delta \mathbf{V} = 0 \tag{121a,b}$$

with D given by relation (54). Hence, through their decrease and using substitution

$$\overline{V} = V - U \tag{122}$$

we find

$$\frac{\partial \overline{\mathbf{V}}}{\partial t} + \overline{\mathbf{V}} \cdot \nabla \overline{\mathbf{V}} = 2\mathbf{U} \cdot \nabla U + 2D\Delta U + D\Delta \overline{\mathbf{V}} - \nabla U \tag{123}$$

Taking into account that the fractal term, $2\mathbf{U} \cdot \nabla U + 2D\Delta U$ intervenes as a pressure (for details see the kinetic significance of fractal potential Q (Bohm, D., 1952)) then we can admit the relation

$$2\mathbf{U} \cdot \nabla U + 2D\Delta U = -2\left(-\frac{\mathbf{U}^2}{2} - D\nabla \cdot \mathbf{U}\right) = -2\nabla\left(\frac{Q}{m_0}\right) = -\frac{\nabla p}{\rho} \tag{124}$$

then equation (123) takes the usual form

$$\frac{\partial \mathbf{V}}{\partial t} + \overline{\mathbf{V}} \cdot \nabla \overline{\mathbf{V}} = -\frac{\nabla p}{\rho} - \nabla U + D\Delta \overline{\mathbf{V}} \tag{125}$$

In particular, if $\nabla U = \mathbf{g}$ is a gravitational accelaration (125) becomes

$$\frac{\partial \overline{\mathbf{V}}}{\partial t} + \overline{\mathbf{V}}.\nabla \overline{\mathbf{V}} = -\frac{\nabla p}{\rho} - \mathbf{g} + D\Delta \overline{\mathbf{V}} \tag{126}$$

Energy conservation law

Energy conservation law, ε in the case of movements on fractal curves of fractal dimension D_F is written under the form

$$\frac{\hat{d}\varepsilon}{dt} = \frac{d\varepsilon}{dt} + \hat{\mathbf{V}} \cdot \nabla \varepsilon - iD\Delta \varepsilon = 0 \tag{127}$$

or, still, by separating the real part from the imaginary one

$$\frac{\partial \varepsilon}{\partial t} + \mathbf{V} \cdot \nabla \varepsilon = 0, \quad -\mathbf{U} \cdot \nabla \varepsilon = D\Delta \varepsilon \tag{128}$$

Hence, through addition and taking into account relation (122), we obtain the expression

$$\frac{\partial \varepsilon}{\partial t} + \overline{\mathbf{V}} \cdot \nabla \varepsilon = D\Delta \varepsilon \tag{129}$$

In particular, for $\varepsilon = 2m_0 D\Omega$ with Ω the wave pulsation (for movements on Peano curves with $D_F = 2$ at Compton scale $\varepsilon = \hbar\Omega$) the previous relation becomes

$$\frac{\partial \Omega}{\partial t} + \overline{\mathbf{V}} \cdot \nabla \Omega = D\Delta \Omega \tag{130}$$

Lorenz type "mechanism"

For an incompressible fractal fluid, the balance equations of the "impulse" -see (126), of the energy -see (129) and "mass" – see (46) with $\rho = const.$ and $\nabla \cdot \mathbf{U} = 0$ become

$$\frac{\partial \overline{\mathbf{V}}}{\partial t} + \overline{\mathbf{V}} \cdot \nabla \overline{\mathbf{V}} = -\frac{\nabla p}{\rho} - \mathbf{g} + D\Delta \overline{\mathbf{V}}$$

$$\frac{\partial \varepsilon}{\partial t} + \overline{\mathbf{V}} \cdot \nabla \varepsilon = D\Delta \varepsilon \tag{131a-c}$$

$$\nabla \overline{\mathbf{V}} = 0$$

Let us take into account the following simplyfing hypothesis:

i. constant density, $\rho = \rho_0 = const.$ excepting the balance equation of the impulse where density is disturbed according to relation

$$\rho = \rho_0 + \delta\rho \tag{132}$$

ii.   the energy "expansion" is a linear one

$$\rho = \rho_0\left[1 - \alpha\left(\varepsilon - \varepsilon_0\right)\right] \tag{133}$$

with α the energy "dilatation" constant.
Under such circumstances, system (131) becomes

$$\rho_0\left(\frac{\partial\overline{\mathbf{V}}}{\partial t} + \overline{\mathbf{V}}\cdot\nabla\overline{\mathbf{V}}\right) + \nabla p = \left(\rho_0 + \delta\rho\right)\mathbf{g} + \rho_0 D\Delta\overline{\mathbf{V}}$$

$$\frac{\partial\varepsilon}{\partial t} + \overline{\mathbf{V}}\cdot\nabla\varepsilon = D\Delta\varepsilon \tag{134a-c}$$

$$\nabla\overline{\mathbf{V}} = 0$$

In order to study the dynamics of system (134), our description closely follows the approach in (Bârzu, A. et al, 2003).
The convection in the fractal fluid takes place when the ascending force that results from energy "dilatation" overcomes the viscous forces. Then we can define the Rayleigh number

$$R = \frac{|F_{asc}|}{|F_{visc}|} \approx \frac{\left|\dfrac{\delta\rho g}{\rho_0}\right|}{\left|D\Delta\overline{\mathbf{V}}\right|} \tag{135}$$

The variation of the density satisfies through (133) the relation

$$\frac{\delta\rho}{\rho_0} \approx \alpha\Delta\varepsilon \tag{136}$$

and the "energy" balance equation (134c) implies

$$\overline{V} \approx \frac{D}{d} \tag{137}$$

where d is the thickness of the fractal fluid level. Substituting (136) and (137) in (135) we obtain Rayleigh's number under the form

$$R = \frac{\alpha\beta g d^4}{D^2} \tag{138}$$

where $\beta = \Delta\varepsilon/d < 0$ is the energy gradient between the superior and inferior frontiers of fluid layer. In the case of convection, Rayleigh's number plays the role of control parameter and takes place for

$$R > R_{critic}$$

In general, R is controlled through the gradient β of the energy.
As reference state, let us choose the stationary rest state $\left(\overline{\mathbf{V}} = 0\right)$, for which equations (134a-c) take the form

$$\begin{cases} \nabla p_S = -\rho_S g\hat{z} = -\rho_0[1-\alpha(\varepsilon_S - \varepsilon_0)]g\hat{z} \\ \Delta\varepsilon_S = 0 \end{cases} \qquad (139a,b)$$

where $\hat{z}$ represents the versor of vertical direction. We take into account that pressure and ε vary only in vertical direction due to the considered symmetry. For ε the conditions on the frontier are

$$\varepsilon(x,y,0) = \varepsilon_0, \quad \varepsilon(x,y,d) = \varepsilon_1 \qquad (140a,b)$$

Integrating equation (139b) with these conditions on the frontier, it will follow that in the reference rest state, the profile of ε on vertical direction is linear.

$$\varepsilon_S = \varepsilon_0 - \beta z \qquad (141)$$

Substituting (141) in (139) and integrating, we obtain

$$p_S(z) = p_0 - \rho_0 g\left(1 + \frac{\alpha\beta z}{2}\right)z \qquad (142)$$

The features of the system in this state do not depend on coefficient D that appears in balance equations.

We study now the stability of the reference state using the method of small perturbations (Bârzu, A. et al, 2003). The perturbed state is characterized by

$$\begin{cases} \varepsilon = \varepsilon_S(z) + \theta(r,t) \\ \rho = \rho_S(z) + \delta\rho(r,t) \\ p = p_S(z) + \partial p(r,t) \\ \overline{\mathbf{V}} = \delta\overline{\mathbf{V}}(r,t) = (u,v,w) \end{cases} \qquad (143a-d)$$

One can notice that the perturbations are time and position functions. Substituting (143) in equations (134) and taking into account (141) and (142) the following equations for perturbations (in linear approximation) are obtained:

$$\nabla \cdot \delta\overline{\mathbf{V}} = 0$$

$$\frac{\partial\theta}{\partial t} = \beta w + D\nabla^2\theta \qquad (144a-d)$$

$$\frac{\partial\delta\overline{\mathbf{V}}}{\partial t} = -\frac{1}{\rho_0}\nabla\delta p + D\nabla^2\delta\overline{\mathbf{V}} + g\alpha\theta\hat{z}$$

We introduce adimensional variables $\overline{r}', t', \theta', \delta\overline{V}', \delta p'$ through the changes

$$\mathbf{r}' = \frac{\mathbf{r}}{d}; \quad t' = \frac{t}{d^2/D}; \quad \theta' = \frac{\theta}{\left(\dfrac{D^2}{g\alpha d^3}\right)}; \quad \delta\overline{\mathbf{V}}' = \frac{\delta\overline{\mathbf{V}}}{D/d}; \quad \delta p' = \frac{\delta p}{\left(\rho_0\dfrac{D^2}{d^2}\right)}$$

Replacing these changes and renouncing, for simplicity, at the prime symbol, the adimensional perturbations satisfy the equations

$$\frac{\partial \overline{\mathbf{V}}}{\partial t} + \overline{\mathbf{V}} \cdot \nabla \overline{\mathbf{V}} = -\nabla p + \theta \hat{z} + \nabla^2 \overline{\mathbf{V}}$$

$$\frac{\partial \overline{\theta}}{\partial t} + (\overline{\mathbf{V}} \cdot \nabla)\theta = Rw + \nabla^2 \theta \tag{145}$$

$$\nabla \cdot \overline{\mathbf{V}} = 0$$

where R is Rayleigh's number.

For R>R$_C$, the reference state becomes unstable, and the convection "patterns" appear. We consider them as being parallel therefore the speed vector will be always perpendicular to their axis. We assume the patterns parallel to the y axis, i.e., the speed component along this direction is zero.

The incompressibility condition becomes

$$u_x + w_z = 0 \tag{146}$$

Equation (146) is satisfied if and only if

$$u = -\psi_z; \; w = \psi_x \tag{147}$$

where $\psi(x,y,z)$ defines Lagrange's current function. The speed field must satisfy the conditions on frontiers (the inferior and superior surfaces)

$$w|_{z=\pm 1/2} = 0 \tag{148}$$

If the frontiers are considered free (the superficial tension forces are neglected), the "shear" component of the pressure tensor is annulated

$$\frac{\partial u}{\partial z}\Big|_{z=\pm 1/2} = 0 \tag{149}$$

Using Lagrange's function, $\psi(x,y,z)$ the limit conditions (148) and (149) become

$$\Psi_x|_{z=\pm 1/2} = 0$$

$$\Psi_{zz}|_{z=\pm 1/2} = 0$$

Let us choose ψ with the form

$$\psi(x,z,t) = \psi_1(t)\cos(\pi z)\sin(qx)$$

According to (147), the components of the speed field are

$$\begin{cases} u = \pi \Psi_1(t)\sin(\pi z)\sin(qx) \\ w = q\Psi_1(t)\cos(\pi z)\cos(qx) \end{cases}$$

The impulse conservation equation (for equation (145)) for directions x and z becomes

$$\begin{aligned} (u_t + uu_x + wu_z) &= -p_x + \Delta u \\ (w_t + uw_x + ww_z) &= -p_z + \Delta w + \theta \end{aligned} \tag{150a,b}$$

We derive (150 a) according to z and (150) according to x. One finds

$$\left[u_{tz} + \frac{\partial}{\partial z}\left(uu_x + wu_z\right) \right] = -p_{xz} + \frac{\partial}{\partial z}\left(\Delta u\right)$$

$$\left[w_{tz} + \frac{\partial}{\partial x}\left(uw_x + ww_z\right) \right] = -p_{zx} + \frac{\partial}{\partial x}\left(\Delta w\right) + \frac{\partial \theta}{\partial x}$$

Through the sum we obtain

$$\left[-\left(\Delta \Psi\right)_t + \frac{\partial}{\partial z}\left(uu_x + wu_z\right) - \frac{\partial}{\partial x}\left(uw_x + ww_z\right) \right] = -\Delta^2 \Psi - \theta_x \tag{151}$$

The value ε being fixed on the two frontiers, we shall have

$$\theta|_{z=\pm 1/2} = 0$$

We consider θ having the form

$$\theta(x,z,t) = \theta_1(t)\cos(\pi z)\cos(qx) + \theta_2(t)\sin(2\pi z) \tag{152}$$

If we consider in (151) the expressions for u, w, θ and ψ it follows that

$$\dot{\psi}_1 = \frac{q\theta_1}{\pi^2 + q^2} - (\pi^2 + q^2)\psi_1 \tag{153}$$

The balance equation for the energy becomes

$$\dot{\theta}_1 = -\pi q\psi_1\theta_2 + qR\psi_1 - (\pi^2 + q^2)\theta_1$$
$$\dot{\theta}_2 = \frac{1}{2}\pi q\psi_1\theta_1 - 4\pi^2\theta_2 \tag{154}$$

In (153) and (154) we change the variables

$$t' = (\pi^2 + q^2)t; \quad X = \frac{\pi q}{\sqrt{2}(\pi^2 + q^2)}\psi_1$$

$$Y = \frac{\pi q^2}{\sqrt{2}(\pi^2 + q^2)^3}\theta_1; \quad Z = \frac{\pi q^2}{(\pi^2 + q^2)^3}\theta_2$$

We obtain the Lorenz type system

$$\dot{X} = (Y - X)$$
$$\dot{Y} = -XZ + rX - Y \tag{155}$$
$$\dot{Z} = XY - bZ$$

where

$$r = \frac{q^2}{(\pi^2 + q^2)^3}R, \quad b = \frac{4\pi^2}{\pi^2 + q^2}$$

The Lorenz system

$$\dot{X} = \sigma(Y - X)$$
$$\dot{Y} = -XZ + rX - Y$$
$$\dot{Z} = XY - bZ$$

reduces to (155) for $\sigma \equiv 1$.

Characteristics of Lorenz type system. Transitions towards chaos.

We consider the evolution equations of Lorenz type system (155) with the notation

$$\dot{x} = (y - x)$$
$$\dot{y} = rx - y - xz \qquad\qquad (156)$$
$$\dot{z} = xy - bz$$

The system is a dissipative one, since the divergence (for details see (Bărzu, A. et al, 2003))

$$\nabla \cdot F = \frac{\partial F_x}{\partial x} + \frac{\partial F_y}{\partial y} + \frac{\partial F_z}{\partial z} = -2 - b < 0$$

since b>0.

Therefore, the phase volume exponentially diminishes in time, as the system tends towards the atractor. For any value of the control parameter r, the system (156) admits as a fixed point the origin

$$x_0 = y_0 = z_0 = 0 \qquad\qquad (157)$$

The characteristic equation is

$$\begin{vmatrix} -1 - \omega & 1 & 0 \\ r - z_0 & -1 - \omega & -x_0 \\ y_0 & x_0 & -b - \omega \end{vmatrix} = 0 \qquad\qquad (158)$$

For the fixed point (157), it takes the form

$$\begin{vmatrix} -1 - \omega & 1 & 0 \\ r & -1 - \omega & 0 \\ 0 & 0 & -b - \omega \end{vmatrix} = 0$$

from where we find

$$(b + \omega)\left[\omega^2 + 2\omega - (r - 1)\right] = 0 \qquad\qquad (159)$$

Since parameters b and r are positive ones, it follows that the first eigenvalue $\omega_1 = -b$ is negative for any values of the parameters. The other two eigenvalues ω_2 and ω_3 satisfy the relations

$$\begin{cases} \omega_2 + \omega_3 = -2 < 0 \\ \omega_2 \omega_3 = -(r - 1) \end{cases} \qquad\qquad (160)$$

According to (160), if $0 < r < 1$ the sum of the two eigenvalues is negative and the product is positive. Therefore, all the eigenvalues are negative and the origin is a stable node. For $r > 1$, according to (160), the origin becomes unstable and two new fixed points appear in a fork bifurcation. These points are noted with C^+ and C^- which corresponds to patterns

$$(C^+)\begin{cases} x_0 = y_0 = \sqrt{b(r-1)} \\ z_0 = r-1 \end{cases}, \quad (C^-)\begin{cases} x_0 = y_0 = -\sqrt{b(r-1)} \\ z_0 = r-1 \end{cases} \tag{161}$$

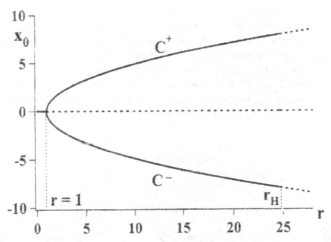

Fig. 1. (according to (Bărzu, A. et al, 2003))

Let us study their stability. Replacing the values that correspond to the branch (C^+) in (158), the characteristic equation becomes

$$\begin{vmatrix} -1-\omega & 1 & 0 \\ 1 & -1-\omega & -\sqrt{b(r-1)} \\ \sqrt{b(r-1)} & \sqrt{b(r-1)} & -b-\omega \end{vmatrix} = 0$$

from where it follows that

$$\omega^3 + \omega^2(b+2) + \omega b(1+r) + 2b(r-1) = 0 \tag{162}$$

If the fixed points (161) will bear a Hopf bifurcation, for a value of control parameter $r_H > 1$, there will be two complex conjugated purely imaginary eigenvalues. Replacing $\omega = i\beta$ in (162) we obtain

$$-i\beta^3 - \beta^2(b+2) + i\beta b(1+r) + 2b(r-1) = 0 \tag{163}$$

Separating the real part from the imaginary one in (163) we obtain the system

$$-\beta^3 + \beta b(1+r) = 0$$
$$-\beta^2(b+2) + 2b(r-1) = 0 \tag{164a,b}$$

From equation (164a) it follows that $\beta^2 = b(1+r)$. Replacing this value in equation (164), Hopf bifurcation takes place in

$$r_H = -\frac{b+4}{b} \tag{165}$$

Considering that $r_H > 1$ the condition for b results

$$b \leq 4 \tag{166}$$

For this value of the control parameter, the two fixed points C^+ and C^- lose their stability in a subcritical Hopf bifurcation. Beyond the bifurcation point all the periodical orbits are unstable and the system has a chaotic behavior. Figures 2a-c to 8a-c show the trajectories, the time evolutions, the phase portraits and the Fourier transform for the different values of the parameters. It follows that when the value of the parameter r increases, there is a complicated succession of chaotic regimes with certain periodicity windows. The limit cycle appears through a reverse subarmonic cascade and loses stability through intermittent transition towards a new chaotic window.

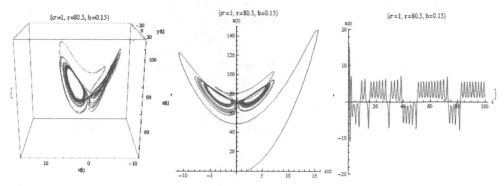

Fig. 2. a) Trajectory b) time evolution c) phase pattern for r=80, b=0.15

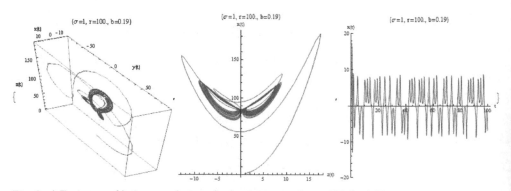

Fig. 3. a) Trajectory b) time evolution c) phase pattern for r=100, b=0.19

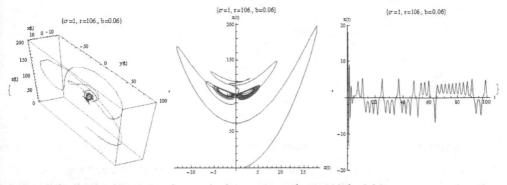

Fig. 4. a) Trajectory b) time evolution c) phase pattern for r=100, b=0.06

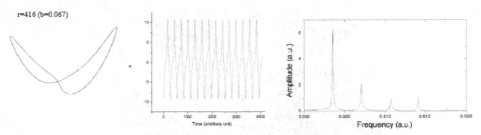

Fig. 5. a) Time evolution b) phase portrait c) the Fourier transform for r=416, b=0.067

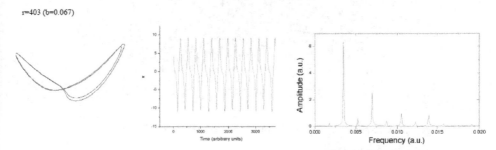

Fig. 6. a) Time evolution b) phase portrait c) the Fourier transform for r=403, b=0.067

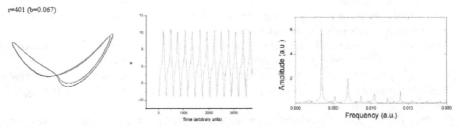

Fig. 7. a) Time evolution b) phase portrait c) the Fourier transform for r=401, b=0.067

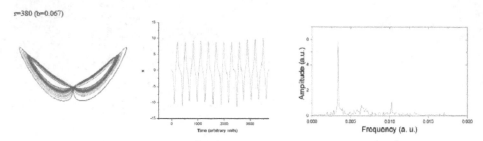

Fig. 8. a) Time evolution b) phase portrait c) the Fourier transform for r=380, b=0.067

In Fig.9 we present the map of the Lyapunov exponent with the value $\sigma = 1$ (the co-ordinates of the light points represent the pairs of values $(x,y) = (b,r)$ for which the probability of entering in a chaotic regime is very high.

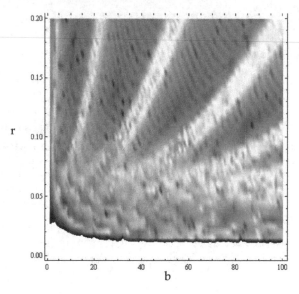

Fig. 9. The Lyapunov exponent map for value $\sigma = 1$ of the Lorenz system

Correspondences with quantum mechanics

The previous analysis states the following:

i.   a model of a physical object can be imagined. This model is built from a Madelung type fluid limited by two carcases that are submitted to an energy "gradient", from the inferior carcase towards the superior one;

ii.   for small energy gradients, i.e. R<R$_C$ the reference state is a stable one. The ascending force resulting from energy "dilatation" is much smaller than the dissipative one.

iii. for energy gradients that impose restriction R>R$_C$ the reference state becomes unstable through the generation of convective type "rolls". The ascensional force is bigger than the dissipative one;

iv.   the increase of energy gradient destroys the convective type "patterns" and induces turbulence;

v. this behavior of fractal fluid can correspond to a Lorenz type "mechanism": limit cycles the convective type "rolls", intermitences ("jumps" between limit cycles) with the "destroy" of the convective type "rolls", chaos with "turbulence" of the convective type state etc.;

vi. the stability of solutions corresponds to the dominant undulatory feature, the wave-corpuscle duality can be correlated with the Lorenz type mechanism: self-organization of the structure through the generation of convective type "rolls" implies the wave-corpuscle transition, while the "jumps" among limit cycles, i.e. the intermittences induce a critical state that corresponds to chaos transition, thus ensuring the dominance of corpuscular effect.

15. Conclusions

Finally we can display the conclusions of this chapter as follows:

-   a critical analisys of the hydrodinamic model of Madelung and of the double solution theory of de Broglie's theory of double solution was performed – departing from here, we built a fractal approximation of motion;
-   we got the equation of motion of the physical object in the fractal approximation and the Eulerian case was studied;
-   the flowing regimes of a rotational fractal fluid were studied;
-   we studied the irotational regime of a fractal fluid and the incorporation of the particle into the associated wave by generating a Schrödinger equation;
-   the extended hydrodinamic model of scale relativity was built and the role of the fractal potential in the process of incorporation of the particle into the wave, specified;
-   we indicated the mechanisms of wave-particle duality by their in phase coherences;
-   we studied the wave-particle duality by stationary flow regimes of a fractal fluid which is coherent in phase, and by non-stationary flow regimes of an incoherent fractal fluid by means of a „polarization" type mechanism;
-   considering the particle as a singularity in the wave, we showed that its incorporation into the associated wave resulted in Einstein's equations in vacuum - contrary, its non-incorporation led to the second quantification;
-   we established a relation between the informational energy and the fractal potential of the complex speed field - it resulted that the generation of forces implies the maximum of the information energy principle;
-   we showed that a particle model in a fractal approximation of motion induced a Lorenz type mechanism.

16. References

Agop, M.; Mazilu, N. (1989). *Fundamente ale fizicii moderne*, Ed. Junimea, Iasi

Agop, M.; Chicoş, L.; Nica, P.; Harabagiu A. (2008). *Euler's fluids and non-differentiable space-time*, Far East Journal of Dynamical systems 10, 1, 93-106.

Agop, M.; Harabagiu, A.; Nica, P.(2008). *Wave-Particle duality through a hydrodynamic modul of the fractal space time theory*, Acta physica Polonica A, 113, 6, 1557-1574

Agop, M.; Colotin, M.; Păun, V. (2009). *Haoticitate, fractalitate şi câmpuri*, Elemente de teorie a fractalilor I. Gottlieb şi C. Mociuţchi, paginile 12-46, Editura ArsLonga Iaşi

Agop, M.; Mazilu, N. (2010). *La răscrucea teoriilor. Între Newton și Einstein – Universul Barbilian*, Ed. Ars Longa, Iași

Albert D.Z..(1994). *Bohm's alternative to quantum mechanics*. Scientific American, 270:32-39

Barbilian, D.(1935). *Apolare und Uberpolare Simplexe*, Mathematica (Cluj), Vol. 11, pp.1-24, (retipărit în opera matematică vol I)

Barbilian, D. (1935). *Die von Einer Quantika Induzierte Riemannsche Metrik*, Comptes rendus de l'Academie Roumaine de Sciences, vol. 2, p.198 (retipărit în Opera Matematica vol. I)

Barbilian, D. (1938). *Riemannsche Raum Cubischer Binarformen*, Comptes rendus de l'Academie Roumanie des Sciences, vol. 2, pg. 345, (retipărit în Opera Matematica vol. I)

Barbilian, D. (1971). *Algebră elementară*, în Opera didactică vol II, Ed. Tehnică București

Barnsley, M. (1988). *Fractals Everywhere. Deterministic Fractal Geometry*, Boston

Bârzu, A.; Bourceanu, G.; Onel, L. (2003). *Dinamica neliniară*, Editura Matrix-Rom, București

Bell J.S. (1987). *Speakable and unspeakable in quantum mechanics*. Cambridge University Press, Cambridge

Benoit Deveand Ed., (2007) *Physics of Semiconductor microcavities from fundamentals to nanoscale decretes*, Wiley-VCH Verlag GmbH Weinheim Germany

Berge, P. ; Pomeau Y. & Vidal Ch.(1984). *L'Ordre dans le chaos*, Hermann

Berndl K.; Dörr D.; Goldstein S.; Peruzzi G. and Zanchi N. (1993) *Existance of Trajectories for Bohmian Mechanics*. International Journal of Theoretical Physics, 32: 2245-2251

Berndl K.; Dörr D.; Goldstein S. and Zanchi N.(1994). *Selfadjointness and the Existance of Deterministic Trajectories in Quantum Theory*, In On three Levels: Micro-, Meso-, and Macroscopic Approches in Physics, (NATO ASI Series B: Physics, Volume 324, Plenum, New-York) pp. 429-434

Bittner E.R.,(2000). *Quantum Tunneling dynamics using hydrodynamic trajectories*, Journal of Chimical - Physics, 112, 9703

Bohm, D. (1952). *A Suggested Interpretation of Quantum Theory in Terms of „Hidden" Variables I*, Phys. Rev. 85 166-179

Bohm D.; Hiley B.J. (1993). *The Undivided Universe: An Ontological Interpretation of Quantum Theory*. Routledge and Kegan Paul, London.

Bohm D. (1952). *A suggested interpretation of quantum theory in terms of „hidden variables": Part II*. Physical Review, 85: 180-193

Bohm D. (1953). *Proof that probability density approaches $|\psi^2|$ in causal interpretation of quantum theory*, Physical Review, 89: 458-466

de Broglie L.(1956). *Un tentative d'interprétation causale et non linéaire de la Mécanique ondulatoire: la theorie de la double solution*, Gauthier-Villars, Paris

de Broglie L. (1957). *La theoree de la Mesure on Mécanique ondulatoire*, Gauthier-Villars, Paris

de Broglie L. (1959). *L'interprétation de la Mécanique ondulatoire*, J. Phys. Rad, 20, 963

de Broglie L. (1963). *Étude critique des bases de l'interprétation actuelle de la Mécanique ondulatoire*, Gauthier-Villars, Paris

de Broglie L. (1964). *La Thermodynamique de la particule isolée (Thermodynamique cachée des particules)*, Gauthier-Villars, Paris

de Broglie L. (1980). *Certitudinile și incertitudinile științei*, Editura Politică, București

Budei, L. (2000). *Modele cu fractali. Aplicații în arhitectura mediului*, Editura Univ. "Gh. Asachi", Iași

Chaline, J.; Nottale, L. ; Grou, P. (2000). *Les arbres d'evolution: Univers S,* Vie, Societes, Edition Hachette

Chaline, J. ; Nottale, L. ; Grou, P. (2009). *Des fleurs pour Scrödinger: La relativite d'echelle et ses applications* Editure Ellipses Marketing

Chioroiu, V. ; Munteanu, L. ; Ştiucă, P. ; Donescu, Ş. (2005). *Introducere în nanomecanică,* Editura Academiei Române, Bucureşti

Ciuti C. ; Camsotto I. (2005). *Quantum fluids effects and parametric instabilities in microcavities,* Physica Status Solidi B, 242, 11, 2224

Dörr D. ; Goldstein S. ; Zanghi N. (1992). *Quantum Mechanics, Randomness and deterministic Reality.* Physics Letters A, 172 : 6-12

Dörr D.; Goldstein S. and Zanchi N. (1993). *A Global Equilibrium as the Foundation of Quantum Randomness,* Foundations of Physics, 23: 712-738

Ernst, F.J. (1968). *New formulation of the Axially Symemetric Gravitational Fielf Problem I,* Phys Rev. 167, 1175, New formulation of the Axially Symmetric Gravitational Fielf Problem II, Phys Rev. 168, p.1415

Ernst, F.J. (1971). *Exterior Algebraic Derivation of Einstein Field Equation Employing a Generalized Basis,* J. Math. Phys. , 12, 2395

Feder, J. & Aharony, A. (Eds) (1990). *Fractals in Physics* North - Holland, Amsterdam

Felsager, B. (1981). *Geometry, Particles and fields.* Odense Univ. Press

Ferry, D. K.; Goodnick, S. M. (2001). *Transport in Nanostructures,* Cambridge University Press

Gouyet JF. (1992). *Physique et Structures Fractals,* Masson Paris

Green, M. B.; Schwarz, J. H.; Witten, E. (1998). *Superstring Theory* vol I, II. Cambridge University, Press, Cambridge

Grössing G. (2008). *Diffusion waves in sub-quantum thermodynamics: Resolution of Einstein's "Particle-in-a-box" objection,* (in press) http://arxiv.org/abs/0806.4462

Halbwacs, F. (1960). *Theorie relativiste des fluids a spin,* Gauthier-Villars, Paris

Harabagiu, A.; Agop, M. (2005). *Hydrodyamic model of scale relativity theory,* Buletinul Institutului Politehnic Iaşi, tomul LI (LV) Fasc. 3-4, 77-82, secţia Matematică, Mecanică teoretică, Fizică

Harabagiu, A.; Niculescu, O.; Colotin, M.; Bibere, T. D.; Gottlieb, I.; Agop, M. (2009). *Particle in a box by means of fractal hydrodynamic model,* Romanian Reports in Physics, 61, 3, 395-400

Harabagiu A. ; Magop, D. ; Agop, M. (2010). *Fractalitate şi mecanică cuantică,* Editura Ars Longa Iaşi

Heck, A. & Perdang, JM. (Eds) (1991). *Applying Fractals in Astronomy;* Springer Verlag

Holland P.R. (1993). *The Quantum Theory of Motion.* Cambridge University Press, Cambridge

Imry, Y. (2002). *Introduction to Mesoscopic Physics,* Oxford University Press, Oxford

Madelung R. (1927). Zs f. Phys.40, 322

Mandelis A.; Nicolaides L.; Chen Y. (2001). *Structure and the reflectionless/refractionless nature of parabolic diffusion-wave fields,* Phys. Rev. Lett. 87, 020801

Mandelis A. (2000). *Diffusion waves and their uses,* Phys. Today 53, 29

Mazilu N.; Agop M. (1994). *Fizica procesului de măsură,* Ed. Ştefan Procopiu, Iaşi

Le Mehante A. (1990). *Les Geometries Fractales,* Hermes, Paris

Mihăileanu, N. (1972). *Geometrie analitică, proiectivă şi diferenţială,* Complemente Editura Didactică şi Pedagogică, Bucureşti

Munceleanu, C.V.; Magop D.; Marin C. ; Agop, M (2010). *Modele fractale în fizica polimerilor*, Ars Longa

El Naschie, MS ; Rössler, OE ; Prigogine, I. (Eds.) (1995) *Quantum mechanics, diffusion and chaotic fractals*, Oxford: Elsevier

Nelson, E. (1985). *Quantum Fluctuations*, Princeton University Press, Princeton, New York

Nottale, L.; Schneider J. (1984). *Fractals and non-standard analysis*; J. Math. Phys. 25,12, 96-300

Nottale, L. (1989). *Fractals and the quantum theory of space-time* Int. J. Mod. Phys. A4 50 47-117

Nottalle, L. (1993). *Fractal Space-Time and Microphysics. Towards a Theory of Scale Relativity*, World Scientific

Nottale, L. (1996). *Scale relativity and fractal space-time: Applications to quantum physics, cosmology and chaotic systems.* Chaos, Solitons & Fractals 7 :877

Nottale, L. (2004). *The theory of scale relativity: Nondifferentiable geometry, Fractal space-time and Quantum Mechanics*, Computing Anticipatory systems: CASYS' 03-33 Sixth International Confference, AIP Confference Proceedings vol 718, pag. 68-95

Peres, A. (1993). *Quantum teory: Concepts and methods*, Klauwer Acad. Publ., Boston

Poole, C. P.; Farach, K. A.; Creswick, R. (1995). *Superconductivity*, San Diego, Academic Press

Popescu, S. (2004). *Probleme actuale ale fizicii sistemelor autoorganizate*, Editura Tehnopress, Iasi

Ruelle D.; Takens, F. (1971). *On the Nature of Turbulence*, Commun. Math. Phys., 20, 167, 23, 343

Ruelle, D. (1975). *Strange Attractors*, The mathematical Intelligencer 2, 126

Sakurai J.J.; San Fu Taun (1994). *Modern Qunatum Mechanics*, Addison-Wesley, Reading, MA

Țițeica, S. (1984). *Mecanică cuantică*, Editura Academiei, București

Weibel, P.; Ord G.; Rössler, OE (Eds) (2005). *Space time physics and fractality*, Festschroft in honer of Mohamad El Naschie Vienna, New York: Springer

5

Approximate Solutions of the Dirac Equation for the Rosen-Morse Potential in the Presence of the Spin-Orbit and Pseudo-Orbit Centrifugal Terms

Kayode John Oyewumi
Theoretical Physics Section, Physics Department,
University of Ilorin, Ilorin
Nigeria

1. Introduction

In quantum mechanics, it is well known that the exact solutions play fundamental role, this is because, these solutions usually contain all the necessary information about the quantum mechanical model under investigation. In recent years, there has been a renewed interest in obtaining the solutions of the Dirac equations for some typical potentials under special cases of spin symmetry and pseudo-spin symmetry (Arima et al., 1969; Hecht and Adler, 1969).

The idea about spin symmetry and pseudo-spin symmetry with the nuclear shell model has been introduced in 1969 by Arima et al. (1969) & Hecht and Adler (1969). This idea has been widely used in explaining a number of phenomena in nuclear physics and related areas. Spin and pseudo-spin symmetric concepts have been used in the studies of certain aspects of deformed and exotic nuclei (Meng & Ring, 1996; Ginocchio, 1997; Ginocchio & Madland, 1998; Alberto et al., 2001; 2002; Lisboa et al., 2004a; 2004b; 2004c; Guo et al., 2005a; 2005b; Guo & Fang, 2006; Ginocchio, 2004; Ginocchio, 2005a; 2005b).

Spin symmetry (SS) is relevant to meson with one heavy quark, which is being used to explain the absence of quark spin orbit splitting (spin doublets) observed in heavy-light quark mesons (Page et al., 2001). On the other hand, pseudo-spin symmetry (PSS) concept has been successfully used to explain different phenomena in nuclear structure including deformation, superdeformation, identical bands, exotic nuclei and degeneracies of some shell model orbitals in nuclei (pseudo-spin doublets)(Arima, et al., 1969; Hecht & Adler, 1969; Meng & Ring, 1996; Ginocchio, 1997; Troltenier et al., 1994; Meng, et al., 1999; Stuchbery, 1999; 2002). Within this framework also, Ginocchio deduced that a Dirac Hamiltonian with scalar S(r) and vector $V(r)$ harmonic oscillator potentials when $V(r) = S(r)$ possesses a spin symmetry (SS) as well as a $U(3)$ symmetry, whereas a Dirac Hamiltonian for the case of $V(r) + S(r) = 0$ or $V(r) = -S(r)$ possesses a pseudo-spin symmetry and a pseudo-$U(3)$ symmetry (Ginocchio, 1997; 2004; 2005a; 2005b). As introduced in nuclear theory, the PSS refers to a quasi-degeneracy of the single-nucleon doublets which can be characterized with the non-relativistic quantum mechanics $(n, \ell, j = \ell + \frac{1}{2})$ and $(n - 1, \ell + 2, j = \ell + \frac{3}{2})$, where n, ℓ and j are the single-nucleon radial, orbital and total angular momentum quantum numbers for a single particle, respectively (Arima et al., 1969; Hecht & Adler, 1969; Ginocchio, 2004;

2005a; 2005b; Page et al., 2001). The total angular momentum is given as $j = \bar{\ell} + \bar{s}$, where $\bar{\ell} = \ell + 1$ is a pseudo- angular momentum and $\bar{s} = \frac{1}{2}$ is a pseudo-spin angular momentum. Meng et al., (1998) deduced that in real nuclei, the PSS is only an approximation and the quality of approximation depends on the pseudo-centrifugal potential and pseudo-spin orbital potential. The orbital and pseudo-orbital angular momentum quantum numbers for SS ℓ and PSS $\bar{\ell}$ refer to the upper-and lower-spinor components (for instance, $F_{n,\kappa}(r)$ and $G_{n,\kappa}(r)$, respectively.

Ginocchio (1997); (1999); (2004); (2005a); (2005b) and Meng et al., (1998) showed that SS occurs when the difference between the vector potential $V(r)$ and scalar potential $S(r)$ in the Dirac Hamiltonian is a constant (that is, $\Delta(r) = V(r) - S(r)$) and PSS occurs when the sum of two potential is a constant (that is, $\Sigma(r) = V(r) + S(r)$).

A large number of investigations have been carried out on the SS and PSS by solving the Dirac equation with various methods (Alberto et al., 2001; 2002; Lisboa et al., 2004a; 2004b; 2004c; Ginocchio, 2005a; 2005b; Xu et al., 2008; Guo et al., 2005a; 2005b; de Castro et al., 2006; Wei and Dong, 2009; Zhang, 2009; Zhang et al., 2009a; Setare & Nazari, 2009; Ginocchio, 1999; Soylu et al., 2007; 2008a; 2008b; Berkdermir, 2006; 2009; Berkdemir & Sever, 2009; Xu & Zhu, 2006; Jia et al., 2006; Zhang et al., 2009a; Zhang et al., 2008; Aydoğdu, 2009; Aydoğdu & Sever, 2009; Wei and Dong, 2008; Jia et al., 2009a; 2009b; Guo et al., 2007).

Some of these potentials are exactly solvable, these include: harmonic potential (Lisboa et al., 2004a; 2004b; 2004c; Ginocchio, 1999; 2005a; 2005b; Guo et al., 2005a; 2005b; de Castro et al., 2006; Akcay & Tezcan, 2009), Coulomb potential (Akcay, 2007; 2009), pseudoharmonic potential (Aydoğdu, 2009; Aydoğdu & Sever, 2009; Aydoğdu & Sever, 2010a), Mie-type potential (Aydoğdu, 2009; Aydoğdu & Sever, 2010b).

Also, for the $\bar{s}$-wave with zero pseudo-orbital angular momentum $\bar{\ell} = 0$ and spin-orbit quantum number $\kappa = 1$, exact analytical solutions have been obtained for some potentials with different methods, such as: Woods-Saxon potential (Aydoğdu, 2009; Guo & Sheng, 2005; Aydoğdu & Sever, 2010c), Eckart potential (Jia et al., 2006), Pöschl-Teller potential (Jia et al, 2009b), Rosen-Morse potential (Oyewumi & Akoshile, 2010), trigonometric Scarf potential (Wei et al., 2010).

However, exact analytical solution for any $\ell-$ states are possible only in a few instances. it is important to mention that most of these potentials can not be solved exactly for $\ell \neq 1(\kappa \neq -1)$ or $\bar{\ell} \neq 0(\kappa \neq 1)$ state, hence, a kind of approximation to the (pseudo or) - centrifugal term is necessary (Pekeris-type approximation) (Ikhdair, 2010; 2011; Ikhdair et al., 2011; Xu et al., 2008; Jia et al., 2009a; 2009b; Wei and Dong, 2009; Zhang et al., 2009b; Soylu et al., 2007; 2008a; 2008b; Zhang et al., 2008; Aydoğdu and Sever, 2010c; Aydoğdu and Sever, 2010d; Bayrak and Boztosun, 2007; Pekeris, 1934; Greene and Aldrich, 1976; Wei and Dong, 2010a; 2010b; 2010c).

With this kind of approximation to the (pseudo or) - centrifugal term, the SS and PSS problems have been solved using different methods to obtain the approximate solutions: AIM (Soylu et al., 2007; 2008a; 2008b; Aydoğdu & Sever, 2010c; Bayrak & Boztosun, 2007; Hamzavi et al., 2010c), Nikiforov- Uvarov method (Aydoğdu & Sever, 2010c; Hamzavi et al., 2010a; 2010b; Berkdemir, 2006; 2009; Berkdemir & Sever, 2009; Ikhdair, 2010; 2011; Ikhdair et al., 2011), functional analysis method (Xu et al., 2008; Wei & Dong, 2010d), SUSY and functional analysis (Jia et al., 2006; 2009a; 2009b; Wei & Dong, 2009; Zhang et al., 2009b; Setare & Nazari, 2009; Wei & Dong, 2010a; 2010b; 2010c). Therefore, by applying a Pekeris-type approximation to the (pseudo or) - centrifugal-like term, the relativistic bound state solutions can be obtained in the framework of the PSS and SS concepts.

In this study, the Rosen-Morse potential is considered, due to the important applications of
in atomic, chemical and molecular Physics as well (Rosen & Morse, 1932). This potential is
very useful in describing interatomic interaction of the linear molecules. The Rosen-Morse
potential is given as

$$V(r) = -V_1 \mathrm{sech}^2 \alpha r + V_2 \tanh \alpha r, \tag{1}$$

where V_1 and V_2 are the depth of the potential and α is the range of the potential, respectively.
Thus, our aim is to employ the newly improved approximation scheme (or Pekeris-type
approximation scheme) in order to obtain the PSS and SS solutions of the Dirac equations
for the Rosen-Morse potential with the centrifugal term. This potential has been studied by
various researchers in different applications (Rosen & Morse, 1932; Yi et al., 2004; Taşkin, 2009;
Oyewumi & Akoshile and reference therein, 2010; Ikhdair, 2010; Ibrahim et al., 2011; Amani
et al., 2011). In the light of this study, standard function analysis approach will be used (Yi et
al, 2004; Taşkin, 2009).
In this chapter, Section 2 contains, the basic equations for the upper- and lower- component of
the Dirac spinors. In Section 3, the approximate analytical solutions of the Dirac equation with
the Rosen-Morse potential with arbitrary κ under pseudospin and spin symmetry conditions
are obtained by means of the standard function analysis approach. Also, the solutions of
some special cases are obtained. The bound state solutions of the relativistic equations
(Klein-Gordon and Dirac) with the equally mixed Rosen-Morse potentials for any ℓ or κ are
contained in Section 4. Section 5 contains contains the conclusions.

2. Basic Equations for the upper- and lower-components of the Dirac spinors

In the case of spherically symmetric potential, the Dirac equation for fermionic massive
spin$-\frac{1}{2}$ particles interacting with the arbitrary scalar potential $S(r)$ and the time-component
$V(r)$ of a four-vector potential can be expressed as (Greiner, 2000; Wei & Dong, 2009; 2010a;
2010b; 2010c; 2010d; Ikhdair, 2010; 2011; Oyewumi & Akoshile, 2010; Ikhdair et al., 2011):

$$\left[c\vec{\alpha}.\vec{P} + \beta[Mc^2 + S(\vec{r})] + V(\vec{r}) - E \right] \psi_{n\kappa}(\vec{r}) = 0, \tag{2}$$

where E is the relativistic energy of the system, M is the mass of a particle, $\vec{P} = -i\hbar\nabla$ is the
momentum operator. $\vec{\alpha}$ and β are 4×4 Dirac matrices, given as

$$\vec{\alpha} = \begin{pmatrix} 0 & \sigma_i \\ \sigma_i & 0 \end{pmatrix}, \quad \beta = \begin{pmatrix} I & 0 \\ 0 & -I \end{pmatrix}, \tag{3}$$

where I is the 2×2 identity matrix and $\sigma_i (i = 1, 2, 3)$ are the vector Pauli matrices.
Following the procedure stated in (Greiner, 2000; Wei & Dong 2009; 2010a; 2010b; 2010c;
2010d; Ikhdair, 2010; 2011; Ikhdair et al., 2011), the spinor wave functions can be written
using the Pauli-Dirac representation as:

$$\psi_{n\kappa}(\vec{r}) = \frac{1}{r} \begin{bmatrix} F_{n\kappa}(r) \, Y_{jm}^{\ell}(\theta, \phi) \\ iG_{n\kappa}(r) \, Y_{jm}^{\bar{\ell}}(\theta, \phi) \end{bmatrix}; \; \kappa = \pm(j + \frac{1}{2}), \tag{4}$$

where $F_{n\kappa}(r)$ and $G_{n\kappa}(r)$ are the radial wave functions of the upper and lower spinors
components, respectively. $Y_{jm}^{\ell}(\theta, \phi)$ and $Y_{jm}^{\bar{\ell}}$ are the spherical harmonic functions coupled

to the total angular momentum j and its projection m on the z-axis. The orbital and pseudo-orbital angular momentum quantum numbers for SS (ℓ) and PSS ($\tilde{\ell}$) refer to the upper $(F_{n\kappa}(r))$ and lower $(G_{n\kappa}(r))$ spinor components, respectively, for which $\ell(\ell+1) = \kappa(\kappa+1)$ and $\tilde{\ell}(\tilde{\ell}+1) = \kappa(\kappa-1)$. For the relationship between the quantum number κ to the quantum numbers for SS (ℓ) and PSS ($\tilde{\ell}$) (Ikhdair, 2010; 2011; Ikhdair et al., 2011; Jia et al., 2009a; 2009b; Xu et al., 2008; Wei & Dong, 2009; Ginocchio, 2004; Zhang et al., 2009b; Setare & Nazari, 2009). For comprehensive reviews, see Ginocchio (1997) and (2005b).

On substituting equation (4) into equation (2), the two-coupled second-order ordinary differential equations for the upper and lower components of the Dirac wave function are obtained as follows:

$$\left(\frac{d}{dr} + \frac{\kappa}{r}\right) F_{n\kappa}(r) = \left[Mc^2 + E_{n\kappa} - \Delta(r)\right] G_{n\kappa},\tag{5}$$

$$\left(\frac{d}{dr} - \frac{\kappa}{r}\right) G_{n\kappa}(r) = \left[Mc^2 - E_{n\kappa} + \Sigma(r)\right] F_{n\kappa}.\tag{6}$$

Eliminating $F_{n\kappa}(r)$ and $G_{n\kappa}(r)$ from equations (5) and (6), the following two Schrödinger-like differential equations for the upper and lower radial spinors components are obtained, respectively as:

$$\left\{-\frac{d^2}{dr^2} + \frac{\kappa(\kappa+1)}{r^2} + \frac{1}{\hbar^2 c^2}\left[Mc^2 + E_{n\kappa} - \Delta(r)\right]\left[Mc^2 - E_{n\kappa} + \Sigma(r)\right]\right\} F_{n\kappa}(r)$$

$$= \frac{\frac{d\Delta(r)}{dr}\left(\frac{d}{dr} + \frac{\kappa}{r}\right)}{\left[Mc^2 + E_{n\kappa} - \Delta(r)\right]} F_{n\kappa}(r),\tag{7}$$

$$\left\{-\frac{d^2}{dr^2} + \frac{\kappa(\kappa-1)}{r^2} + \frac{1}{\hbar^2 c^2}\left[Mc^2 + E_{n\kappa} - \Delta(r)\right]\left[Mc^2 - E_{n\kappa} + \Sigma(r)\right]\right\} G_{n\kappa}(r)$$

$$= -\frac{\frac{d\Sigma(r)}{dr}\left(\frac{d}{dr} - \frac{\kappa}{r}\right)}{\left[Mc^2 - E_{n\kappa} + \Sigma(r)\right]} G_{n\kappa}(r),\tag{8}$$

where $\Delta(r) = V(r) - S(r)$ and $\Sigma(r) = V(r) + S(r)$ are the difference and the sum of the potentials $V(r)$ and $S(r)$, respectively.

In the presence of the SS, that is, the difference potential $\Delta(r) = V(r) - S(r) = C_s =$ constant or $\frac{d\Delta(r)}{dr} = 0$, then, equation (7) reduces into

$$\left\{-\frac{d^2}{dr^2} + \frac{\kappa(\kappa+1)}{r^2} + \frac{1}{\hbar^2 c^2}\left[Mc^2 + E_{n\kappa} - C_s\right]\Sigma(r)\right\} F_{n\kappa}(r)$$

$$= \left[E_{n\kappa}^2 - M^2 c^4 + C_s(Mc^2 - E_{n\kappa})\right] F_{n\kappa}(r),\tag{9}$$

where $\kappa(\kappa+1) = \ell(\ell+1)$, $\kappa = \begin{cases} \ell, & \text{for } \kappa < 0 \\ -(\ell+1), & \text{for } \kappa > 0 \end{cases}$. The SS energy eigenvalues depend on n and κ, for $\ell \neq 0$, the states with $j = \ell \pm \frac{1}{2}$ are degenerate. Then, the lower component

Approximate Solutions of the Dirac Equation for the Rosen-Morse Potential in the Presence of the Spin-Orbit and
Pseudo-Orbit Centrifugal Terms

137

$G_{n\kappa}(r)$ of the Dirac spinor is obtained as

$$G_{n,\kappa}(r) = \frac{1}{Mc^2 + E_{n\kappa} - C_s}\left[\frac{d}{dr} + \frac{\kappa}{r}\right]F_{n\kappa}(r), \tag{10}$$

where $E_{n\kappa} + Mc^2 \neq 0$, only real positive energy state exist when $C_s = 0$ (Guo & Sheng, 2005; Ikhdair, 2010; Ikhdair et al., 2011).

Also, under the PSS condition, that is, the sum potential $\Sigma(r) = V(r) + S(r) = C_{ps}$ constant or $\frac{d\Sigma(r)}{dr} = 0$, then, equation (7) becomes

$$\left\{-\frac{d^2}{dr^2} + \frac{\kappa(\kappa - 1)}{r^2} - \frac{1}{\hbar^2 c^2}\left[Mc^2 - E_{n\kappa} + C_{ps}\right]\Delta(r)\right\}G_{n\kappa}(r)$$

$$= \left[E_{n\kappa}^2 - M^2c^4 + C_{ps}(Mc^2 - E_{n\kappa})\right]G_{n\kappa}(r), \tag{11}$$

and the upper component $F_{n\kappa}(r)$ is obtained as

$$\Gamma_{n,\kappa}(r) = \frac{1}{Mc^2 - E_{n\kappa} + C_{ps}}\left[\frac{d}{dr} - \frac{\kappa}{r}\right]G_{n\kappa}(r), \tag{12}$$

where $E_{n\kappa} - Mc^2 \neq 0$, only real negative energy state exist when $C_{ps} = 0$. Also, κ is related to the pseudo-orbital angular quantum number $\bar{\ell}$ as $\kappa(\kappa - 1) = \bar{\ell}(\bar{\ell} + 1)$, $\kappa = \begin{cases} -\bar{\ell}, & \text{for } \kappa < 0 \\ (\bar{\ell} + 1), & \text{for } \kappa > 0 \end{cases}$, which implies that $j = \bar{\ell} \pm \frac{1}{2}$ are degenerate for $\bar{\ell} \neq 0$ (Guo & Sheng, 2005; Ikhdair, 2010; Ikhdair et al., 2011). It is required that the upper and lower spinor components must satisfy the following boundary conditions $F_{n\kappa}(0) = G_{n\kappa}(0) = 0$ and $F_{n\kappa}(\infty) = G_{n\kappa}(\infty) = 0$ for bound state solutions.

Exact solutions of equations (9) and (11) with the Rosen-Morse potential (1) can be obtained only for the s−wave ($\kappa = 0, -1$) and ($\kappa = 0, 1$) due to the spin-orbit (or pseudo) centrifugal term $\frac{\kappa(\kappa+1)}{r^2}$ (or $\frac{\kappa(\kappa-1)}{r^2}$). Therefore, a newly improved approximation in dealing with the spin-orbit (or pseudo) centrifugal term to obtain the approximate solutions for the Rosen-Morse is adopted.

This type of approximation, (Pekeris-type) approximation can be traced back to Pekeris (1934), and for short-range potential, Greene & Aldrich (1976) proposed a good approximation to the centrifugal term $(1/r^2)$. The idea about the use of approximation to centrifugal (or pseudo centrifugal) term has received much attention and considerable interest due to its wide range of applications (Wei & Dong, 2010a; 2010b; 2010c; 2010d; Aydoğdu & Sever, 2010; Zhang et al., 2009b; Jia et al., 2009a; 2009b; Lu, 2005; Ikhdair, 2010; Ikhdair et al., 2011). We adopt the centrifugal (or pseudo centrifugal) approximation introduced by Lu (2005) for values of κ that are not large and vibrations of the small amplitude about the minimum. This approximation to the centrifugal or (pseudo centrifugal) term near the minimum point $r = r_0$ introduced by Lu (2005) is given as follows:

$$\frac{1}{r^2} \approx \frac{1}{r_0^2}\left[c_0 + c_1\left(\frac{-e^{-2\alpha r}}{1 + e^{-2\alpha r}}\right) + c_2\left(\frac{-e^{-2\alpha r}}{1 + e^{-2\alpha r}}\right)^2\right], \tag{13}$$

where

$$C_0 = 1 - \left(\frac{1+e^{-2\alpha r_0}}{2\alpha r_0}\right)^2 \left(\frac{8\alpha r_0}{1+e^{-2\alpha r_0}} - (3+2\alpha r_0)\right),$$

$$C_1 = -2(e^{2\alpha r_0}+1)\left[3\left(\frac{1+e^{-2\alpha r_0}}{2\alpha r_0}\right) - (3+2\alpha r_0)\left(\frac{1+e^{-2\alpha r_0}}{2\alpha r_0}\right)\right],$$

$$C_2 = (e^{2\alpha r_0}+1)^2 \left(\frac{1+e^{-2\alpha r_0}}{2\alpha r_0}\right)^2 \left[(3+2\alpha r_0) - \left(\frac{4\alpha r_0}{1+e^{-2\alpha r_0}}\right)\right], \tag{14}$$

other higher terms are neglected.

3. Bound state solutions of the Dirac equation with the Rosen-Morse potential with arbitrary κ

3.1 Spin symmetry solutions of the Dirac equation with the Rosen-Morse potential with arbitrary κ

In equation (9), we adopt the choice of $\Sigma(r) = 2V(r) \to V(r)$ as earlier illustrated by Alhaidari et al. (2006), which enables us to reduce the resulting solutions into their non-relativistic limits under appropriate transformations, that is,

$$\Sigma(r) = -4V_1 \frac{e^{-2\alpha r}}{(1+e^{-2a\alpha r})^2} + V_2 \frac{(1-e^{-2\alpha r})}{(1+e^{-2a\alpha r})}. \tag{15}$$

Using the centrifugal term approximation in equation (13) and introducing a new variable of the form $z = e^{-2\alpha r}$ in equation (9), the following equation for the upper component spinor $F_{n\kappa}(r)$ is obtained as:

$$z^2 \frac{d^2}{dz^2}F_{n\kappa}(z) + z\frac{d}{dz}F_{n\kappa}(z) + \frac{1}{4\alpha^2}\left\{\frac{1}{\hbar^2 c^2}\left[E_{n\kappa}^2 - M^2 c^4 + C_s(Mc^2 - E_{n\kappa})\right]\right\}F_{n\kappa}(z)$$

$$-\frac{\kappa(\kappa+1)}{4\alpha^2}\left\{\frac{1}{r_0^2}\left[C_0 + C_1\frac{z}{1-z} + C_2\frac{z^2}{(1-z)^2}\right] - \frac{4\tilde{V}_1 z}{(1-z)^2} - \tilde{V}_2 - \frac{2\tilde{V}_2 z}{(1-z)}\right\}F_{n\kappa}(z), \tag{16}$$

where

$$\tilde{V}_1 = \frac{V_1}{\hbar^2 c^2}[Mc^2 + E_{n\kappa} - C_s] \text{ and } \tilde{V}_2 = \frac{V_2}{\hbar^2 c^2}[Mc^2 + E_{n\kappa} - C_s]. \tag{17}$$

The upper component spinor $F_{n\kappa}(z)$ has to satisfy the boundary conditions, $F_{n\kappa}(z) = 0$ at $z \to 0$ $(r \to \infty)$ and $F_{n\kappa}(z) = 1$ at $z \to 1$ $(r \to 0)$. Then, the function $F_{nk}(z)$ can be written as

$$F_{n\kappa}(z) = (1-z)^{1+q} z^\beta f_{n\kappa}(z), \tag{18}$$

where

$$q = \frac{1}{2}\left[-1 + \sqrt{1 + \frac{\kappa(\kappa+1)C_2}{\alpha^2 r_0^2} + \frac{4\tilde{V}_1}{\alpha^2}}\right] \tag{19}$$

and

$$-\beta^2 = \frac{1}{4\alpha^2}\left\{\frac{1}{\hbar^2 c^2}\left[E_{n\kappa}^2 - M^2 c^4 + C_s(Mc^2 - E_{n\kappa})\right] - \frac{\kappa(\kappa+1)}{r_0^2}C_0 - \tilde{V}_2\right\}. \tag{20}$$

Approximate Solutions of the Dirac Equation for the Rosen-Morse Potential in the Presence of the Spin-Orbit and Pseudo-Orbit Centrifugal Terms

139

On substituting equation (18) into equation (16) with equations (17), (19) and (20) , the second-order differential equation is obtained as

$$z(1-z)\frac{d^2}{dz^2}f_{n\kappa}(z) + [(2\beta+1) - (2q+2\beta+3)z]\frac{d}{dz}f_{n\kappa}(z)$$

$$- \left[(2\beta+1)(1+q) + \frac{\tilde{V}_2+2\tilde{V}_1}{2\alpha^2} + \frac{\kappa(\kappa+1)C_1}{4\alpha^2 r_0^2}\right]f_{n\kappa}(z), \tag{21}$$

whose solutions are the hypergeometric functions (Gradshteyn & Ryzhik, 2007), its general form can be expressed as

$$f_{n\kappa}(z) = A\,{}_2F_1(a,b;c;z) + Bz^{1-c}\,{}_2F_1(a-c+1,b-c+1;2-c;z), \tag{22}$$

in which the first term can be expressed as:

$$_2F_1(a,b;c;z) = \frac{\Gamma(c)}{\Gamma(a)\Gamma(b)}\sum_{k=0}^{\infty}\frac{\Gamma(a+k)\Gamma(b+k)z^k}{\Gamma(c+k)k!}, \tag{23}$$

where

$$a = 1+q+\beta-\gamma$$
$$b = 1+q+\beta+\gamma$$
$$c = 1+2\beta$$

$$\gamma = \sqrt{\beta^2 - \frac{(\tilde{V}_2+2\tilde{V}_1)}{2\alpha^2} + q(1+q) - \frac{\kappa(\kappa+1)C_1}{4\alpha^2 r_0^2}}. \tag{24}$$

The hypergeometric function $f_{n\kappa}(z)$ can be reduced to polynomial of degree n, whenever either a or b equals to a negative integer $-n$. This implies that the hypergeometric function $f_{n\kappa}(z)$ given by equation (23) can only be finite everywhere unless

$$a = 1+q+\beta-\gamma = -n; \quad n = 0,\,1,\,2,\,3,\,\ldots\,. \tag{25}$$

Using equations (17), (19) and (20) in equation (25), an explicit expression for the energy eigenvalues of the Dirac equation with the Rosen-Morse potential under the spin symmetry condition is obtained as:

$$(Mc^2 + E_{n\kappa} - C_s)(Mc^2 - E_{n\kappa} + V_2) = -\frac{\kappa(\kappa+1)C_0}{r_0^2}\hbar^2 c^2$$

$$+4\alpha^2\hbar^2 c^2\left[\frac{\frac{(C_2-C_1)}{4\alpha^2 r_0^2}\kappa(\kappa+1) - \frac{(Mc^2+E_{n\kappa}-C_s)V_2}{2\alpha^2\hbar^2 c^2}}{2(n+q+1)} - \frac{(n+q+1)}{2}\right]^2. \tag{26}$$

It is observed that, the spin symmetric limit leads to quadratic energy eigenvalues. Hence, the solution of equation (26) consists of positive and negative energy eigenvalues for each n and κ. In 2005, Ginocchio has shown that there are only positive energy eigenvalues and no bound negative energy eigenvalues exist in the spin limit. Therefore, in the spin limit, only positive energy eigenvalues are chosen for the spin symmetric limit.

Using equations (18) to (25), the radial upper component spinor can be obtained as

$$F_{n\kappa}(r) = N_{n\kappa}(1 + e^{-2\alpha r})^{1+q}(-e^{-2\alpha r})^{\beta} \, _2F_1(-n, n + 2(\beta + q + 1); 2\beta + 1; -e^{-2\alpha r})$$

$$= N_{n\kappa}\frac{n!\Gamma(2\beta + 1)}{\Gamma(n + \beta + 1)}(1 + e^{-2\alpha r})^{1+q}(-e^{-2\alpha r})^{\beta}P_n^{(2\beta, \, 2q+1)}(1 + 2e^{-2\alpha r}), \tag{27}$$

$N_{n\kappa}$ is the normalization constant which can be determined by the condition that $\int_0^{\infty} | F_{n\kappa}(r) |^2 \, dr = 1$.

By making use of the equation (23) and the following integral (see formula (7.512.12) in Gradshteyn & Ryzhik (2007)):

$$\int_0^1 (1-x)^{\mu-1}x^{\nu-1} \, _pF_q(a_1, .., a_p; \, b_1, ..b_q; \, ax)dx = \frac{\Gamma(\mu)\Gamma(\nu)}{\Gamma(\mu+\nu)} \, _{p+1}F_{q+1}(\nu, \, a_1, .., a_p; \, \mu+\nu, \, b_1, ..b_q; a) \,, \tag{28}$$

which is valid for $Re\mu > 0$, $Re\nu > 0$, $p \leq q + 1$, if $p = q + 1$, then $| \, q < 1 \, |$, this leads to

$$N_{n\kappa} = \left[\frac{\Gamma(2q + 3)\Gamma(2\beta + 1)}{2\alpha\Gamma(n)}\sum_{k=0}^{\infty}\frac{(-1)^k \, (n + 2(1 + \beta + q))_k \, \Gamma(n + k)}{k!(k + 2\beta)!\Gamma\left(k + 2(\beta + q + \frac{3}{2})\right)}A_{n\kappa}\right]^{-1/2}, \tag{29}$$

where $A_{n\kappa} = \, _3F_2(2\beta + k, -n, \, n + 2(1 + \beta + q); k + 2(\beta + q + \frac{3}{2}); 2\beta + 1; 1)$ and $(x)_a = \frac{\Gamma(x+a)}{\Gamma(x)}$ (Pochhammer symbol). In order to find the lower component spinor, the recurrence relation of the hypergeometric function (Gradshteyn & Ryzhik, 2007)

$$\frac{d}{d\xi}[\, _2F_1 \, (a, \, b, \, c; \xi)] = \left(\frac{ab}{c}\right)\frac{d}{d\xi} \, _2F_1 \, (a + 1, \, b + 1, \, c + 1; \xi) \,, \tag{30}$$

is used to evaluate equation (10) and this is obtained as

$$G_{n\kappa}(r) = \frac{N_{n\kappa}(1 + e^{-2\alpha r})^{1+q}(-e^{-2\alpha r})^{\beta}}{[Mc^2 + E_{n\kappa} - C_s]}\left[-2\alpha\beta - \frac{2\alpha e^{-2\alpha r}}{1 + e^{-2\alpha r}} + \frac{\kappa}{r}\right]$$

$$\times \, _2F_1(-n, n + 2(\beta + q + 1); 2\beta + 1; -e^{-2\alpha r}) +$$

$$\frac{N_{n\kappa}(1 + e^{-2\alpha r})^{1+q}(-e^{-2\alpha r})^{\beta+1}}{[Mc^2 + E_{n\kappa} - C_s]}\left\{\frac{2\alpha n \, [n + 2(\beta + q + 1)]}{(2\beta + 1)}\right\}$$

$$\times \, _2F_1(-n + 1, n + 2(\beta + q + \frac{3}{2}); 2(\beta + 1); -e^{-2\alpha r}). \tag{31}$$

3.2 Pseudopin symmetry solutions of the Dirac equation with the Rosen-Morse potential with arbitrary κ

In the case of pseudospin symmetry, that is, the difference as in equation (11). $\frac{d\Sigma(r)}{dr} = 0$ or $\Sigma(r) = \text{Constant} = C_{ps}$, and taking into consideration the choice of $\Delta(r) = 2V(r) \rightarrow V(r)$ as earlier illustrated by Alhaidari et al. (2006). Then,

Approximate Solutions of the Dirac Equation for the Rosen-Morse Potential in the Presence of the Spin-Orbit and
Pseudo-Orbit Centrifugal Terms

141

$$\Delta(r) = -4V_1 \frac{e^{-2\alpha r}}{(1 + e^{-2a\alpha r})^2} + V_2 \frac{(1 - e^{-2\alpha r})}{(1 + e^{-2a\alpha r})}. \tag{32}$$

With the pseudo-centrifugal approximation in equation (13) and substituting $z = -e^{-2\alpha r}$, then, the following equation for the lower component spinor $G_{n\kappa}(r)$ is obtained as:

$$z^2 \frac{d^2}{dz^2} G_{n\kappa}(z) + z \frac{d}{dz} G_{n\kappa}(z) + \frac{1}{4\alpha^2} \left\{ \frac{1}{\hbar^2 c^2} \left[E_{n\kappa}^2 - M^2 c^4 - C_{ps}(Mc^2 + E_{n\kappa}) \right] \right\} G_{n\kappa}(z)$$

$$-\frac{\kappa(\kappa-1)}{4\alpha^2} \left\{ \frac{1}{r_0^2} \left[C_0 + C_1 \frac{z}{1-z} + C_2 \frac{z^2}{(1-z)^2} \right] - \frac{4\tilde{V}_3 z}{(1-z)^2} - \tilde{V}_4 - \frac{2\tilde{V}_4 z}{(1-z)} \right\} G_{n\kappa}(z), \tag{33}$$

where

$$\tilde{V}_3 = \frac{V_1}{\hbar^2 c^2} [Mc^2 - E_{n\kappa} + C_{ps}] \text{ and } \tilde{V}_4 = \frac{V_2}{\hbar^2 c^2} [Mc^2 - E_{n\kappa} + C_{ps}]. \tag{34}$$

With boundary conditions in the previous subsection, then, writing the function $G_{n\kappa}(z)$ as

$$G_{n\kappa}(z) = (1-z)^{1+\bar{q}} z^{\bar{\beta}} g_{n\kappa}(z), \tag{35}$$

where

$$\bar{q} = \frac{1}{2} \left[-1 + \sqrt{1 + \frac{\kappa(\kappa-1)C_2}{\alpha^2 r_0^2} - \frac{4\tilde{V}_3}{\alpha^2}} \right] \tag{36}$$

and

$$-\bar{\beta}^2 = \frac{1}{4\alpha^2} \left\{ \frac{1}{\hbar^2 c^2} \left[E_{n\kappa}^2 - M^2 c^4 - C_{ps}(Mc^2 + E_{n\kappa}) \right] - \frac{\kappa(\kappa-1)}{r_0^2} C_0 + \tilde{V}_4 \right\}. \tag{37}$$

On substituting equation (35) into equation (33) and using equations (34), (36) and (37), equation (33) becomes

$$z(1-z) \frac{d^2}{dz^2} g_{n\kappa}(z) + \left[(2\bar{\beta}+1) - (2\bar{q} + 2\bar{\beta} + 3)z \right] \frac{d}{dz} g_{n\kappa}(z)$$

$$- \left[(2\bar{\beta}+1)(1+\bar{q}) - \frac{\tilde{V}_4 + 2\tilde{V}_3}{2\alpha^2} + \frac{\kappa(\kappa-1)C_1}{4\alpha^2 r_0^2} \right] g_{n\kappa}(z), \tag{38}$$

whose solutions are the hypergeometric functions (Gradshteyn & Ryzhik, 2007), its general form can be expressed as

$$f_{n\kappa}(z) = A \, {}_2F_1(a, b; c; z) + B z^{1-c} \, {}_2F_1(a - c + 1, b - c + 1; 2 - c; z), \tag{39}$$

in which the first term can be expressed as:

$$ {}_2F_1(a, b; c; z) = \frac{\Gamma(c)}{\Gamma(a)\Gamma(b)} \sum_{k=0}^{\infty} \frac{\Gamma(a+k)\Gamma(b+k)z^k}{\Gamma(c+k)k!}, \tag{40}$$

where

$$a = 1 + \bar{q} + \bar{\beta} - \bar{\gamma}$$

$$b = 1 + \bar{q} + \bar{\beta} + \bar{\gamma}$$

$$c = 1 + 2\bar{\beta}$$

$$\bar{\gamma} = \sqrt{\bar{\beta}^2 + \frac{(\tilde{V}_2 + 2\tilde{V}_3)}{2\alpha^2} + \bar{q}(1 + \bar{q}) - \frac{\kappa(\kappa - 1)C_1}{4\alpha^2 r_0^2}}. \qquad (41)$$

Also, in the similar fashion as obtained in the case of the spin symmetry condition, an explicit expression for the energy eigenvalues of the Dirac equation with the Rosen-Morse potential under the pseudospin symmetry is obtained as:

$$(Mc^2 - E_{n\kappa} + C_{ps})(Mc^2 + E_{n\kappa} - V_2) = -\frac{\kappa(\kappa - 1)C_0}{r_0^2}\hbar^2 c^2$$

$$+ 4\alpha^2 \hbar^2 c^2 \left[\frac{\frac{(C_2 - C_1)}{4\alpha^2 r_0^2}\kappa(\kappa - 1) + \frac{(Mc^2 - E_{n\kappa} + C_{ps})V_2}{2\alpha^2 \hbar^2 c^2}}{2(n + \bar{q} + 1)} - \frac{(n + \bar{q} + 1)}{2} \right]^2. \qquad (42)$$

It is observed that, the pseudospin symmetric limit leads to quadratic energy eigenvalues. Therefore, the solution of equation (42) consists of positive and negative energy eigenvalues for each n and κ. Since, it has been shown that there are only negative energy eigenvalues and no bound positive energy eigenvalues exist in the pseudospin limit (Ginocchio, 2005). Therefore, in the pseudospin limit, only negative energy eigenvalues are chosen.

The radial lower component spinor can be obtained by considering equations (35)-(41) as

$$G_{n\kappa}(r) = \bar{N}_{n\kappa}(1 + e^{-2\alpha r})^{1+\bar{q}}(-e^{-2\alpha r})^{\bar{\beta}} \, _2F_1(-n, n + 2(\bar{\beta} + \bar{q} + 1); 2\bar{\beta} + 1; -e^{-2\alpha r})$$

$$= \bar{N}_{n\kappa} \frac{n!\Gamma(2\bar{\beta} + 1)}{\Gamma(n + \bar{\beta} + 1)}(1 + e^{-2\alpha r})^{1+\bar{q}}(-e^{-2\alpha r})^{\bar{\beta}} P_n^{(2\bar{\beta}, \, 2\bar{q}+1)}(1 + 2e^{-2\alpha r}) \qquad (43)$$

$\bar{N}_{n\kappa}$ is the normalization constant which can be determined by the condition that $\int_0^\infty |G_{n\kappa}(r)|^2 \, dr = 1$ and by making use of the equations (23) and (28), we have

$$\bar{N}_{n\kappa} = \left[\frac{\Gamma(2\bar{q} + 3)\Gamma(2\bar{\beta} + 1)}{2\alpha\Gamma(n)} \sum_{k=0}^\infty \frac{(-1)^k (n + 2(1 + \bar{\beta} + \bar{q}))_k \Gamma(n + k)}{k!(k + 2\bar{\beta})!\Gamma\left(k + 2(\bar{\beta} + \bar{q} + \frac{3}{2})\right)} \bar{A}_{n\kappa} \right]^{-1/2}, \qquad (44)$$

where $\bar{A}_{n\kappa} = \, _3F_2(2\bar{\beta} + k, -n, n + 2(1 + \bar{\beta} + \bar{q}); k + 2(\bar{\beta} + \bar{q} + \frac{3}{2}); 2\bar{\beta} + 1; 1)$ and $(x)_a = \frac{\Gamma(x+a)}{\Gamma(x)}$ (Pochhammer symbol).

Approximate Solutions of the Dirac Equation for the Rosen-Morse Potential in the Presence of the Spin-Orbit and Pseudo-Orbit Centrifugal Terms

143

Similarly, by using equation (12) $F_{n\kappa}(r)$ can also be obtained as

$$F_{n\kappa}(r) = \frac{\overline{N}_{n\kappa}(1 + e^{-2\alpha r})^{1+\bar{q}}(-e^{-2\alpha r})^{\bar{\beta}}}{[Mc^2 - E_{n\kappa} + C_{ps}]} \left[-2\alpha\bar{\beta} - \frac{2\alpha e^{-2\alpha r}}{1 + e^{-2\alpha r}} - \frac{\kappa}{r} \right]$$
$$\times {_2F_1}(-n, n + 2(\bar{\beta} + \bar{q} + 1); 2\bar{\beta} + 1; -e^{-2\alpha r}) +$$
$$\frac{\overline{N}_{n\kappa}(1 + e^{-2\alpha r})^{1+\bar{q}}(-e^{-2\alpha r})^{\bar{\beta}+1}}{[Mc^2 - E_{n\kappa} + C_{ps}]} \left\{ \frac{2\alpha n \left[n + 2(\bar{\beta} + \bar{q} + 1)\right]}{(2\bar{\beta} + 1)} \right\}$$
$$\times {_2F_1}(-n + 1, n + 2(\bar{\beta} + \bar{q} + \tfrac{3}{2}); 2(\bar{\beta} + 1); -e^{-2\alpha r}). \tag{45}$$

It is pertinent to note that, the negative energy solution for the pseudospin symmetry can be obtained directly from the positive energy solution of the spin symmetry using the parameter mapping (Berkdemir & Cheng, 2009; Ikhdair, 2010):

$$F_{n\kappa}(r) \leftrightarrow G_{n\kappa}(r), V(r) \to -V(r), \text{ (or } V_1 \to -V_1 \text{ and } V_2 \to -V_2), E_{n\kappa} \to -E_{n\kappa} \text{ and } C_s \to -C_{ps}.$$

3.3 Remarks
In this work, solutions of some special cases are studied:

3.3.1 s-wave solutions:
Our results include any arbitrary κ values, therefore, there is need to investigate if our results will give similar results for s-wave for the spin symmetry when $\kappa = -1$ or $\ell = 0$ and for the pseudospin when $\kappa = 1$ or $\bar{\ell} = 0$.
For the SS, $\kappa = -1$ (or $\ell = 0$) in equation (26) gives

$$(Mc^2 + E_{n,-1} - C_s)(Mc^2 - E_{n,-1} + V_2) = 4\alpha^2\hbar^2 c^2 \left[\frac{\frac{(Mc^2 + E_{n,-1} - C_s)V_2}{2\alpha^2\hbar^2 c^2}}{2(n + q_1 + 1)} - \frac{(n + q_1 + 1)}{2} \right]^2, \tag{46}$$

where

$$q_1 = \frac{1}{2}\left[-1 + \sqrt{1 + \frac{4V_1[Mc^2 + E_{n,-1} - C_s]}{\alpha^2\hbar^2 c^2}} \right]. \tag{47}$$

For the PSS, $\kappa = 1$ (or $\bar{\ell} = 0$) in equation (42) gives

$$(Mc^2 - E_{n,1} + C_{ps})(Mc^2 + E_{n,1} - V_2) = 4\alpha^2\hbar^2 c^2 \left[\frac{\frac{(Mc^2 - E_{n,1} + C_{ps})V_2}{2\alpha^2\hbar^2 c^2}}{2(n + \bar{q}_1 + 1)} - \frac{(n + \bar{q}_1 + 1)}{2} \right]^2, \tag{48}$$

where

$$\bar{q}_1 = \frac{1}{2}\left[-1 + \sqrt{1 - \frac{4V_1[Mc^2 - E_{n,1} + C_{ps}]}{\alpha^2\hbar^2 c^2}} \right]. \tag{49}$$

The corresponding upper and lower component spinors for the SS and PSS can be obtained also. The above solutions are identical with the results obtained by Oyewumi & Akoshile (2010) and Ikhdair (2010).

3.3.2 Solutions for the standard Eckart potential:

By setting $V_1 = -V_1$ and $V_2 = -V_2$ in equation (1), we have the standard Eckart potential. The energy eigenvalues for the SS and the PSS are given, respectively as:

$$(Mc^2 + E_{n\kappa} - C_s)(Mc^2 - E_{n\kappa} - V_2) = -\frac{\kappa(\kappa+1)C_0}{r_0^2}\hbar^2 c^2$$

$$+4\alpha^2\hbar^2 c^2\left[\frac{\frac{(C_2-C_1)}{4\alpha^2 r_0^2}\kappa(\kappa+1) + \frac{(Mc^2+E_{n\kappa}-C_s)V_2}{2\alpha^2\hbar^2 c^2}}{2(n+q_2+1)} - \frac{(n+q_2+1)}{2}\right]^2 \tag{50}$$

and

$$(Mc^2 - E_{n\kappa} + C_{ps})(Mc^2 + E_{n\kappa} + V_2) = -\frac{\kappa(\kappa-1)C_0}{r_0^2}\hbar^2 c^2$$

$$+4\alpha^2\hbar^2 c^2\left[\frac{\frac{(C_2-C_1)}{4\alpha^2 r_0^2}\kappa(\kappa-1) - \frac{(Mc^2-E_{n\kappa}+C_{ps})V_2}{2\alpha^2\hbar^2 c^2}}{2(n+\bar{q}_2+1)} - \frac{(n+\bar{q}_2+1)}{2}\right]^2, \tag{51}$$

where q_2 and $\bar{q}_2$ are obtained, respectively as:

$$q_2 = \frac{1}{2}\left[-1+\sqrt{1+\frac{\kappa(\kappa+1)C_2}{\alpha^2 r_0^2} - \frac{4V_1[Mc^2+E_{n\kappa}-C_s]}{\alpha^2\hbar^2 c^2}}\right]$$

$$\bar{q}_2 = \frac{1}{2}\left[-1+\sqrt{1+\frac{\kappa(\kappa-1)C_2}{\alpha^2 r_0^2} + \frac{4V_1[Mc^2-E_{n\kappa}+C_{ps}]}{\alpha^2\hbar^2 c^2}}\right]. \tag{52}$$

The corresponding upper and lower component spinors for the SS and the PSS can easily be obtained from equations (27), (31), (43) and (45).

3.3.3 Solutions of the PT-Symmetric Rosen-Morse potential:

The choice of $V_2 = iV_2$ in equation (1) gives the PT-Symmetric Rosen-Morse potential (Jia et al., 2002; Yi et al., 2004; Taşkin, 2009; Oyewumi & Akoshile, 2010; Ikhdair, 2010):

$$V(r) = -V_1\mathrm{sech}^2\alpha r + iV_2\tanh\alpha r. \tag{53}$$

For a given potential $V(r)$, if $V(-r) = V^*(r)$ (or $V(\eta - r) = V^*(r)$) exists, then, the potential $V(r)$ is said to be PT-Symmetric. Here, P denotes the parity operator (space reflection, $P : r \to -r$, or $r \to \eta - r$) and T denotes the time reversal operator ($T : i \to -i$).

For the case of the SS and the PSS solutions of this PT-Symmetric version of the Rosen-Morse potential, the energy eigenvalue equations are:

$$(Mc^2 + E_{n\kappa} - C_s)(Mc^2 - E_{n\kappa} + iV_2) = -\frac{\kappa(\kappa+1)C_0}{r_0^2}\hbar^2 c^2$$

$$+4\alpha^2\hbar^2 c^2\left[\frac{\frac{(C_2-C_1)}{4\alpha^2 r_0^2}\kappa(\kappa+1) - i\frac{(Mc^2+E_{n\kappa}-C_s)V_2}{2\alpha^2\hbar^2 c^2}}{2(n+q+1)} - \frac{(n+q+1)}{2}\right]^2 \tag{54}$$

Approximate Solutions of the Dirac Equation for the Rosen-Morse Potential in the Presence of the Spin-Orbit and Pseudo-Orbit Centrifugal Terms

145

and

$$(Mc^2 - E_{n\kappa} + C_{ps})(Mc^2 + E_{n\kappa} - iV_2) = -\frac{\kappa(\kappa-1)C_0}{r_0^2}\hbar^2 c^2$$

$$+4\alpha^2\hbar^2 c^2 \left[\frac{\frac{(C_2-C_1)}{4\alpha^2 r_0^2}\kappa(\kappa-1) + i\frac{(Mc^2-E_{n\kappa}+C_{ps})V_2}{2\alpha^2\hbar^2 c^2}}{2(n+\bar{q}+1)} - \frac{(n+\bar{q}+1)}{2} \right]^2, \quad (55)$$

respectively. q and $\bar{q}$ have their usual values as in equations (19) and (36), the corresponding upper and lower component spinors for the SS and the PSS can be obtained directly from equations (27), (31), (43) and (45).

3.3.4 Solutions of the reflectionless-type potential:

If we choose $V_2 = 0$ and $V_1 = \frac{1}{2}\xi(\xi+1)$ in equation (1), then equation (1) becomes the reflectionless-type potential (Grosche & Steiner, 1995; 1998; Zhao et al., 2005):

$$V(r) = -\xi(\xi+1)\text{sech}^2\alpha r, \quad (56)$$

where ξ is an integer, that is, $\xi = 1, 2, 3, \ldots$.

For the SS solutions of the reflectionless-type potential, the energy eigenvalues, the upper and the lower component spinors are obtained, respectively as:

$$(Mc^2 + E_{n\kappa} - C_s)(Mc^2 - E_{n\kappa}) = -\frac{\kappa(\kappa+1)C_0}{r_0^2}\hbar^2 c^2 + \alpha^2\hbar^2 c^2 \left[\frac{\frac{(C_2-C_1)}{4\alpha^2 r_0^2}\kappa(\kappa+1)}{2(n+q_3+1)} - \frac{(n+q_3+1)}{2} \right]^2, \quad (57)$$

$$F_{n\kappa}(r) = N_{n\kappa}(1+e^{-2\alpha r})^{1+q_3}(-e^{-2\alpha r})^{\beta_3} {}_2F_1(-n, n+2(\beta_3+q_3+1); 2\beta_3+1; -e^{-2\alpha r})$$

$$= N_{n\kappa}\frac{n!\Gamma(2\beta_3+1)}{\Gamma(n+\beta_3+1)}(1+e^{-2\alpha r})^{1+q_3}(-e^{-2\alpha r})^{\beta_3} P_n^{(2\beta_3, 2q_3+1)}(1+2e^{-2\alpha r}) \quad (58)$$

and

$$G_{n\kappa}(r) = \frac{N_{n\kappa}(1+e^{-2\alpha r})^{1+q_3}(-e^{-2\alpha r})^{\beta_3}}{[Mc^2 + E_{n\kappa} - C_s]} \left[-2\alpha\beta_3 - \frac{2\alpha e^{-2\alpha r}}{1+e^{-2\alpha r}} + \frac{\kappa}{r} \right]$$

$$\times {}_2F_1(-n, n+2(\beta_3+q_3+1); 2\beta_3+1; -e^{-2\alpha r}) +$$

$$\frac{N_{n\kappa}(1+e^{-2\alpha r})^{1+q_3}(-e^{-2\alpha r})^{\beta_3+1}}{[Mc^2 + E_{n\kappa} - C_s]} \left\{ \frac{2\alpha n[n+2(\beta_3+q_3+1)]}{(2\beta_3+1)} \right\}$$

$$\times {}_2F_1(-n+1, n+2(\beta_3+q_3+\tfrac{3}{2}); 2(\beta_3+1); -e^{-2\alpha r}), \quad (59)$$

where

$$q_3 = \frac{1}{2}\left[-1 + \sqrt{1 + \frac{\kappa(\kappa+1)C_2}{\alpha^2 r_0^2} + \frac{2\xi(\xi+1)[Mc^2 + E_{n\kappa} - C_s]}{\alpha^2\hbar^2 c^2}} \right] \quad (60)$$

and

$$\beta_3 = \sqrt{\frac{\kappa(\kappa+1)}{4\alpha^2 r_0^2} C_0 - \frac{1}{4\alpha^2 \hbar^2 c^2} \left[E_{n\kappa}^2 - M^2 c^4 + C_s(Mc^2 - E_{n\kappa}) \right]}. \tag{61}$$

For the PSS solutions of the reflectionless-type potential, the energy eigenvalues, the upper and the lower component spinors are obtained, respectively as:

$$(Mc^2 - E_{n\kappa} + C_{ps})(Mc^2 + E_{n\kappa}) = -\frac{\kappa(\kappa-1)C_0}{r_0^2}\hbar^2 c^2 + \alpha^2 \hbar^2 c^2 \left[\frac{\frac{(C_2 - C_1)}{\alpha^2 r_0^2}\kappa(\kappa-1)}{2(n+\bar{q}_3+1)} - \frac{(n+\bar{q}_3+1)}{2} \right]^2, \tag{62}$$

$$G_{n\kappa}(r) = \overline{N}_{n\kappa}(1+e^{-2\alpha r})^{1+\bar{q}_3}(-e^{-2\alpha r})^{\bar{\beta}_3} \, {}_2F_1(-n, n+2(\bar{\beta}_3 + \bar{q}_3 + 1); 2\bar{\beta}_3 + 1; -e^{-2\alpha r})$$

$$= \overline{N}_{n\kappa} \frac{n!\Gamma(2\bar{\beta}_3+1)}{\Gamma(n+\bar{\beta}_3+1)}(1+e^{-2\alpha r})^{1+\bar{q}_3}(-e^{-2\alpha r})^{\bar{\beta}_3} P_n^{(2\bar{\beta}_3, \, 2\bar{q}_3+1)}(1+2e^{-2\alpha r}) \tag{63}$$

and

$$F_{n\kappa}(r) = \frac{\overline{N}_{n\kappa}(1+e^{-2\alpha r})^{1+\bar{q}_3}(-e^{-2\alpha r})^{\bar{\beta}_3}}{[Mc^2 - E_{n\kappa} + C_{ps}]}\left[-2\alpha\bar{\beta}_3 - \frac{2\alpha e^{-2\alpha r}}{1+e^{-2\alpha r}} - \frac{\kappa}{r} \right]$$

$$\times {}_2F_1(-n, n+2(\bar{\beta}_3 + \bar{q}_3 + 1); 2\bar{\beta}_3 + 1; -e^{-2\alpha r}) +$$

$$\frac{\overline{N}_{n\kappa}(1+e^{-2\alpha r})^{1+\bar{q}_3}(-e^{-2\alpha r})^{\bar{\beta}_3+1}}{[Mc^2 - E_{n\kappa} + C_{ps}]}\left\{ \frac{2\alpha n \left[n+2(\bar{\beta}_3 + \bar{q}_3 + 1) \right]}{(2\bar{\beta}_3 + 1)} \right\}$$

$$\times {}_2F_1(-n+1, n+2(\bar{\beta}_3 + \bar{q}_3 + \tfrac{3}{2}); 2(\bar{\beta}_3 + 1); -e^{-2\alpha r}), \tag{64}$$

where

$$\bar{q}_3 = \frac{1}{2}\left[-1 + \sqrt{1 + \frac{\kappa(\kappa-1)C_2}{\alpha^2 r_0^2} - \frac{2\xi(\xi+1)[Mc^2 - E_{n\kappa} + C_{ps}]}{\alpha^2 \hbar^2 c^2}} \right] \tag{65}$$

and

$$\bar{\beta}_3 = \sqrt{\frac{\kappa(\kappa-1)}{4\alpha^2 r_0^2} C_0 - \frac{1}{4\alpha^2 \hbar^2 c^2}\left[E_{n\kappa}^2 - M^2 c^4 - C_{ps}(Mc^2 + E_{n\kappa}) \right]}. \tag{66}$$

3.3.5 Solutions of the non-relativistic limit

The approximate solutions of the Schrödinger equation for the Rosen-Morse potential including the centrifugal term can be obtained from our work. This can be done by equating $C_s = 0, S(r) = V(r) = \Sigma(r)$ in equations (26) and (27). By using the following appropriate transformations suggested by Ikhdair (2010):

$$\frac{(Mc^2 + E_{n\kappa})}{\hbar^2 c^2} \to \frac{2\mu}{\hbar^2}$$

$$Mc^2 - E_{n\kappa} \to -E_{n\ell} \tag{67}$$

$$\kappa \to \ell,$$

the non-relativistic limit of the energy equation and the associated wave functions, respectively become:

$$E_{n\ell} = V_2 + \frac{\ell(\ell+1)\hbar^2 C_0}{2\mu r_0^2} - \frac{\hbar^2 c^2}{2\mu} \left[\frac{(n+1)^2 + (2n+1)q_0 + \frac{\ell(\ell+1)C_1}{4\alpha^2 r_0^2} + \frac{\mu}{\alpha^2 \hbar^2}(2V_1 + V_2)}{(n+q_0+1)} \right]^2,$$

(68)

and

$$F_{n\ell}(r) = N_{n\ell}(1 + e^{-2\alpha r})^{1+q_0}(-e^{-2\alpha r})^{\beta_0} \, _2F_1(-n, n + 2(\beta_0 + q_0 + 1); 2\beta_0 + 1; -e^{-2\alpha r})$$

$$= N_{n\ell} \frac{n!\Gamma(2\beta_0 + 1)}{\Gamma(n + \beta_0 + 1)} (1 + e^{-2\alpha r})^{1+q_0}(-e^{-2\alpha r})^{\beta_0} P_n^{(2\beta_0, \, 2q_0 + 1)}(1 - 2z),$$

(69)

where

$$q_0 = \frac{1}{2} \left[-1 + \sqrt{1 + \frac{\ell(\ell+1)C_2}{\alpha^2 r_0^2} + \frac{8\mu V_1}{\alpha^2 \hbar^2}} \right]$$

(70)

and

$$\beta_0 = \sqrt{\frac{\ell(\ell+1)}{4\alpha^2 r_0^2} C_0 + \frac{\mu V_2}{2\alpha^2 \hbar^2} - \frac{\mu E_{n\ell}}{2\alpha^2 \hbar^2}}.$$

(71)

By using the appropriate transformations suggested by Ikhdair (2010), the non-relativistic limit of energy equation and the associated wave functions of the Schrödinger equation for the Rosen-Morse potential are recovered completely. These results are identical with the results of Ikhdair (2010), Taşkin (2009)(note that Taşkin (2009) used $\hbar = \mu = 1$ in his calculations).

4. The relativistic bound state solutions of the Rosen-Morse potential with the centrifugal term

The Klein-Gordon and the Dirac equations describe relativistic particles with zero or integer and 1/2 integral spins, respectively (Landau & Lifshift 1999; Merzbacher, 1998; Greiner, 2000; Alhaidari et al., 2006; Dong, 2007). However, the exact solutions are only possible for a few simple systems such as the hydrogen atom, the harmonic oscillator, Kratzer potential and pseudoharmonic potential.

In the following specific examples, Soylu et al. (2008c) obtained the s-wave solutions of the Klein-Gordon equation with equal scalar and vector Rosen-Morse potential by using the asymptotic iteration method. Also, Yi et al. (2004) obtained the energy equation and the corresponding wave functions of the Klein-Gordon equation for the Rosen-Morse-type potential by using standard and functional method.

For the approximate solutions of the Schrödinger equation for the Rosen-Morse potential with the centrifugal term, that is $\ell \neq 0$ or $\kappa = 1$, with the standard function analysis method, Taşkin (2009) has used the newly improved Pekeris-type approximation introduced by Lu (2005). In addition, Ikot and Akpabio (2010) solved this same problem by using the Nikiforov-Uvarov method, they used the approximation scheme introduced by Jia et al. (2009a, 2009b) and Xu et al. (2010).

In the recent years, some researchers have used the Pekeris-type approximation scheme for the centrifugal term to solve the relativistic equations to obtain the ℓ or $\kappa-$ wave energy equations and the associated wave functions of some potentials. These include: Morse potential (Bayrak et al., 2010), hyperbolical potential (Wei & Liu, 2008), Manning-Rosen potential (Wei & Dong, 2010), Deng-Fan oscillator (Dong, 2011).

In the context of the standard function analysis approach, the approximate bound state solutions of the arbitrary ℓ-state Klein-Gordon and κ-state Dirac equations for the equally mixed Rosen-Morse potential will be obtained by introducing a newly improved approximation scheme to the centrifugal term.

4.1 Approximate bound state solutions of the Klein-Gordon equation for the Rosen-Morse potential for $\ell \neq 0$

The time-independent Klein-Gordon equation with the scalar $S(r)$ and vector $V(r)$ potentials is given as (Landau & Lifshift, 1999; Merzbacher, 1998; Greiner, 2000; Alhaidari et al., 2006):

$$\left\{ -\hbar^2 c^2 \nabla^2 + \left[Mc^2 + S(r) \right]^2 - [E - V(r)]^2 \right\} \psi(r,\theta,\phi) = 0, \tag{72}$$

where $M, \hbar$ and c are the rest mass of the spin-0 particle, Planck's constant and velocity of the light, respectively. For spherical symmetrical scalar and vector potentials, putting

$$\psi_{n,\ell,m}(r,\theta,\phi) = \frac{1}{r} U_{n,\ell}(r) Y_{\ell,m}(\theta,\phi), \tag{73}$$

where $Y_{\ell,m}(\theta,\phi)$ is the spherical harmonic function, we obtain the radial Klein-Gordon equation as

$$U''_{n,\ell}(r) + \frac{1}{\hbar^2 c^2} \left\{ E^2 - M^2 c^4 - 2 \left[EV(r) + Mc^2 S(r) \right] + \left[V^2(r) - S^2(r) \right] - \frac{\ell(\ell+1)\hbar^2 c^2}{r^2} \right\} U_{n,\ell}(r) = 0. \tag{74}$$

We are considering the case when the scalar and vector potentials are equal (that is, $S(r) = V(r)$), coupled with the resulting simplification in the solution of the relativistic problems as discussed by Alhaidari et al., 2006, we have

$$U''_{n,\ell}(r) + \frac{1}{\hbar^2 c^2} \left\{ E^2 - M^2 c^4 - \left[E + Mc^2 \right] V(r) - \frac{\ell(\ell+1)\hbar^2 c^2}{r^2} \right\} U_{n,\ell}(r) = 0. \tag{75}$$

This equation cannot be solved analytically for the Rosen-Morse potential with $\ell \neq 0$, unless, we introduce the approximation scheme (earlier discussed in this chapter) to the centrifugal term. With this approximation scheme, and the potential in (1) together with the transformation $z = -e^{-2\alpha r}$ in equation (75), we have

$$z^2 U''_{n,\ell}(z) + z U'_{n,\ell}(z)$$

$$+ \left[\frac{E^2 - M^2 c^4}{4\alpha^2 \hbar^2 c^2} - \frac{\tilde{V}_5}{\alpha^2} \frac{z}{(1-z)^2} - \frac{\tilde{V}_6}{4\alpha^2} \frac{(1+z)}{(1-z)} - \frac{\ell(\ell+1)C_0}{4\alpha^2 r_0^2} - \frac{\ell(\ell+1)C_1}{4\alpha^2 r_0^2} \frac{z}{(1-z)} - \frac{\ell(\ell+1)C_2}{4\alpha^2 r_0^2} \frac{z^2}{(1-z)^2} \right] U_{n,\ell}(z) = 0, \tag{76}$$

where

$$\tilde{V}_5 = \frac{V_1}{\hbar^2 c^2}[E + Mc^2]$$

$$\tilde{V}_6 = \frac{V_2}{\hbar^2 c^2}[E + Mc^2]. \tag{77}$$

In the similar manner, the energy equation of the arbitrary ℓ-state Klein-Gordon equation with equal scalar and vector potentials of the Rosen-Morse potential is obtained as follows:

$$(E_{n,\ell}^2 - M^2 c^4) = (E_{n,\ell} + Mc^2)V_2 + \frac{\ell(\ell+1)C_0}{r_0^2}\hbar^2 c^2$$

$$-4\alpha^2 \hbar^2 c^2 \left[\frac{\frac{(C_2 - C_1)}{4\alpha^2 r_0^2}\ell(\ell+1) - \frac{(E_{n,\ell} + Mc^2)V_2}{2\alpha^2 \hbar^2 c^2}}{2(n+\delta_1+1)} - \frac{(n+\delta_1+1)}{2} \right]^2, \tag{78}$$

where

$$\delta_1 - \frac{1}{2}\left[-1 + \sqrt{1 \mid \frac{\ell(\ell+1)C_2}{\alpha^2 r_0^2} + \frac{4(E_{n,\ell} + Mc^2)}{\alpha^2 \hbar^2 c^2}} \right]. \tag{79}$$

The associated wave function can be expressed as

$$U_{n,\ell}(r) = N_{n,\ell}(1 + e^{-2\alpha r})^{1+\delta_1}(-e^{-2\alpha r})^{\xi_1} {}_2F_1(-n, n + 2(\xi_1 + \delta_1 + 1); 2\xi_1 + 1; -e^{-2\alpha r})$$

$$= N_{n,\ell}\frac{n!\Gamma(2\xi_1 + 1)}{\Gamma(n + \xi_1 + 1)}(1 + e^{-2\alpha r})^{1+\delta_1}(-e^{-2\alpha r})^{\xi_1} P_n^{(2\xi_1, 2\delta_1 + 1)}(1 + 2e^{-2\alpha r}) \tag{80}$$

where

$$\xi_1 = \sqrt{\frac{\ell(\ell+1)C_0}{4\alpha^2 r_0^2} + \frac{V_2(E_{n,\ell} + Mc^2)}{4\alpha^2 \hbar^2 c^2} - \frac{E^2 - M^2 c^4}{4\alpha^2 \hbar^2 c^2}} \tag{81}$$

and $N_{n,\ell}$ is the normalization constant which can easily be determined in the usual manner.

4.2 Approximate bound state solutions of the Dirac equation for the Rosen-Morse potential for any κ

In this subsection, we consider equations (2), (3) and (4), and on re-writing equations (5) and (6) for the case of equal scalar and vector, i. e. $V(r) = S(r)$, we have the following two coupled differential equations:

$$\left(\frac{d}{dr} - \frac{\kappa}{r} \right) F_{n\kappa}(r) = \left[Mc^2 + E_{n\kappa} \right] G_{n\kappa}, \tag{82}$$

$$\left(\frac{d}{dr} + \frac{\kappa}{r} \right) G_{n\kappa}(r) = \left[Mc^2 - E_{n\kappa} \right] F_{n\kappa}. \tag{83}$$

With the substitution of equation (82) into equation (83) and taking into consideration the suggestion of Alhaidari et al., (2006), a Schrödinger-like equation for the arbitrary spin-orbit coupling quantum number κ is obtained as

$$\frac{d^2 F_{n\kappa}(r)}{dr^2} + \frac{1}{\hbar^2 c^2}\left\{ [E_{n\kappa}^2 - M^2 c^4] - [Mc^2 + E_{n\kappa}]V(r) - \frac{\hbar^2 c^2 \kappa(\kappa-1)}{r^2} \right\} F_{n\kappa}(r) = 0. \tag{84}$$

Here, it is observed that equation (84) is identical with equation (75). Therefore, the energy equation of the Dirac equation with the equally mixed Rosen-Morse potential for arbitrary $\kappa-$state is obtained as

$$(E_{n\kappa}^2 - M^2c^4) = (E_{n\kappa} + Mc^2)V_2 + \frac{\kappa(\kappa-1)C_0}{r_0^2}\hbar^2c^2$$

$$-4\alpha^2\hbar^2c^2 \left[\frac{\frac{(C_2-C_1)}{4\alpha^2 r_0^2}\kappa(\kappa-1) - \frac{(E_{n\kappa}+Mc^2)V_2}{2\alpha^2\hbar^2c^2}}{2(n+\delta_2+1)} - \frac{(n+\delta_2+1)}{2} \right]^2, \tag{85}$$

where

$$\delta_2 = \frac{1}{2}\left[-1 + \sqrt{1 + \frac{\kappa(\kappa-1)C_2}{\alpha^2 r_0^2} + \frac{4(E_{n\kappa}+Mc^2)V_1}{\alpha^2\hbar^2c^2}} \right]. \tag{86}$$

The associated upper component spinor $F_{n\kappa}(r)$ is obtained as

$$F_{n\kappa}(r) = N_{n\kappa}(1+e^{-2\alpha r})^{1+\delta_2}(-e^{-2\alpha r})^{\xi_2} \,_2F_1(-n, n+2(\xi_2+\delta_2+1); 2\xi_2+1; -e^{-2\alpha r})$$

$$= N_{n\kappa}\frac{n!\Gamma(2\xi_2+1)}{\Gamma(n+\xi_2+1)}(1+e^{-2\alpha r})^{1+\delta_2}(-e^{-2\alpha r})^{\xi_2}P_n^{(2\xi_2,\,2\delta_2+1)}(1+2e^{-2\alpha r}). \tag{87}$$

On substituting equation (87) into equation (82) and by using the recurrence relation of the hypergeometric function in equation (30), the lower component spinor $G_{n\kappa}(r)$ can be obtained as

$$G_{n\kappa}(r) = \frac{N_{n\kappa}(1+e^{-2\alpha r})^{1+\delta_2}(-e^{-2\alpha r})^{\xi_2}}{[E_{n\kappa}+Mc^2]}\left[-2\alpha\xi_2 - \frac{2\alpha e^{-2\alpha r}}{1+e^{-2\alpha r}} - \frac{\kappa}{r} \right]$$

$$\times \,_2F_1(-n, n+2(\xi_2+\delta_2+1); 2\xi_2+1; -e^{-2\alpha r}) +$$

$$\frac{N_{n\kappa}(1+e^{-2\alpha r})^{1+\delta_2}(-e^{-2\alpha r})^{\xi_2+1}}{[E_{n\kappa}+Mc^2]}\left\{ \frac{2\alpha n\,[n+2(\xi_2+\delta_2+1)]}{(2\xi_2+1)} \right\}$$

$$\times \,_2F_1(-n+1, n+2(\xi_2+\delta_2+\tfrac{3}{2}); 2(\xi_2+1); -e^{-2\alpha r}), \tag{88}$$

where

$$\xi_2 = \sqrt{\frac{\kappa(\kappa-1)C_0}{4\alpha^2 r_0^2} + \frac{V_2(E_{n\kappa}+Mc^2)}{4\alpha^2\hbar^2c^2} - \frac{E^2-M^2c^4}{4\alpha^2\hbar^2c^2}} \tag{89}$$

and $N_{n\kappa}$ is the normalization constant which can easily be determined in the usual manner. Substitution of $F_{n\kappa}(r)$ and $G_{n\kappa}(r)$ into equation (5) gives the bound state spinors of the Dirac equation with the equally mixed Rosen-Morse potential for the arbitrary spin-orbit coupling quantum number κ. In the similar manner, approximate solutions can be obtained when $S(r) = -V(r)$.

5. Conclusions

The approximate analytical solutions of the Dirac equation with the Rosen-Morse potential with arbitrary κ under the pseudospin and spin symmetry conditions have been studied, the standard function analysis approach has been adopted. The Pekeris-type approximation scheme (a newly improved approximation scheme) has been used for the centrifugal (or pseudo centrifugal) term in order to solve for any values of κ.

Under the PSS and SS conditions, the energy equations, the upper- and the lower-component spinors for the Rosen-Morse potential for any κ have been obtained. The solutions of some special cases are also considered and the energy equations with their associated spinors for the PSS and SS are obtained, these include:

(i) the s-state solution,

(ii) the standard Eckart potential,

(iii) the PT-Symmetric Rosen-Morse potential,

(iv) the reflectionless-type potential,

(v) the non-relativistic limit.

Also, in the context of the standard function analysis approach, the approximate bound state solutions of the arbitrary ℓ-state Klein-Gordon and κ-state Dirac equations for the equally mixed Rosen-Morse potential are obtained by introducing a newly improved approximation scheme to the centrifugal term. The approximate analytical solutions with the Dirac-Rosen-Morse potential for any κ or ℓ have been obtained. The upper- and lower-component spinors have been expressed in terms of the hypergeometric functions (or Jacobi polynomials). The approximate analytical solutions obtained in this study are the same with other results available in the literature.

6. Acknowledgments

The author thanks his host Prof. K. D. Sen of the School of Chemistry, University of Hyderabad, India during his TWAS-UNESCO Associate research visit where part of this work was done. Also, he thanks his host Prof. M. N.Hounkonnou (the President of the ICMPA-UNESCO Chair), University of Abomey-Calavi, Republic of Benin where this work has been finalized. Thanks to Ms. Maja Bozicevic for her patience. He acknowledges the University of Ilorin for granting him leave. eJDS (ICTP) is acknowledged. Also, he appreciates the efforts of Profs. Ginocchio, J. N., Dong, S.H., Wei, G.F., Taşkin, F., Grosche, C., Berkdemir, C., Zhang, M. C., Hamzavi and his collaborators for communicating their works to me.

7. References

[1] Akcay, H. (2007). The Dirac oscillator with a Coulomb-like tensor potential, *Journal of Physics A: Mathematical & Theoretical* Vol. 40: 6427 - 6432.

[2] Akcay, H. (2009). Dirac equation with scalar and vector quadratic potentials and Coulomb-like tensor potential, *Physics Letters A* Vol. 373: 616 - 620.

[3] Akcay, H. & Tezcan, C. (2009). Exact solutions of the Dirac equation with harmonic oscillator potential including a Coulomb-like tensor potential, *International Journal of Modern Physics C* Vol. 20 (No. 6): 931 - 940.

[4] Alberto, P., Fiolhais, M., Malheiro, M., Delfino, A. & Chiapparini, M. (2001). Isospin asymmetry in the pseudospin dynamical symmetry, *Physical Review Letters* Vol. 86 (No. 22): 5015 - 5018.

[5] Alberto, P., Fiolhais, M., Malheiro, M., Delfino, A. & Chiapparini, M. (2002). Pseudospin symmetry as a relativstic dynamical symmetry in the nucleus, *Physical Review C* Vol. 65: 034307-1 - 034307-9.

[6]   Alhaidari, A. D. , Bahlouli, H. & Al-Hasan, A. (2006). Dirac and Klein-Gordon equations with equal scalar and vector potentials, *Physics Letters A*   Vol. 349: 87 - 97.

[7]   Amani, A. R. , Moghrimoazzen, M. A. , Ghorbanpour, H. and Barzegaran, S. (2011). The ladder operators of Rosen-Morse Potential with Centrifugal term by Factorization Method, *African Journal of Mathematical Physics* Vol. 10: 31 - 37.

[8]   Arima, A. , Harvey, M. & Shimizu, K. (1969). Pseudo LS coupling and pseudo SU3 coupling schemes, *Physics Letters B* Vol. 30: 517 – 522.

[9]   Aydoğdu, O. & Sever, R. (2009). Solution of the Dirac equation for pseudoharmonic potential by using the Nikiforov-Uvarov method, *Physica Scripta*   Vol. 80: 015001-1 - 015001-6.

[10] Aydoğdu, O. (2009). Pseudospin symmetry and its applications, *Ph. D. thesis, Middle East Technical University*, Turkey.

[11] Aydoğdu, O. & Sever, R. (2010a). Exact pseudospin symmetric solution of the Dirac equation for pseudoharmonic potential in the presence of tensor potential, *Few-Body system* Vol. 47: 193 - 200.

[12] Aydoğdu, O. & Sever, R. (2010b). Exact solution of the Dirac equation with the Mie-type potential under the pseudospin and spin symmetry limit, *Annals of Physics* Vol. 325: 373 - 383.

[13] Aydoğdu, O. & Sever, R. (2010c). Pseudospin and spin symmetry in the Dirac equation with Woods-Saxon potential and Tensor potential, *European Physical Journal A* Vol. 43: 73 - 81.

[14] Aydoğdu, O. & Sever, R. (2010d). Pseudospin and spin symmetry for the ring-shaped generlized Hulthén potential, *International Journal of Modern Physics A* Vol. 25 (No. 21): 4067 - 4079.

[15] Bayrak, O. & Boztosun, I. (2007). The pseudospin symmetric solution of the Morse potential for any κ state, *Journal of Mathematical & Theoretical* Vol. 40: 11119 - 11127.

[16] Bayrak, O., Soylu, A. & Boztosun, I. (2010). The relativistic treatment of spin-0 particles under the rotating Morse oscillator, *Journal of Mathematical Physics* Vol. 51: 112301-1 - 112301-6.

[17] Berkdemir, C. (2006). Pseudospin symmetry in the relativistic Morse potential includingthe spin-orbit coupling term *Nuclear Physics A* Vol. 770: 32 - 39.

[18] Berkdemir, C. & Cheng, Y. F. (2009). On the exact solutions of the Dirac equation with a novel angle-dependent potential, *Physica Scripta* Vol. 79: 035003-1 - 035005-6.

[19] Berkdemir, C. (2009). Erratum to "Pseudospin symmetry in the relativistic Morse potential including the spin-orbit coupling term" [*Nuclear Physics A* Vol. 770: 32 - 39; (2006)], *Nuclear Physics A* Vol. 821: 262 - 263.

[20] Berkdemir, C. & Sever, R. (2009). Pseudospin symmetry solution of the Dirac equation with an angle-dependent potential, *Journal of Physics A: Mathematical & Theoretical* Vol. 41: 0453302-1 - 0453302-11.

[21] de Castro, A. S., Alberto, P., Lisboa, R. & Malheiro, M. (2006). Relating pseudospin and spin symmetries through charge conjugation and chiral transformations: The case of the relativistic harmonic oscillator, *Physical Review C* Vol. 73: 054309-1 - 054309-13.

[22] Dong, S. H. (2011). Relativistic treatment of spinless particles subject to a rotating Deng-Fan oscillator, *Communication in Theoretical Physics* Vol. 55 (No. 6): 969 - 971.

[23] Dong, S. H. (2007): *Factorization Method in Quantum Mechanics (Fundamental Theories of Physics)*, Springer, Netherlands.

[24] Ginocchio, J. N. (1997). Pseudospin as a relativistic symmetry, *Physical Review Letters* Vol. 78: 436 - 439; A relativistic symmetry in nuclei, *Physics Reports* Vol. 315: 231 - 240.

[25] Ginocchio, J. N. & Madland, D. G.(1998). Pseudospin symmetry and relativistic single-nucleon wave functions, *Physical Review C* Vol. 57 (No. 3): 1167 - 1173.

[26] Ginocchio, J. N. (1999). A relativistic symmetry in nuclei, *Physics Reports* Vol. 315: 231 - 240.

[27] Ginocchio, J. N. (2004). Relativistic harmonic oscillator with spin symmetry, *Physical Review C* Vol. 69: 034318-1 - 034318-8.

[28] Ginocchio, J. N. (2005a). Relativistic symmetries in nuclei and hadrons, *Physics Reports* Vol. 414 (No. 4-5): 165 - 261.

[29] Ginocchio, J. N. (2005b). $U(3)$ and Pseudo-$U(3)$ Symmetry of the Relativistic Harmonic Oscillator, *Physical Review Letters* Vol. 95: 252501-1 - 252501-3.

[30] Gradshteyn, I.S. & Ryzhik, I. M. (2007). *Table of Integrals, Series, and Products*, Academic Press, New York.

[31] Greene, R. L. & Aldrich, C. (1976). Variational Wave Functions for a Screened Coulomb Potential, *Physics Review A* Vol. 14: 2363 - 2366.

[32] Greiner, W. (2000). *Relativistic Quantum Mechanics: Wave equations*, Springer, Berlin.

[33] Groschc, C. & Steiner, F. (1995). How to solve path integrals in quantum mechanics , *Journal of Mathematical Physics* Vol. 36 (No. 5): 2354 - 2386; (1998). *Handbook of Feynman path integrals*, Springer-Vcrlag, Berlin.

[34] Guo, J. Y. & Sheng, Z. Q. (2005). Solution of the Dirac equation for the Woods-Saxon potential with spin and pseudospin symmetry, *Physics Letters A'* Vol. 338: 90 - 96

[35] Guo, J. Y., Wang, R. D. & Fang, X. Z. (2005a). Pseudospin symmetry in the resonant states of nuclei, *Physical Review C* Vol. 72: 054319-1 - 054319-8.

[36] Guo, J. Y., Fang, X. Z. & Xu, F. X. (2005b). Pseudospin symmetry in the relativistic harmonic oscillator *Nuclear Physics A* Vol.757: 411 - 421.

[37] Guo, J. Y. and Fang, X. Z. (2006). Isospin dependence of pseudospin symmetry in nuclear resonant states, *Physical Review C* Vol. 74: 024320-1 - 024320-8.

[38] Guo, J. Y., Zhou, F., Guo, L. F. & Zhou, J. H. (2007). Exact solution of the continuous states for generalized asymmetrical Hartmann potentials under the condition of pseudospin symmetry, *International Journal of Modern Physics* Vol. 22 (No. 26): 4825 - 4832.

[39] Hamzavi, M., Rajabi, A. A. & Hassanabadi, H. (2010a). Exact spin and pseudospin symmetry solutions of the Dirac equation for Mie-type potential including a Coulomb-like tensor potential, *Few-Body System* Vol. 48: 171 - 183.

[40] Hamzavi, M., Hassanabadi, H. & Rajabi, A. A. (2010b). Exact solution of Dirac equation for Mie-type potential by using the Nikiforov-Uvarov method under the pseudospin and spin symmetry limit, *Modern Physics Letters A* Vol. 25 (No. 28): 2447 - 2456.

[41] Hamzavi, M., Rajabi, A. A. & Hassanabadi, H. (2010c). Exact pseudospin symmetry solution of the Dirac equation for spatially-dependent mass Coulomb potential including a Coulomb-like tensor interaction via asymptotic iteration method, *Physics Letters A* Vol. 374: 4303 - 4307.

[42] Hecht, K. T. & Adler, A. (1969). Generalized seniority for favored $J \neq 0$ pair in mixed configurations, *Nuclear Physics A* Vol. 137: 129 - 143.

[43] Ibrahim, T. T., Oyewumi, K. J. & Wyngaardt, S. M. (2011). Approximate solution of N-dimensional relativistic wave equations, submitted for publication.

[44] Ikhdair, S. M.(2010). Approximate solutions of the Dirac equation for the Rosen-Morse potential including the spin-orbit centrifugal term, *Journal of Mathematical Physics* Vol. 51: 023525-1 - 023525-16.

[45] Ikhdair, S. M.(2011). An approximate κ state solutions of the Dirac equation for the generalized Morse potential under spin and pseudospin symmetry, *Journal of Mathematical Physics* Vol. 52: 052303-1 - 052303-22.

[46] Ikhdair, S. M., Berkdermir, C. & Sever, R. (2011). Spin and pseudospin symmetry along with orbital dependency of the Dirac-Hulthén problem, *Applied Mathematics and Computation* Vol. 217: 9019 - 9032.

[47] Ikot, A. N. & Akpabio, L. E. (2010). Approximate solution of the Schrödinger equation with Rosen-Morse potential including the centrifugal term, *Applied Physics Research* Vol. 2 (No. 2): 202 - 208.

[48] Jia, C. S., Guo, J. P. & Pend, X. L. (2006). Exact solution of the Dirac-Eckart problem with spin and pseudospin symmetry, *Journal of Physics A: Mathematical & General* Vol. 39: 7737 - 7744.

[49] Jia, C. S., Li, S. C., Li, Y. & Sun, L. T. (2002). Pseudo-Hermitian potential models with PT symmetry, *Physics Letters A* Vol. 300: 115 -121.

[50] Jia, C. S., Chen, T. & Cui, L. G. (2009b). Approximate analytical solutions of the Dirac equation with the generalized pöschl-Teller potential including the pseudo-centrifugal term, *Physics Letters A* Vol. 373: 1621 - 1626.

[51] Jia, C. S., Liu, J. Y. , Wang, P. Q. & Lin, X. (2009b). Approximate analytical solutions of the Dirac equation with the hyperbolic potential in the presence of the spin symmetry and pseudo-spin symmetry, *International Journal of Theoretical Physics* Vol. 48: 2633 - 2643.

[52] Landau, L. D. and Lifshitz, E. M. (1999): *Quantum Mechanics (Non-relativistic Theory): Course of Theoretical Physics*, Butterworth-Heinemann, Oxford.

[53] Lisboa, R., Malheiro, M. & Alberto, P. (2004a). The nuclear pseudospin symmetry along an isotopic chain, *Brazilian Journal of Physics* Vol. 34 (No. 1A): 293 - 296.

[54] Lisboa, R., Malheiro, M., de Castro, A. S., Alberto, P. & Fiolhais, M. (2004b). Pseudospin symmetry and the relativistic harmonic oscillator, *Physical Review C* Vol. 69: 024319-1 - 024319-15.

[55] Lisboa, R., Malheiro, M., de Castro, A. S., Alberto, P. & Fiolhais, M. (2004c).Harmonic oscillator and nuclear pseudospin,*Proceedings of American Institute of Physics*, IX Hadron Physics & VII Relativistic Aspects of Nuclear Physics: A joint meeting on QCD & QGP in the edited by M. E. Bracco, M. Chiapparini, E. Ferreira & T. Kodama, AIP, New York, pp. 569 - 571.

[56] Lu, J. (2005). Analytic quantum mechanics of diatomic molecules with empirical potentials, *Physics Scripta* Vol. 72: 349 - 352.

[57] Merzbacher, E. (1998). *Quantum Mechanics*, John Wiley, New York.

[58] Meng, J. & Ring, P. (1996). Relativistic Hartree-Bogoliubov description of the neutron halo in [11]Li, *Physical Review Letters* Vol. 77: 3963 - 3966.

[59] Meng, J., Sugarwara-Tanabe, K., Yamaji, S. & Arima, A. (1998). Pseudospin symmetry in relativistiv mean field theory, *Physical Review C* Vol. 58: R628 - R631.

[60] Meng, J., Sugarwara-Tanabe, K., Yamaji, S. & Arima, A. (1999). Pseudospin symmetry in Zr and Sn isotopes from the proton drip line to the neutron drip line, *Physical Review C* Vol.59: 154 - 163.

Approximate Solutions of the Dirac Equation for the Rosen-Morse Potential in the Presence of the Spin-Orbit and
Pseudo-Orbit Centrifugal Terms

155

[61] Oyewumi, K. J. & Akoshile, C. A. (2010). Bound-state solutions of the Dirac-Rosen-Morse potential with spin and pseudospin symmetry, *European Physical Journal A* Vol. 45: 311 - 318.

[62] Page, P. R., Goldman, T. & Ginocchio, J. N. (2001). Relativistic symmetry supresses Quark spin-orbit splitting, *Physical Review Letters* Vol. 86: 204 - 207.

[63] Pekeris, C. L. (1934). The rotation-vibration coupling in diatomic molecules, *Physical Review* Vol. 45: 98 - 103.

[64] Rosen, N. & Morse, P. M. (1932). On the Vibrations of Polyatomic Molecules, *Physics Review* Vol. 42 (No. 2): 210 - 217.

[65] Setare, M. R. & Nazari, Z. (2009). Solution of Dirac equations with five-parameter exponent-type potential, *Acta Physica Polonica B* Vol. 40 (No. 10): 2809 - 2824.

[66] Soylu, A, Bayrak, O. and Boztosun, I. (2007). An approximate solution of the Dirac-Hultén problem with pseudospin and spin symmetry for any κ state, *Journal Mathematical Physics* Vol. 48: 082302-1 - 082302-9.

[67] Soylu, Bayrak, O. and Boztosun, I. (2008a). κ state solutions of the Dirac equationfor the Eckart potential with pseudospin and spin symmetry, *Journal of Physic. A: Mathematical & Theoretical* Vol. 41: 065308-1 - 065308-8.

[68] Soylu, Bayrak, O. and Boztosun, I. (2008b). Kappa statesolutionsof Dirac-Hulténand Dirac-Eckart problems with pseudospin and spin symmetry, *Proceedings of American Institute of Physics*, Nuclear Physics and Astrophysics: From Stable beams to Exotic Nuclei edited by I. Boztosun & A. B. Balantekin, AIP, New York, pp. 322 - 325.

[69] Soylu, A., Bayrak, O. & Boztosun, I. (2008c). Exact solutions of Klein-Gordon equationwith scalar and vector Rosen-Morse type potentials, *Chinese Physics Letters* Vol. 25 (No. 8): 2754 - 2757.

[70] Stuchbery, A. E. (1999). Magnetic behaviour in the pseudo-Nilsson model, *Journal of Physics G* Vol. 25: 611 - 616.

[71] Stuchbery, A. E. (2002). Magnetic properties of rotational states in the pseudo-Nilsson model, *Nuclear Physics A* Vol. 700: 83 - 116.

[72] Taşkin, F. (2009). Approximate solutions of the Dirac equation for the Manning-Rosen potential including the spin-orbit coupling term, *International Journal of Theoretical Physics* Vol. 48: 1142 - 1149.

[73] Troltenier, D., Nazarewicz, W., Szymanski, Z., & Draayer, J. P. (1994). On the validity of the pseudo-spin concept for axially symmetric deformed nuclei, *Nuclear Physics A* Vol. 567: 591 - 610

[74] Wei, G. F. & Dong, S. H. (2008). Approximately analytical solutions of the Manning-Rosen potential with the spin-orbit coupling term and spin symmetry, *Physics Letters A* Vol. 373: 49 - 53.

[75] Wei, G. F. & Liu, X. Y. (2008). The relativistic bound states of the hyperbolic potential with the centrifugal term *Physica Scripta*, Vol. 78: 065009-1 - 065009-5

[76] Wei, G. F. & Dong, S. H. (2009). Algebraic approach to pseudospin symmetry for the Dirac equation with scalar and vector modified Pöschl-Teller potentials, *European Physical Letters* Vol. 87: 40004-p1 - 40004-p6.

[77] Wei, G. F. & Dong, S. H. (2010a). Pseudospin symmetry in the relativistic Manning-Rosen Potential to the pseudo-centrifugal term, *Physics Letters B* Vol. 686: 288 - 292.

[78] Wei, G. F. & Dong, S. H. (2010b). Pseudospin symmetry for modified Rosen-Morse potential including a Pekeris-type approximation to the pseudo-centrifugal term, *European Physical Journal A* Vol. 46: 207 - 212.

[79] Wei, G. F. & Dong, S. H. (2010c). A novel algebraic approach to spin symmetry for Dirac equation with scalar and vector second Pöschl-Teller potentials, *European Physical Journal A* Vol. 43: 185 - 190.

[80] Wei, G. F. & Dong, S. H. (2010d). Spin symmetry in the relativistic symmetrical well potential including a proper approximation to the spin-orbit coupling term, *Physica Scripta* Vol. 81: 035009-1 - 035009-5.

[81] Wei, G. F., Duan, X. Y. & Liu, X. Y. (2010). Algebraic approach to spin symmetry for the Dirac equation with the trigonometric Scarf potential, *International Journal of Modern Physics A* Vol. 25(No. 8): 1649 - 1659.

[82] Xu, Q. & Zhu, S. J. (2006). Pseudospin symmetry and spin symmetry in the relativistic Woods-Saxon, *Nuclear Physics A* Vol. 768: 161 -169.

[83] Xu, Y., He, S. & Jia, C. S.(2008). Approximate analytical solutions of the Dirac equationwith the Pöschl-Teller potential including the spin-orbit coupling term, *journal of Physics A: Mathematical & Theoretical* Vol. 41: 0255302-1 - 255302-8.

[84] Xu, Y., He, S. & Jia, C. S. (2010). Approximate analytical solutions of the Klein-Gordon equation with the Pöschl-Teller potential including the centrifugal term' *Physica Scripta*, Vol. 81: 045001-1 - 045001-7

[85] Yi, L. Z., Diao, Y. F., Liu, J.Y. & Jia, C. S. (2004). Bound states of the Klein-Gordon equation with vector and scalar Rosen-Morse potentials, *Physics Letters A* Vol. 333: 212 - 217.

[86] Zhang, M. C. (2009). Pseudospin symmetry and a double ring-shaped spherical harmonic oscillator potential, *Central European Journal of Physics* Vol. 7(No. 4): 768 - 773. Pseudospin symmetry for a ring-shaped non-spherical harmonic oscillator potential, *International Journal of Theoretical Physics*, Vol 48: 2625 - 2632.

[87] Zhang, L. H., Li, X. P. & Jia, C. S. (2008). Analytical approximation to the solution of the Dirac equation with the Eckart potential including the spin-orbit coupling term, *Physics Letters A* Vol. 372: 2201 - 2207.

[88] Zhang, M. C., Huang-Fu, G. Q. & An, B. (2009a). Pseudospin symmetry for a new ring-shaped non-spherical harmonic oscillator potential, *Physica Scripta* Vol. 80: 065018-1 - 065018-5

[89] Zhang, L.H, Li, X. P. & Jia, C. S. (2009b). Approximate analytical solutions of the Dirac equation with the generalized Morse potential model in the presence of the spin symmetry and pseudo-spin symmetry, *Physica Scripta* Vol. 80: 035003-1 - 0035003-6.

[90] Zhao, X. Q., Jia, C. S. & Yang, Q. B. (2005). Bound states of relativistic particles in the generalized symmetrical double-well potential, *Physics Letters A*, Vol. 337: 189 - 196.

6

Quantum Mechanics Entropy and a Quantum Version of the H-Theorem

Paul Bracken
Department of Mathematics,
University of Texas, Edinburg, TX
USA

1. Introduction

Entropy is a fundamental concept which emerged along with other ideas during the development of thermodynamics and statistical mechanics Landau and Lifshitz (1978); Lieb and Yngvason (1999). Entropy has developed foremost out of phenomenological thermodynamical considerations such as the second law of thermodynamics in which it plays a prominent role Wehrl (1978). With the intense interest in the investigation of the physics of matter at the atomic and subatomic quantum levels, it may well be asked whether this concept can emerge out of the study of systems at a more fundamental level. In fact, it may be argued that a correct definition is only possible in the framework of quantum mechanics, whereas in classical mechanics, entropy can only be introduced in a rather limited and artificial way. Entropy relates macroscopic and microscopic aspects of nature, and ultimately determines the behavior of macroscopic systems. It is the intention here to present an introduction to this subject in a readable manner from the quantum point of view. There are many reasons for undertaking this. The intense interest in irreversible thermodynamics Grössing (2008), the statistical mechanics of astrophysical objects Padmanabhan (1990); Pathria (1977), quantum gravity and entropy of black holes Peres & al. (2004), testing quantum mechanics Ballentine (1970) and applications to condensed matter and quantum optics Haroche & al. (2006); Raimond & al. (2001) are just a few areas which are directly or indirectly touched on here.

Let us begin by introducing the concept of entropy from the quantum mechanical perspective, realizing that the purpose is to focus on quantum mechanics in particular. Quantum mechanics makes a clear distinction between observables and states. Observables such as position and momentum are mathematically described by self-adjoint operators in a Hilbert space. States, which are generally mixed, can be described by a density matrix, which is designated by ρ throughout. This operator ρ is Hermitean, has trace one and yields the expectation value of an observable A in the state ρ through the definition

$$\langle A \rangle = Tr(\rho A). \tag{1.1}$$

Entropy is not an observable, so there does not exist an operator with the property that its expectation value in some state would be the entropy. In fact, entropy is a function of state. If the given state is described by the density matrix ρ, its entropy is defined to be

$$S(\rho) = -k_B \, Tr(\rho \log(\rho)).$$ (1.2)

This formula is due to von Neumann von Neumann (1955), and generalizes the classical expression of Boltzmann and Gibbs to the quantum regime. Of course, k_B is Boltzmann's constant, and the natural logarithm is used throughout. If k_B is put equal to one, the entropy becomes dimensionless. Thus, entropy is a well-defined quantity, no matter what size or type of system is considered. It is always greater than or equal to zero, and equal to zero exactly for pure states.

It will be useful to give some interpretation of von Neumann's formula. The discovery for which Boltzmann is remembered is his formula for entropy which appeared in 1877, namely,

$$S = k_B \log(W),$$ (1.3)

This form for S was established by making a connection between a generalization S of thermostatic entropy and the classical H-function. The identification of the constant on the right of (1.3) as Boltzmann's was proposed by Planck. Equation (1.3) taken in conjunction with the H-theorem, interprets the second law, $\Delta S \geq 0$, simply as the tendency of an isolated system to develop from less probable states to more probable states; that is, from small W to large W. Thermostatic equilibrium corresponds to the state in which W attains its maximum value. In fact, equation (1.3) has had far reaching consequences. It led Planck, for example, to his quantum hypothesis, which is that the energy of radiation is quantized, and then from there to the third law of thermodynamics. The H-theorem provided an explanation in mechanical terms of the irreversible approach of macroscopic systems towards equilibrium. By correlating entropy with the H-function and thermodynamic probability, Boltzmann revealed the statistical character of the second law. Of course, Boltzmann was restricted to a classical perspective. The question as to whether the number of microstates makes literal sense classically has been discussed as an objection to his approach. As stated by Pauli Pauli (2000), a microstate of a gas for example is defined as a set of numbers which specify in which cell each atom is located, that is, a number labeling the atom, an index for the cell in which atom s is located and a label for the microstate. The macrostate is uniquely determined by the microstate, however the converse does not hold. For every macrostate .there are very many microstates, as will be discussed. Boltzmann's fundamental hypothesis is then: All microstates are equally probable.

However, as Planck anticipated, in quantum mechanics such a definition immediately makes sense. There is no ambiguity at all, as there is a natural idea of microstate. The number of microstates may be interpreted as the number of pure states with some prescribed expectation values. Suppose there are W different pure states in a system, each occuring with the same probability. Then the entropy is simply $S = \log(W)$. However, the density matrix of the system is given by $\rho = (1/W)\mathcal{P}$, where $\mathcal{P}$ is a W-dimensional projection operator. Thus, the correspondence follows immediately, that is, $\log W = -Tr\,[\rho \log \rho]$.

Each density matrix can be diagonalized Wehrl (1978),

$$\rho = \sum_k p_k \, |k\rangle \langle k|,$$ (1.4)

where $|k\rangle$ is a normalized eigenvector corresponding to the eigenvalue p_k and $|k\rangle\langle k|$ is a projection operator onto $|k\rangle$ with $p_k \geq 0$ and $\sum_k p_k = 1$. Here the coefficients are positive probabilities and not complex amplitudes as in a quantum mechanical superposition. Substituting (1.4) into (1.2) finally yields,

$$S(\rho) = -\sum_k p_k \log(p_k). \tag{1.5}$$

There is a more combinatorial approach Wehrl (1978). This will come up again subsequently when ensembles take the place of a density operator. If N measurements are performed, one will obtain as a result that for large N, the system is found $p_1 N$ times in $|1\rangle$, $p_2 N$ times in state $|2\rangle$ and so on, all having the same weight. By straightforward counting, there results

$$W_N = \frac{N!}{(p_1 N)!(p_2 N)! \cdots}. \tag{1.6}$$

When $N \to \infty$, Stirling's formula can be applied to the logarithm of (1.6) so the entropy is

$$\log \frac{N!}{n_1! n_2! \cdots} = N \log(N) - N - \sum_j (n_j \log n_j - n_j) = -N \sum_j p_j \log(p_j). \tag{1.7}$$

Dividing both sides of (1.7) by N, then as $N \to \infty$ (1.5) is recovered. It should also be noted that (1.5) is of exactly the same form as Shannon entropy, which can be thought of as a measure of unavailable information.

Of course, another way to look at this is to consider N copies of the same Hilbert space, or system, in which there are microstates $|1\rangle \otimes |2\rangle \cdots$ such that $|1\rangle$ occurs $p_1 N$ times, $|2\rangle$ occurs $p_2 N$ times, and so forth. Again (1.6) is the result, and according to Boltzmann's equation, one obtains $\log(W_N)$ for the entropy as in (1.5). In (1.5), S is maximum when all the p_j are equal to $1/N$.

By invoking the constraint $\sum_k p_k = 1$, (1.5) takes the form

$$S = -\sum_{k=1}^{N-1} p_k \log(p_k) - p_N \log(p_N), \tag{1.8}$$

where $p_N = 1 - \sum_{k=1}^{N-1} p_k$, and all other p_k are considered to be independent variables. Differentiating S in (1.8), it is found that

$$\frac{\partial S}{\partial p_k} = -\log(p_k) + \log(p_N).$$

This vanishes of course when $p_k = p_N = N^{-1}$ and this solution is the only extremum of S. To summarize, entropy is a measure of the amount of chaos or lack of information about a system. When one has complete information, that is, a pure state, the entropy is zero. Otherwise, it is greater than zero, and it is bigger the more microstates exist and the smaller their statistical weight.

2. Basic properties of entropy

There are several very important properties of entropy function (1.5) which follow from simple mathematical considerations and are worth introducing at this point Peres (1995).

The first point to make is that the function $S(\mathbf{p})$ is a concave function of its arguments $\mathbf{p} = (p_1, \cdots, p_N)$. For any two probability distributions $\{p_j\}$ and $\{q_j\}$, and any $\lambda \in [0, 1]$, S defined in (1.5) satisfies the following inequality

$$S(\lambda \mathbf{p} + (1 - \lambda \mathbf{q})) \geq \lambda S(\mathbf{p}) + (1 - \lambda)S(\mathbf{q}), \qquad \lambda \in [0, 1]. \tag{2.1}$$

This can be proved by differentiating S twice with respect to λ to obtain,

$$\frac{d^2 S(\lambda \mathbf{p} + (1 - \lambda)\mathbf{q})}{d\lambda^2} = -\sum_j \frac{(p_j - q_j)^2}{\lambda p_j + (1 - \lambda)q_j} \leq 0. \tag{2.2}$$

This is a sufficient condition for a function to be concave. Equality holds only when $p_j = q_j$, for all j. The physical meaning of inequality (2.1) is that mixing different probability distributions can only increase uniformity.

If N is the maximum number of different outcomes obtainable in a test of a given quantum system, then any test that has exactly N different outcomes is called a maximal test, called T here. Suppose the probabilities p_m for the outcomes of a maximal test T which can be performed on that system are given. It can be shown that this entropy never decreases if it is elected to perform a different maximal test. The other test may be performed either instead of T, or after it, if test T is repeatable.

To prove this statement, suppose the probabilities for test T are $\{p_m\}$ and those for a subsequent test are related to the $\{p_m\}$ by means of a doubly stochastic matrix $P_{\mu m}$. This is a matrix which satisfies $\sum_\mu P_{\mu m} = 1$ and $\sum_m P_{\mu m} = 1$. In this event,

$$q_\mu = \sum_m P_{\mu m} p_m$$

are the probabilities for the subsequent test. The new entropy is shown to satisfy the inequality $S(\mathbf{q}) \geq S(\mathbf{p})$. To prove this statement, form the difference of these entropies based on (1.5),

$$\sum_m p_m \log(p_m) - \sum_\mu q_\mu \log(q_\mu) = \sum_m p_m (\log(p_m) - \sum_\mu P_{\mu m} \log(q_\mu))$$

$$= \sum_{m\mu} p_m (P_{\mu m} \log(p_m) - P_{\mu m} \log(q_\mu))$$

$$= \sum_{m\mu} p_m P_{\mu m} \log(\frac{p_m}{q_\mu}).$$

In the second line, $\sum_\mu P_{\mu m} = 1$ has been substituted to get this result. Using the inequality $\log x \geq 1 - x^{-1}$, where equality holds when $x = 1$, and the fact that S has a negative sign, it follows that

$$S(\mathbf{q}) - S(\mathbf{p}) \geq \sum_{m\mu} p_m P_{\mu m}(1 - \frac{q_\mu}{p_m}) = \sum_{m\mu}(p_m P_{\mu m} - q_\mu P_{\mu m}) = \sum_\mu (q_\mu - q_\mu \sum_m P_{\mu m}) = 0. \tag{2.3}$$

The equality sign holds if and only if $P_{\mu m}$ is a permutation matrix, so the sets are identical. After a given preparation whose result is represented by a density matrix ρ, different tests correspond to different sets of probabilities, and therefore to different entropies. The entropy of a preparation can be defined as the lowest value attained by (1.5) for any complete test performed after that preparation. The optimal test which minimizes S is shown to be the one that corresponds to the orthonormal basis v_μ given by the eigenvectors of the density matrix ρ

$$\rho\, v_\mu = w_\mu v_\mu.$$

In this basis, ρ is diagonal and the eigenvalues w_μ satisfy $0 \le w_\mu \le 1$ and $\sum_\mu w_\mu = 1$. A basic postulate of quantum mechanics asserts that the density matrix ρ completely specifies the statistical properties of physical systems that were subjected to a given preparation. All the statistical predictions that can be obtained from (1.1) for an operator are the same as if we had an ordinary classical mixture, with a fraction w_μ of the systems with certainty in the state v_μ. Therefore, if the maximal test corresponding to the basis v_μ is designed to be repeatable, the probabilities w_μ remain unchanged and entropy S remains constant. The choice of any other test can only increase the entropy, as in the preceding result. This proves that the optimal test, which minimizes the entropy, is the one corresponding to the basis that diagonalizes the density matrix.

The entropic properties of composite systems obey numerous inequalities as well. Let $\{v_m\}$ and $\{e_\mu\}$ be two orthonormal basis sets for the same physical system. Let $\rho = \sum w_m |v_m\rangle \langle v_m|$ and $\sigma = \sum \omega_\mu |e_\mu\rangle \langle e_\mu|$ be two different density matrices. Their relative entropy $S(\sigma|\rho)$ is defined to be

$$S(\sigma|\rho) = Tr[\rho(\log \rho - \log \sigma)]. \tag{2.4}$$

Let us evaluate $S(\sigma|\rho)$ in (2.3) in the $|v_\mu\rangle$ basis where ρ is diagonal. The diagonal elements of $\log \sigma$ are

$$(\log \sigma)_{mm} = \langle v_m, \sum_\mu \log \omega_\mu |e_\mu\rangle \langle e_\mu|v_m\rangle = \sum_\mu \log \omega_\mu |\langle e_\mu, v_m\rangle|^2 = \sum_\mu \log \omega_\mu \, P_{\mu m}. \tag{2.5}$$

The matrix $P_{\mu m}$ is doubly stochastic, so as in (2.3) we have

$$S(\sigma|\rho) = \sum_m w_m (\log w_m - \sum_\mu P_{\mu m} \log \omega_\mu) = \sum_{\mu m} w_m P_{\mu m} \log\left(\frac{w_m}{\omega_\mu}\right) \ge 0. \tag{2.6}$$

Equality holds in (2.6) if and only if $\sigma = \rho$.

Inequality (2.6) can be used to prove a subadditivity inequality. Consider a composite system, with density matrix ρ, then the reduced density matrices of the subsystems are called ρ_1 and ρ_2. Then matrices ρ, ρ_1 and ρ_2 satisfy,

$$S(\rho) \le S(\rho_1) + S(\rho_2). \tag{2.7}$$

This inequality implies that a pair of correlated systems involves more information than the two systems separately.

To prove this, suppose that w_m, ω_μ and $W_{m\mu} = w_m \omega_\mu$ are the eigenvalues of ρ_1, ρ_2 and $\rho_1 \otimes \rho_2$, respectively, then

$$\sum_m w_m \log w_m + \sum_\mu \omega_\mu \log \omega_\mu = \sum_{m\mu} W_{m\mu} \log W_{m\mu}.$$

This has the equivalent form,

$$S(\rho_1) + S(\rho_2) = S(\rho_1 \otimes \rho_2).$$

Consider now the relative entropy

$$S(\rho_1 \otimes \rho_2 | \rho) = Tr[\rho(\log \rho - \log \rho_1 \otimes \rho_2)] = Tr[\rho(\log \rho - \log \rho_1 - \log \rho_2)].$$

It has just been shown that relative entropy is nonnegative, so it follows from this that

$$Tr(\rho \log \rho) \geq Tr(\rho \log \rho_1) + Tr(\rho \log \rho_2).$$

Since $Tr(\rho \log \rho_1) = \sum_{m\mu n\nu} \rho_{m\mu,n\nu} (\log \rho_1)_{nm} \delta_{\nu\mu} = Tr(\rho_1 \log \rho_1)$, and similarly for $Tr(\rho \log \rho_2)$, it follows that

$$Tr(\rho \log \rho) \geq Tr(\rho_1 \log \rho_1) + Tr(\rho_2 \log \rho_2).$$

Now using (1.2), it follows that

$$-S(\rho) \geq -S(\rho_1) - S(\rho_2).$$

Multiplying both sides by minus one, (2.7) follows.

3. Entanglement and entropy

The superposition principle applied to composite systems leads to the introduction of the concept of entanglement Mintet & al. (2005); Raimond & al. (2001), and provides an important application for the density matrix. A very simple composite object is a bipartite quantum system S which is composed of two parts A and B. The states of A and B belong to two separate Hilbert spaces called $\mathcal{H}_A$ and $\mathcal{H}_B$ which are spanned by the bases $|i_A\rangle$ and $|i_B\rangle$, and may be discrete or continuous. If A and B are prepared independently of each other and are not coupled together at some point, S is described by the tensor product $|\psi_S\rangle = |\psi_A\rangle \otimes |\psi_B\rangle$. Each subsystem is described by a well-defined wave function. Any manipulation of one part leaves the measurement prediction for the other part unchanged. System S can also be prepared by measuring joint observables, which act simultaneously on A and B. Even if S has been prepared by measuring separate observables, A and B can become coupled by means of an interaction Hamiltonian. In this instance, it is generally impossible to write the global state $|\psi_S\rangle$ as a product of partial states associated to each component of S.

This is what the expression *quantum entanglement* means. The superposition principle is at the heart of the most intriguing features of the microscopic world. A quantum system may exist in a linear superposition of different eigenstates of an observable, suspended between different classical realities, as when one says a particle can be at two positions at the same time. It seems to be impossible to get a classical intuitive representation of superpositions. When the superposition principle is applied to composite systems, it leads to the concept of entanglement. Moreover, as Bell has shown, entanglement cannot be consistent with any local theory containing hidden variables.

Even if the state S cannot be factorized according to the superposition principle, it can be expressed as a sum of product states $|i_A\rangle \otimes |\mu_B\rangle$, which make up a basis of the global Hilbert space, $\mathcal{H}_S$. Consequently, an entangled state can be expressed as

$$|\psi_S\rangle = \sum_{i,\mu} \alpha_{i\mu}|i_A\rangle \otimes |\mu_B\rangle \neq |\psi_A\rangle \otimes |\psi_B\rangle, \tag{3.1}$$

where the $\alpha_{i\mu}$ are complex amplitudes. The states $|\psi_S\rangle$ contain information not only about the results of measurements on A and B separately, but also on correlations between these measurements. In an entangled state, each part loses its quantum identity. The quantum content of the global state is intricately interwoven between the parts. Often it is the case that there is interest in carrying out measurements on one part without looking at another part. For example, what is the probability of finding a result when measuring observable O_A attached to subsystem A, without worrying about B. The complete wave function $|\psi_S\rangle$ can be used to predict the experimental outcomes of the measurement of $O_A \otimes 1_B$. This can also be done by introducing the density operator ρ_S of a system described by the quantum state $|\psi_S\rangle$, which is just the projector

$$\rho_S = |\psi_S\rangle\langle\psi_S|. \tag{3.2}$$

It has the same information content as $|\psi_S\rangle$, and for all predictions on S, all quantum rules can be expressed in such a fashion; for example, the expectation values of an observable O_S of S is found by (1.1). The probability of finding the system in $|i\rangle$ after a measurement corresponding to the operator $\rho_i = |i\rangle\langle i|$ is given by $|\langle i|\psi_S\rangle|^2$ in the quantum description and $Tr(\rho_i\rho_S)$ in terms of the density matrix.

The density operator approach is very advantageous for describing one subsystem, A, without looking at B. A partial density operator ρ_A can be determined which has all the predictive information about A alone, by tracing ρ_S over the subspace of B

$$\rho_A = Tr_B(\rho_S) = \sum_{i,i',\mu} \alpha_{i\mu}\alpha^*_{i'\mu}|i_A\rangle\langle i_A|. \tag{3.3}$$

Thus, the probability of finding A in state $|j_A\rangle$ is found by computing the expectation value of the projector $\rho_j = |j_A\rangle\langle j_A|$, which is $\pi_j = Tr\,(\rho_A\rho_j)$. Predictions on A can be done without considering B. The information content of ρ_A is smaller than in ρ_S, since correlations between A and B are omitted. To say that A and B are entangled is equivalent to saying that ρ_A and ρ_B are not projectors on a quantum state. There is however a basis in $\mathcal{H}_A$ in which ρ_A is diagonal. Let us call it $|j_A\rangle$, so that ρ_A is given by

$$\rho_A = \sum_j \lambda_j|j_A\rangle\langle j_A|. \tag{3.4}$$

In (3.4), λ_j are positive or zero eigenvalues which sum to one. By neglecting B, there is acquired only a statistical knowledge of state A, with a probability λ_j of finding it in $|j_A\rangle$.

It is possible to express the state for S in a representation which displays the entanglement. The superposition (3.1) claims nothing as to whether the state can be factored. To put this property in evidence, choose a basis in $\mathcal{H}_A$, called $|j_A\rangle$ in which ρ_A is diagonal. Then (3.1) is written

$$|\psi_S\rangle = \sum_j |j_A\rangle|\tilde{j}_B\rangle, \tag{3.5}$$

where state $|\tilde{j}_B\rangle$ is given by

$$|\tilde{j}_B\rangle = \sum_\mu \alpha_{j\mu}|\mu_B\rangle. \tag{3.6}$$

The $|\tilde{j}_B\rangle$ are mirroring in $\mathcal{H}_B$ the basis of orthonormal states in $\mathcal{H}_A$ in which ρ_A is diagonal. These mirror states are also orthogonal to each other as can be seen by expressing the fact that ρ is diagonal

$$\langle j_A|\rho_A|j'_A\rangle = \lambda_j\delta_{jj'} = \langle \tilde{j}_B|\tilde{j}_B\rangle.$$

At this point, the mirror state can be normalized by means of the transformation $|\hat{j}_B\rangle = |j_B\rangle/\sqrt{\lambda_j}$ giving rise to the Schmidt expansion,

$$|\psi_S\rangle = \sum_j \sqrt{\lambda_j}|j_A\rangle|\hat{j}_B\rangle. \tag{3.7}$$

The sum over a basis of product mirror states exhibits clearly the entanglement between A and B. The symmetry of this expression shows that ρ_A and ρ_B have the same eigenvalues. Any pure entangled state of a bipartite system can be expressed in this way.

Now a measure of the degree of entanglement can be defined using the density matrix. As the λ_j become more spread out over many non-zero values, more information is lost by concentrating on one system and disregarding correlations between A and B. This loss of mutual information can be linked to the degree of entanglement. This information loss could be measured by calculating the von Neumann entropy of A or B from (1.5)

$$S_A = S_B = -\sum_j \lambda_j\log(\lambda_j) = -Tr(\rho_A\log\rho_A) = -Tr(\rho_B\log\rho_B). \tag{3.8}$$

This is the entropy of entanglement $S_e = S_A = S_B$, and it expresses quantitatively the degree of disorder in our knowledge of the partial density matrices of the two parts of the entangled system S.

If the system is separable, then one λ_j is non-zero and $S_e = 0$, so maximum information on the states of both parts obtains. As soon as two λ_j are non-zero, S_e becomes strictly positive and A and B are entangled. The maximum entropy, hence maximum entanglement obtains when the λ_j are equally distributed among the A and B subspaces. It is maximal and equal to $\log N_A$, when ρ_A is proportional to $\mathbf{1}_A$, that is $\rho_A = \mathbf{1}_A/N_A$. In a maximally entangled state, local measurements performed on one part of the system are not predictable at all. What can be predicted are the correlations between the measurements performed on both parts. For example, consider a bipartite system in which one part has dimension two. There are only two λ-values in the Schmidt expansion, and satisfy $\lambda_1 + \lambda_2 = 1$. Then from (1.5), the entropy when $\lambda_1 \in (0,1)$ is,

$$S_e = -\lambda_1\log(\lambda_1) - (1-\lambda_1)\log(1-\lambda_1). \tag{3.9}$$

The degree of entanglement is equal to zero when $\lambda_1 = 0$ or 1 and passes through a maximum at $\lambda_1 = 1/2$ at which $S_e = 1$. The degree of entanglement measured by the von Neumann entropy is invariant under local unitary transformations acting on A or B separately, a direct consequence of the invariance of the spectrum of the partial density operators.

Consider the case of a two-level system with states $|0\rangle$ and $|1\rangle$, where the density matrix is a two-by-two hermitean matrix given by

$$\rho_A = \begin{pmatrix} \rho_{00} & \rho_{01} \\ \rho_{10} & \rho_{11} \end{pmatrix}. \tag{3.10}$$

The entropy can be calculated for this system. Its positive diagonal terms are the probabilities of finding the system in $|0\rangle$ or $|1\rangle$ and they sum to one. The nondiagonal terms satisfy $\rho_{01} = \rho_{10}^*$ and are zero for a statistical mixture of $|0\rangle$ and $|1\rangle$. Since ρ_A is a positive operator

$$|\rho_{10}| = |\rho_{01}| \leq \sqrt{\rho_{00}\rho_{11}}.$$

is satisfied, and the upper bound is reached for pure states.
The density matrix ρ_A can be expanded with real coefficients onto the operator basis made up of the identity matrix $\mathbf{I}$ and the Pauli matrices σ_i

$$\rho_A = \frac{1}{2}(\mathbf{I} + \mathbf{R} \cdot \boldsymbol{\sigma}), \tag{3.11}$$

where $\mathbf{R} = (u, v, w)$ is three-dimensional and $\boldsymbol{\sigma} = (\sigma_x, \sigma_y, \sigma_z)$. The components of $\mathbf{R}$ are linked to the elements of the density matrix as follows

$$u = \rho_{10} + \rho_{01}, \qquad v = i(\rho_{01} - \rho_{10}), \qquad w = \rho_{00} - \rho_{11}.$$

The modulus R of $\mathbf{R}$ satisfies $R \leq 1$, equality holding only for pure states. This follows from $Tr(\rho_A^2) \leq 1$. If nonlinear functions of an observable A are defined as $f(A) = \sum f(a_k)|e_k\rangle\langle e_k|$, the von Neumann entropy of ρ is

$$S = -\frac{1+R}{2}\log(\frac{1+R}{2}) - \frac{1-R}{2}\log(\frac{1-R}{2}). \tag{3.12}$$

To each density matrix ρ_A, the end of the vector $\mathbf{R}$ can be located on the surface of a sphere. The surface of the sphere $R = 1$ is the set of pure states with $S = 0$. The statistical mixtures correspond to inside the sphere $R < 1$. The closer the point to the center, the larger the von Neumann entropy. The center of the sphere corresponds to the totally unpolarized maximum entropy state.
Any mixed state can be represented in an infinite number of ways as a statistical mixture of two pure states, since any $\mathbf{P}$ with its end inside the sphere can be expressed as a vector sum of a $\mathbf{P}_1$ and $\mathbf{P}_2$ whose ends are at the intersection of the sphere with an arbitrary line passing by the extreme end of $\mathbf{P}$, so one can write $\mathbf{P} = \lambda \mathbf{P}_1 + (1 - \lambda)\mathbf{P}_2$ for $0 < \lambda < 1$. The density matrix which is a linear function of $\mathbf{P}$ is then a weighted sum of the projectors on the pure states $|u_1\rangle$ and $|u_2\rangle$ corresponding to $\mathbf{P}_1$ and $\mathbf{P}_2$,

$$\rho_A = \frac{1}{2}[\mathbf{I} + \lambda \mathbf{P}_1 \cdot \boldsymbol{\sigma} + (1 - \lambda)\mathbf{P}_2 \cdot \boldsymbol{\sigma}] = \lambda|u_1\rangle\langle u_1| + (1 - \lambda)|u_2\rangle\langle u_2|. \tag{3.13}$$

Thus, there exists an ambiguity of representation of the density operator which, if $P \neq 0$, can be lifted by including the condition that $|u_1\rangle$ and $|u_2\rangle$ be orthogonal.
Before finishing, it is worth discussing the following application, which seems to have very important ramifications. A violation of the second law arises if nonlinear modifications are

introduced into Schrödinger's equation Weinberg (1989). A nonlinear Schrödinger equation does not violate the superposition principle in the following sense. The principle asserts that the pure states of a physical system can be represented by rays in a complex linear space, but does not demand that the time evolution obeys a linear equation. Nonlinear variants of Schrödinger's equation can be created with the property that if $u(0)$ evolves to $u(t)$ and $v(0)$ to $v(t)$, the pure state represented by $u(0) + v(0)$ does not evolve into $u(t) + v(t)$, but into some other pure state.

The idea here is to show that such a nonlinear evolution violates the second law of thermodynamics. This is provided the other postulates of quantum mechanics remain as they are, and that the equivalence of the von Neumann entropy to ordinary entropy is maintained. Consider a mixture of quantum systems which are represented by a density matrix

$$\rho = \lambda \Pi_u + (1 - \lambda)\Pi_v, \tag{3.14}$$

where $0 < \lambda < 1$ and Π_u, Π_v are projection operators on the pure states u and v. In matrix form the density matrix is represented as

$$\rho = \begin{pmatrix} \lambda & \lambda\langle v|u\rangle \\ (1-\lambda)\langle u|v\rangle & 1-\lambda \end{pmatrix}.$$

The eigenvalues are found by solving the polynomial $\det(\rho - w\mathbf{1}) = 0$ for the eigenvalues w. Setting $x = |\langle u, v\rangle|^2$, they are given by

$$w_j = \frac{1}{2} \pm [\frac{1}{4} - \lambda(\lambda - 1)(1 - x)]^{1/2}, \qquad j = 1, 2. \tag{3.15}$$

The entropy of this mixture is found by putting w_j into (1.5)

$$S = -w_1 \log(w_1) - w_2 \log(w_2). \tag{3.16}$$

The polynomial $p(\lambda) = 4\lambda(1 - \lambda)$ has range $(0,1)$ when $\lambda \in (0,1)$, so it follows that $s = 4\lambda(1 - \lambda)(1 - x) \in (0,1)$ as well. Setting $f = 1 - s$, then when $s \in (0,1)$, the derivative of (3.16) is given by

$$\frac{\partial S}{\partial x} = -\frac{\lambda(1 - \lambda)}{\sqrt{1 - 4\lambda(1 - \lambda)(1 - x)}} \log\left(\frac{(1 + \sqrt{f})^2}{4\lambda(1 - \lambda)(1 - x)}\right)$$

$$= -\frac{\lambda(1 - \lambda)}{\sqrt{1 - s}} \log\left(\frac{(1 + \sqrt{1 - s})^2}{s}\right) < 0.$$

Consequently, if pure quantum states evolve as $u(0) \to u(t)$ and $v(0) \to v(t)$, the entropy of the mixture ρ shall not decrease provided that $x(t) \leq x(0)$, or in terms of the definition of x, $|\langle u(t), v(t)\rangle|^2 \leq |\langle u(0), v(0)\rangle|^2$. If say $\langle u(0), v(0)\rangle = 0$, then also $\langle u(t), v(t)\rangle = 0$, so orthogonal states remain orthogonal. Consider now a complete orthogonal set u_k. For every v,

$$\sum_k |\langle u_k, v\rangle|^2 = 1.$$

If there exists m such that $|\langle u_m(t), v(t)\rangle|^2 < |\langle u_m(0), v(0)\rangle|^2$, there must also exist some n for which the reverse holds, $|\langle u_n(t), v(t)\rangle|^2 > |\langle u_n(0), v(0)\rangle|^2$. In this event, the entropy of a mixture of u_n and v will spontaneously decrease in a closed system, which is in violation of the second law of thermodynamics. To retain the law, $|\langle u(t), v(t) \rangle|^2 = |\langle u(0), v(0) \rangle|^2$ must hold for every u and v. From Wigner's theorem, the mapping $v(0) \rightarrow v(t)$ is unitary, so Schrödinger's equation must be linear if the other postulates of quantum mechanics remain fixed.

4. Ensemble methods in quantum mechanics

In classical mechanics, one relinquishes the idea of a description of the microscopic mechanical states of trillions of microscopic interacting particles by instead computing averages over a virtual ensemble of systems which replicate the real system. Quantum theory is faced with a similar problem, and the remedy takes the form of the Gibbs ensemble. This last section will take a slightly different track and discusses ensemble theory in quantum mechanics. Two of the main results will be to produce a quantum version of the H-Theorem, and to show how the quantum mechanical canonical ensemble can be formulated.

An astronomical number of states, or of microstates, is usually compatible with a given set of macroscopic parameters defining a macrostate of a thermophysical system. Consequently, a virtual quantum mechanical ensemble of systems is invoked, which is representative of the real physical system. The logical connection between a physical system and ensemble is made by requiring the time average of a mechanical property G of a system in thermodynamic equilibrium equal its ensemble average calculated with respect to an ensemble made up of $N^* \rightarrow \infty$ systems representing the actual system

$$\bar{G} = \langle G \rangle. \tag{4.1}$$

The ensemble average $\langle G \rangle$ is the ordinary mean of G over all the systems of the ensemble. If N_r^* systems are in a state with eigenvalue G_r corresponding to G,

$$N^* \langle G \rangle = \sum_r N_r^* G_r, \tag{4.2}$$

where the sum is over all allowed states.

Adopt as a basic set the states $\psi_{jrm\cdots}$ uniquely identifiable by the quantum numbers $j, r, m, \cdots$ referring to a set of compatible properties. A particular system of the ensemble will not permanently be in one of these states $\psi_{jrm\cdots}$, as there exists only a probability to find a system in any one. Let us compress the basic states to read ψ_{jr} if we let r stand for the entire collection of quantum numbers $r, m, \cdots$. These cannot strictly be eigenstates of the total energy, since a system occupying a particular eigenstate of its total Hamiltonian H at any one moment will remain in this state forever. The state of the real system, which the ensemble is to represent, is a superposition of eigenstates belonging to the same or different values of the energy. To obtain an ensemble where the individual members are to change, we suppose the basic set ψ_{jr} is made up of eigenstates of the unperturbed Hamiltonian H^0. Assume it is possible to write

$$H = H^0 + H^1, \tag{4.3}$$

such that H^1 is a small perturbation added to the unperturbed Hamiltonian H^0, and vary with the physical system considered.

Suppose E_j^0 are the eigenvalues of the unperturbed H^0 and ψ_{jr}^0 the eigenstates corresponding to them, where r again denotes a set of compatible quantum numbers. Introducing H^1 now changes the energy eigenvalues and energy eigenfunctions by an amount E_{jr}^1 and ψ_{jr}^1, which should be very small compared with the unperturbed values. It is precisely the eigenstates ψ_{jr}^0 of H^0 rather than H that are used as basic states for the construction of the ensemble. Since these for the most part will appear in what follows, we continue to omit the superscript for both the eigenfunctions ψ_{jr} and eigenvalues E_{jr} whenever the situation indicates that unperturbed quantities are intended. A perturbed system finding itself initially in any one of the unperturbed states ψ_{jr} does not remain indefinitely in this state, but will continually undergo transitions to other unperturbed states ψ_{ks} due to the action of the perturbation H^1. In analogy with a classical system, a quantum ensemble is described by the number of systems N_{jr}^* in each state ψ_{jr}. The probability P_{jr} of finding a system, selected at random from the ensemble, in the state ψ_{jr} is clearly

$$P_{jr} = \frac{N_{jr}^*}{N^*}. \tag{4.4}$$

The quantities N_{jr}^* must sum up to N^*,

$$\sum_{jr} N_{jr}^* = N^*, \qquad \sum_{jr} P_{jr} = 1. \tag{4.5}$$

An ensemble can be representative of a physical system in thermodynamic equilibrium only in this context if the occupation numbers N_{jr}^* are constants. A more general picture could consider the occupation numbers as functions of time $N_{jr}^* = N_{jr}^*(t)$. The ensemble corresponds to a system removed from equilibrium. Let us ask then how do the N_{jr}^* vary with time.

Quantum mechanics claims the existence of $A_{ks}^{jr}(t)$ which determine the probability of a system in state ψ_{jr} at time zero to be in ψ_{ks} at time t. The final state could correspond to the initial state. Since $N_{jr}^*(0)$ systems are in a state specified by quantum numbers jr at $t = 0$, $A_{ks}^{jr}(t)N_{jr}^*(0)$ systems will make the transition from jr to ks during $(0, t)$ The number of systems in ks at time t will be

$$N_{ks}^*(t) = \sum_{j} \sum_{r} A_{ks}^{jr}(t)\, N_{jr}^*(0). \tag{4.6}$$

The $A_{ks}^{jr}(t)$ must satisfy the condition $\sum_j \sum_r A_{ks}^{jr}(t) = 1$. Multiplying this by $N_{ks}^*(0)$ and subtracting from (4.6) gives

$$N_{ks}^*(t) - N_{ks}^*(0) = \sum_{j} \sum_{r} A_{ks}^{jr}(t)[N_{jr}^*(0) - N_{ks}^*(0)]. \tag{4.7}$$

This is the change in occupation number over $(0, t)$. Dividing (4.7) by N^* and using (4.4) gives

$$P_{ks}(t) - P_{ks}(0) = \sum_{j} \sum_{r} A_{ks}^{jr}[P_{jr}(0) - P_{ks}(0)]. \tag{4.8}$$

A stationary ensemble or one in statistical equilibrium defined as $N_{ks}^*(t) = N_{ks}^*(0)$ for all ks holds when $N_{jr}^*(0) = N_{ks}^*(0)$, at least when $A_{ks}^{jr}(t) \neq 0$. The contribution to the right side of (4.8) comes from an extremely narrow interval $\Delta E = 2\hbar/t$ centered at $E_j = E_k$, as indicated by perturbation theory. In this interval, it can be assumed $P_{jr}(0)$ depends on the j-index weakly enough that we can use $P_{ks}(0)$ in their place, so the term in brackets in (4.8) does not depend on j. The energy spectrum is very nearly continuous for a thermophysical system, so the sum over j can be approximated by an integral over E. This implies an approximation of the form

$$\sum_j A_{ks}^{jr}(t) = tW_{sr}^{(k)}. \tag{4.9}$$

The quantities $W_{sr}^{(k)}$ are time independent provided H^1 is time independent. Consequently, they are nonnegative and depend only on the displayed indices. Substituting (4.9) and $P_{jr}(0) = P_{kr}(0)$ into (4.8) gives

$$\frac{1}{t}[P_{ks}(t) - P_{ks}(0)] = \sum_r W_{sr}^{(k)}[P_{kr}(0) - P_{ks}(0)]. \tag{4.10}$$

In the limit when t becomes arbitrarily small, (4.10) can be approximated by expanding about $t = 0$ on the left to give the final result for the time rate of change of the probability P_{ks},

$$\dot{P}_{ks} = \sum_r W_{sr}^{(k)}[P_{kr}(0) - P_{ks}(0)]. \tag{4.11}$$

This equation was first derived by W. *Pauli*, and will lead to a quantum version of the H-Theorem next. It signifies that of the $N^* P_{kr}(0)$ systems occupying state kr at $t = 0$, $N^* P_{kr}(0)W_{sr}^{(k)}$ will, per unit time, go over to ks. Thus, the $W_{ks}^{(k)}$ are interpreted as transition probabilities per unit time that the system will go from state kr to ks. They must satisfy $W_{sr}^{(k)} \geq 0$ and the symmetry conditions $W_{rs}^{(k)} = W_{sr}^{(k)}$. This is also referred to as the principle of microscopic reversibility.

4.1 A quantum H-theorem
The ensemble which represents a real physical system is determined by the thermodynamic state and environment of the actual system. The virtual ensemble has constituents which must duplicate both aspects. Of great practical interest and the one considered here is the case of isolated systems. An isolated system is characterized not only by a fixed value of the energy E, but also by a definite number of particles and volume V. Under these conditions, a quantum H-theorem can be formulated Yourgrau et al. (1966). Classically the error with which the energy of the real system can be specified can be theoretically reduced to zero. However, quantum theory claims there is a residual error specified by the uncertainty relation. All members of the ensemble cannot be said then to occupy eigenstates belonging to the same energy. It must be assumed the systems are distributed over energy levels lying within a finite range, ΔE. The following restrictions on the occupation numbers of the ensemble are imposed for an isolated system

$$N_{jr}^* \neq 0, \quad E_j \in I_{\Delta E} = (E - \frac{1}{2}\Delta E, E + \frac{1}{2}\Delta E), \quad N_{jr}^* = 0, \quad E_j \notin I_{\Delta E}. \tag{4.12}$$

It will be shown that the ensemble specified by (4.12) exhibits a one-directional development in time ending ultimately in equilibrium.

Pauli's equation can be used to obtain the rate of change of the quantum mechanical H-function which is defined to be

$$H^*(t) = \sum_s P_s \log(P_s). \tag{4.13}$$

The summation in (4.13) is extended over the group of states whose energies are approximately E, that is in the interval $I_{\Delta E}$ for example. Now, differentiate (4.13) with respect to t and use the fact that $\sum_s P_s = 1$ to get

$$\dot{H}^* = \sum_s \dot{P}_s \log(P_s) + \sum_s \dot{P}_s = \sum_s \dot{P}_s \log(P_s). \tag{4.14}$$

Now requiring that $\dot{P}_s$ be determined by (4.11), $\dot{H}^*$ in (4.14) becomes

$$\dot{H}^*(t) = \sum_s \sum_r W_{sr}(P_r - P_s) \log(P_s). \tag{4.15}$$

Interchanging r and s and using the symmetry property $W_{rs} = W_{sr}$, this is

$$\dot{H}^*(t) = \sum_r \sum_s W_{rs}(P_s - P_r) \log(P_r) = -\sum_r \sum_s W_{sr}(P_r - P_s) \log(P_r). \tag{4.16}$$

Adding (4.15) and (4.16) yields the following result,

$$\dot{H}^*(t) = -\frac{1}{2} \sum_r \sum_s W_{sr}(P_r - P_s)(\log(P_r) - \log(P_s)). \tag{4.17}$$

Recalling that $W_{sr} \geq 0$ as well as the inequality $(x - y)(\log x - \log y) \geq 0$ for each (r, s), it follows that $(P_r - P_s)(\log(P_r) - \log(P_s)) \geq 0$. Consequently, each term in the sum in (4.17) is either zero or positive, hence $H^*(t)$ decreases monotonically with time,

$$\dot{H}^*(t) \leq 0. \tag{4.18}$$

Equality holds if and only if $P_s = P_r$ for all pairs (r, s) such that $W_{sr} \neq 0$. Thus H^* decreases and statistical equilibrium is reached only when this condition is fulfilled. Originally enunciated by Boltzmann in a classical context, (4.18) constitutes a quantum mechanical version of the H-theorem.

4.2 Quantum mechanical canonical ensemble

Let us devise an ensemble which is representative of a closed isothermal system of given volume, or characterized by definite values of the parameters T, V and N. This approach brings us back to one of the ways entropy was formulated in the introduction, and need not rely on the specification of a density matrix. Suppose there are N^* members of the ensemble each with the same values of V and N as the real system. However, they are not completely isolated from each other, so each is surrounded by a surface that does not permit the flow of particles but is permeable to heat. The collection of systems can be packed into the form of a lattice and the entire construction immersed in a heat reservoir at temperature T until equilibrium is attained. The systems are isothermal such that each is embedded in a heat reservoir composed of the remaining $N^* - 1$.

Once the ensemble is defined, it can be asked which fraction of the N^* systems occupies any particular eigenstate of the unperturbed Hamiltonian of the experimental system. Let us study the ensemble then which is regarded as a large thermophysical system having energy E^*, volume $V^* = N^*V$ and made up of N^*N particles.The quantum states of this large supersystem belonging to energy E^* are to be enumerated. The thermal interaction energy is assumed to be so small that a definite energy eigenstate can be assigned to each individual system at any time. As energy can be exchanged between constituent systems, the eigenstates accessible to them do not pertain to one value of energy. The energy eigenstates of a system are written $E_1, E_2, \cdots, E_j, \cdots$ with $E_{j+1} \geq E_j$. Only one system-state j belongs to energy eigenvalue E_j. An energy eigenstate of the supersystem is completely defined once the energy eigenstate occupied by each system is specified.

It is only needed to stipulate the number N_j^* of systems occupying every system state j. Any set of values of the occupation numbers $N_1^*, N_2^*, \cdots$ define a quantum mechanical distribution. Clearly the W^* supersystem states calculated by

$$W(N_1^*, N_2^*, \cdots) = \frac{N^*!}{N_1^*!N_2^*!\cdots} \tag{4.19}$$

are compatible with a given distribution $N_1^*, N_2^*, \cdots$. Not all sets of N_j^* are admissible. The physically relevant ones satisfy the two constraints

$$\sum_j N_j^* = N^*, \qquad \sum_j N_j^* E_j = E^*. \tag{4.20}$$

The supersystem then consists of a number N^* of fixed but arbitrary systems with a constant energy E^*.

The number of physically possible supersystem states is clearly given as

$$\Omega^*(E^*, N^*) = \sum_C W^*(N_1^*, N_2^*, \cdots), \tag{4.21}$$

where the summation is to be extended over all N_j^* satisfying constraints (4.20). According to the earlier postulate, all allowed quantum states of an isolated system are equiprobable. Consequently, from this principle all states which satisfy (4.20) occur equally often. The probability P^* that a particular distribution $N_1^*, N_2^*, \cdots$ is actualized is the quotient of W^* and Ω^*,

$$P^*(N_1^*, N_2^*, \cdots) = \frac{W^*(N_1^*, N_2^*, \cdots)}{\Omega^*(E^*, N^*)}. \tag{4.22}$$

With respect to P^* in (4.22), the average value of the occupation number N_k^* is given quite simply by

$$\bar{N}_k^* = \sum_k N_k^* P^*(N_1^*, N_2^*, \cdots). \tag{4.23}$$

Substituting P^* into (4.23), it can be written as

$$\Omega^*(E^*, N^*)\bar{N}_k^* = N^* \sum_k \frac{(N^*-1)!}{N_1^*!N_2^*!\cdots(N_k^*-1)!\cdots}. \tag{4.24}$$

To obtain a more useful expression for $\bar{N}_k^*$, the right-hand side can be transformed to a set of primed integers. To this end, define

$$N^{*'} = N^* - 1, \qquad N_k^{*'} = N_k^* - 1, \qquad N_j^{*'} = N_j^*, \qquad j \neq k.$$

Using these, constraints (4.20) get transformed into

$$\sum N_j^{*'} = N_1^* + N_2^* + \cdots + (N_j^* - 1) + \cdots = N^* - 1,$$
$$\sum N_j^{*'} E_j = N_1^* E_1 + \cdots + (N_k^* - 1) E_k + \cdots = E^* - E_k. \tag{4.25}$$

Consequently,

$$\Omega^*(E^*, N^*) = N^* {\sum}' \frac{N^{*'}!}{N_1^{*'}! N_2^{*'}! \cdots}, \tag{4.26}$$

where the prime means the sum extends over all $N_k^{*'}$ which satisfy constraints (4.25). Comparing (4.26) with (4.20), the right-hand side of (4.26) is exactly $N^* \Omega^*(E^* - E_k, N^* - 1)$. Thus dividing by $\Omega^*(E^*, N^*)$, we have

$$\bar{N}_k^* = \frac{N^* \Omega^*(E^* - E_k, N^* - 1)}{\Omega^*(E^*, N^*)}.$$

Dividing this by N^* and taking the logarithm of both sides results in the expression,

$$\log \frac{\bar{N}_k^*}{N^*} = \log \Omega^*(E^* - E_k, N^* - 1) - \log \Omega^*(E^*, N^*). \tag{4.27}$$

The result in (4.27) can be expanded in a Taylor series to first order if we take $N^* \gg 1$ and $E^* \gg E_k$,

$$\log \frac{\bar{N}_k^*}{N^*} = -\frac{\partial \log \Omega^*}{\partial E^*} E_k^* - \frac{\partial \log \Omega^*}{\partial N^*} = -\beta E_k^* - \alpha. \tag{4.28}$$

From the constraint $N^* = \sum_j \bar{N}_j^* = N^* e^{-\alpha} \sum e^{-\beta E_j}$, e^α can be obtained. Replacing this back in (4.28) and exponentiating gives

$$\bar{N}_k^* = N^* \frac{e^{-\beta E_k}}{\sum_k e^{-\beta E_j}}. \tag{4.29}$$

The result in (4.29) gives what the average distribution of systems over system states will be in a supersystem at equilibrium. The instantaneous distribution will fluctuate around this distribution. The relative fluctuations of the occupation numbers for large enough N^* are negligible, so to this accuracy, $\bar{N}_k^*/N^*$ can be equated to P_k. Setting $Z = \sum_j e^{-\beta E_j}$, the instantaneous probability that an arbitrarily chosen system of this supersystem will be in system state k can be summarized as follows

$$P_k = Z^{-1} e^{-\beta E_k}. \tag{4.30}$$

This distribution is the quantum version of the canonical distribution in phase space, and is referred to as the quantum mechanical canonical ensemble. The function Z so defined is called the partition function.

In effect, this formalism has permitted the construction of a type of measuring device. Let us show that the microscopic ideas which have led to these results immediately imply consequences at the macroscopic level. To this end, it will be established what the exact form of the connection between Z and the Helmholtz free energy F actually is. The starting point is the second part of (4.20). Putting $U = E^*/N^*$, it implies

$$U = \sum_j P_j E_j. \tag{4.31}$$

Formula (4.31) is in agreement with the postulate maintaining that the energy U of the physical system must be identified with the ensemble average $\langle E \rangle$ of the energy.

Begin by considering the change dU of the energy U when the experimental system remains closed but undergoes an infinitesimal reversible process. Equation (4.31) implies that

$$dU = \sum_j (E_j \, dP_j + P_j \, dE_j). \tag{4.32}$$

Now (4.30) can be solved for E_j in the form $E_j = -\beta^{-1}(\log Z + \log P_j)$. Consequently, since $\sum_j P_j = 1$, it is found that $\sum_j dP_j = 0$. Combining these it then follows that

$$-\sum_j E_j dP_j = \beta^{-1} \sum_j (\log Z + \log P_j) \, dP_j = \beta^{-1} \sum_j \log P_j \, dP_j = \beta^{-1} d(\sum_j P_j \log P_j). \tag{4.33}$$

Further, with $-\not{d}W$ the work done on the system during the given process, we have that

$$\sum_j P_j \, dE_j = -\not{d}W, \tag{4.34}$$

Combining (4.33) and (4.34), we get the result

$$dU = -\beta^{-1} d(\sum_j P_j \log P_j) - \not{d}W. \tag{4.35}$$

Comparing (4.35) with the first law $dU = \not{d}Q - \not{d}W$, it is asserted that

$$\beta \not{d}Q = -d(\sum_j P_j \log P_j). \tag{4.36}$$

Since the right-hand side of (4.36) is an exact differential, it is concluded that β is an integrating factor for $\not{d}Q$. By the second law of thermodynamics, β must be proportional to T^{-1} and the proportionality constant must be the reciprocal of k_B. With β of this form, when combined with the second law $\not{d}Q = T dS$, we have

$$dS = -k_B d(\sum_j P_j \log P_j). \tag{4.37}$$

This can be easily integrated to give

$$S = -k_B \sum_j P_j \log P_j + C, \tag{4.38}$$

where the integrating constant C is independent of both T and V. In fact the additive property of entropy requires that $C = 0$.

This complicated procedure has returned us in some sense to where we began with (1.5), but by a different route. To get a relation between Z and F, use (4.30), (4.31) and (4.38) to write

$$TS = -k_B T \sum_j P_j \log P_j = k_B T \log Z + \sum_j P_j E_j = k_B T \log Z + U. \tag{4.39}$$

Consequently, since $F = U - TS$, (4.39) implies the following result

$$F = -k_B T \log Z. \tag{4.40}$$

Through the construction of these ensembles at a fundamental quantum level, a formalism has been obtained which will allow us to obtain concrete predictions for many equilibrium thermodynamic properties of a system once the function $Z = Z(T, V, N)$ is known. In fact, it follows from the thermodynamic equation

$$dF = -S\, dT - p\, dV + \mu\, dN, \tag{4.41}$$

where μ is the chemical potential per molecule, that

$$S = -(\frac{\partial F}{\partial T})_{V,N} = k_B \log Z + k_B T (\frac{\partial Z}{\partial T})_{V,N}, \tag{4.42}$$

$$p = -(\frac{\partial F}{\partial V})_{N,T} = k_B T (\frac{\partial Z}{\partial V})_{N,T},$$

$$\mu = (\frac{\partial F}{\partial N})_{T,V} = -k_B T (\frac{\partial \log Z}{\partial N})_{T,V},$$

$$U = F + TS = k_B T^2 (\frac{\partial \log Z}{\partial T})_{V,N}. \tag{4.43}$$

As an application of these results, consider the one-dimensional harmonic oscillator which has quantum mechanical energy eigenvalues given by

$$\epsilon_n = (n + \frac{1}{2})\hbar\omega, \qquad n = 0, 1, 2, \cdots . \tag{4.44}$$

The single-oscillator partition function is given by

$$z(\beta) = \sum_{n=0}^{\infty} e^{-\beta(n+\frac{1}{2})\hbar\omega} = (2\sinh(\frac{1}{2}\beta\hbar\omega))^{-1}.$$

The N-oscillator partition function is then given by

$$Z_N(\beta) = [z(\beta)]^N = (2\sinh(\frac{1}{2}\beta\hbar\omega)]^{-N}. \tag{4.45}$$

The Helmholtz free energy follows from (4.40),

$$F = Nk_B T \log(2\sinh(\frac{1}{2}\beta\hbar\omega)).$$

By means of F, (4.42) and (4.43) imply that $\mu = F/N$, $p = 0$ and the entropy and energy are

$$S = Nk_B[\frac{\beta\hbar\omega}{e^{\beta\hbar\omega} - 1} - \log(1 - e^{-\beta\hbar\omega})], \quad U = N[\frac{1}{2}\hbar\omega + \frac{\hbar\omega}{e^{\beta\hbar\omega} - 1}]. \tag{4.46}$$

5. Conclusions

It has been seen that formulating the concept of entropy at the microscopic level can be closely related to studying the foundations of quantum mechanics. Doing so provides a useful formalism for exploring many complicated phenomena such as entanglement at this level. Moreover, predictions can be established which bridge a gap between the microscopic and the macroscopic realm. There are many other topics which branch out of this introduction to the subject. For example there is a great deal of interest now in the study of the quantization of nonintegrable systems Gutzwiller (1990), which has led to the field of quantum chaos. There are many indications of links in this work between the areas of nonintegrability and the kind of ergodicity assumed in statistical mechanics which should be pursued.

6. References

Ballentine L. (1970). The Statistical Interpretation of Quantum Mechanics, *Reviews of Modern Physics*, 42, 358-381.

Grössing G. (2008). The vacuum fluctuation theorm: Exact Schrödinger equation via nonequilibrium thermodynamics, *Physics Letters*, B 372, 4556-4563.

Gutzwiller M. (2009). Chaos in Classical and Quantum Mechanics, Springer-Verlag, NY.

Haroche S. and Raimond J. (2006). *Exploring the Quantum*, Oxford University Press.

Landau L. and Lifshitz E. (1978). *Statistical Physics*, Pergamon Press.

Lieb E. and Yngvason J. (1999). The physics and mathematics of the second law of thermodynamics, *Physics Reports*, 310, 1-96.

Mintet F., Carvalho A., Kees M. and Buchleitner A. (2005). Measures and dynamics of entangled states, *Physics Reports*, 415, 207-259.

von Neumann J. (1955). *Mathematical Foundations of Quantum Mechanics*, translated by R. T. Beyer, Princeton University Press, Princeton, NJ.

Padmanabhan T. (1990). Statistical Mechanics of self gravitating systems, *Physics Reports*, 188, 285-362.

Pathria R. (1977). *Statistical Mechanics*, Pergamon Press.

Pauli W. (2000). *Statistical Mechanics, Pauli Lectures on Physics*, vol. 4, Dover, Mineola, NY.

Peres A. (1995). *Quantum Theory: Concepts and Methods*, Klewer Academic Publishers, Dordrecht, The Netherlands.

Peres A. and Terno D. (2004). Quantum Information and Relativity Theory, *Reviews of Modern Physics*, 76, 93-123.

Raimond J., Brune M. and Haroche S. (2001). Manipulating quantum entanglement with atoms and photons in a cavity, *Reviews of Modern Physics*, 73, 565-582.

Rajeev S. (2008). Quantization of contact manifolds and thermodynamics, *Annals of Physics*, 323, 768-782.

Wehrl A. (1978). General Properties of Entropy, *Reviews of Modern Physics*, 50, 221-260.

Weinberg S. (1989). Testing Quantum Mechanics, *Annals of Physics*, 194, 336-386.

Yourgrau W., van der Merwe A. and Raw G., (1982). *Treatise on Irreversible and Statistical Thermophysics*, Dover, Mineola, NY.

7

The 'Computational Unified Field Theory' (CUFT): Harmonizing Quantum and Relativistic Models and Beyond

Jonathan Bentwich
Brain Perfection LTD
Israel

1. Introduction

Perhaps the most troubling enigma in modern natural sciences is the principle contradiction that exists between quantum mechanics and Relativity theory (Greene, 2003) ; Indeed, this principle incompatibility between Quantum Mechanics and Relativity Theory propelled Einstein to relentlessly pursuit a 'Unified Field Theory' (Einstein, 1929, 1931, 1951) and subsequently prompted an intensive search for a 'Theory of Everything' (TOE) (Bagger & Lambert, 2007; Elis, 1986; Hawkins, 2002; Polchinski, 2007; Brumfiel, 2006). The principle contradictions that exist between quantum mechanics and relativity theory are:

a. Probabilistic vs. deterministic models of physical reality:

Relativity theory is based on a positivistic model of 'space-time' in which an object or an event possesses clear definitive 'space-time', 'energy-mass' properties and which therefore gives rise to precise predictions regarding the prospective 'behavior' of any such object or event (e.g., given an accurate description of its initial system's state). In contrast, the probabilistic interpretation of quantum mechanics posits the existence of only a 'probability wave function' which describes physical reality in terms of complimentary 'energy-space' or 'temporal-mass' uncertainty wave functions (Born, 1954; Heisenberg, 1927). This means that at any given point in time all we can determine (e.g., at the subatomic quantum level) is the statistical likelihood of a given particle or event to possesses a certain 'spatial-energetic' and 'temporal-mass' complimentary values. Moreover, the only probabilistic nature of quantum mechanics dictates that this statistical uncertainty is almost 'infinite' prior to our measurement of the particle's physical properties and 'collapses' upon our interactive measurement of it into a relatively defined (complimentary) physical state... Hence, quantum mechanics may only provide us with a probabilistic prediction regarding the physical features of any given subatomic event – as opposed to the relativistic positivistic (deterministic) model of physical reality.

b. "Simultaneous-entanglement" vs. "non-simultaneous-causality" features:

quantum and relativistic models also differ in their (a-causal) *'simultaneous-entanglement'* vs. *'non-simultaneous-causal'* features; In Relativity theory the speed of light represents the ultimate constraint imposed upon the transmission of any physical signal (or effect), whereas quantum mechanics advocates the existence of a 'simultaneous-entanglement' of quantum effects (e.g., that are not bound by the speed of light constraint). Hence, whereas the relativistic model is based on strict causality – i.e., which separates between any spatial-temporal 'cause' and 'effect' through the speed of light (non-simultaneous) signal barrier, quantum entanglement allows for 'a-causal' *simultaneous effects* that are independent of any light-speed constraint (Horodecki et al., 2007).

c. Single vs. multiple spatial-temporal modeling:

Finally, whereas Relativity theory focuses on the conceptualization of only a *single* spatial point at any given time instant – i.e., which therefore possesses a well defined spatial position, mass, energy, or temporal measures, quantum mechanics allows for the measurement   (and   conceptualization)   of   multiple   spatial-temporal   points (simultaneously) – giving rise to a (probability) *'wave* function'; Indeed, it is hereby hypothesized that this principle distinction between a *single spatial-temporal* quantum *'particle'* or localized relativistic object (or event) and a *multi- spatial-temporal* quantum *'wave'* (function) may both shed light on some of the key conceptual differences between quantum and relativistic modeling as well as potentially assist us in bridging the apparent gap between these two models of physical reality (based on a conceptually higher-ordered computational framework).

2. The 'Duality Principle': Constraining quantum and relativistic 'Self-Referential Ontological Computational System' (SROCS) paradigms

However, despite these (apparent) principle differences between quantum and relativistic models of physical reality it is hypothesized that both of these theories share a basic *'materialistic-reductionistic'* assumption underlying their basic (theoretical) computational structure: It is suggested that mutual to both quantum and relativistic theoretical models is a fundamental   'Self-Referential-Ontological-Computational-System'   (SROCS)   structure (Bentwich, 2003a, 2003b, 2003c, 2004, 2006) which assumes that it is possible to determine the 'existence' or 'non-existence' of a certain 'y' factor solely based on its *direct physical interaction* $(PR\{x,y\}/di1)$ with another 'x' factor (e.g., at the same 'di1' computational level), thus:

$$SROCS: PR\{x,y\}/di1 \rightarrow ['y' \text{ or } '\neg y]'/di1.$$

But, a strict computational-empirical analysis points out that such (quantum and relativistic) SROCS computational structure may also inevitably lead to *'logical inconsistency'* and inevitable consequent *'computational indeterminacy'* – i.e., a principle inability of the (hypothesized) SROCS computational structure to determine whether the particular 'y' element "exists" or "doesn't exist": Indeed (as will be shown below) such 'logical inconsistency' and subsequent 'computational indeterminacy' occurs in the specific case in which the direct physical interaction between the 'x' and 'y' factors leads to a situation in which the 'y' factor "*doesn't* exist", which is termed: a 'Self-Referential-Ontological-Negative-System', SRONCS)... However, since there exist ample empirical evidence that both quantum and relativistic computational systems *are capable* of determining whether a particular 'y' element (e.g., state/s or value/s) "exists" or "doesn't exist" then this contradicts the SRONCS (above mentioned) inevitable 'computational indeterminacy', thereby calling for a reexamination of the currently assumed quantum and relativistic SROCS/SRONCS computational structure;

Indeed, this analysis (e.g., delineated below) points at the existence of a (new) computational *'Duality Principle'* which asserts that the computation of any hypothetical quantum or relativistic (x,y) relationship/s must take place at a *conceptually higher-ordered computational framework 'D2'* – e.g., that is (in principle) irreducible to any direct (or even indirect) physical interaction between the (quantum or relativistic) 'x' and 'y' factors (but which can nevertheless determine the association between any two given 'x' and 'y' factors) (Bentwich, 2003a, 2003b, 2003c, 2004, 2006a, 2006b).

In the case of Relativity theory, such basic SROCS computational structure pertains to the computation of any spatial-temporal or energy-mass value/s of any given event (or object) – solely based on its *direct physical interaction* with any hypothetical (differential) relativistic observer; We can therefore represent any such (hypothetical) spatial-temporal or energy-mass value/s (of any given event or object) as a particular *'Phenomenon'*: 'P[s-t $(i...n)$, e-m $(i...n)$]'; Therefore, based on the (above) relativistic 'materialistic-reductionistic' assumption whereby the specific value of any (spatial-temporal or energy-mass) 'Phenomenon' value is computed solely based on its direct physical interaction ('di1') with a specific (hypothetical differential) relativistic observer, we obtain the (above mentioned) SROCS computational structure:

$$\text{SROCS: PR\{O-}\textit{diff}\,, \text{P[s-t } (i...n), \text{ e-m } (i...n)] \,\}/\text{di1}$$
$$\rightarrow \{\text{'P[s-t } (i), \text{ e-m } (i)]' \text{ or '}\textit{not } \text{P[s-t } (i), \text{ e-m } (i)]'\}.$$

Hence, according to the above mentioned SROCS computational structure the relativistic SROCS computes the "existence" or "non-existence" of any particular 'Phenomenon' (e.g., specific 'spatial-temporal' or 'energy-mass' 'i' value/s of any given object/event) – solely based upon the direct physical interaction (PR.../di1) between the potential (exhaustive hypothetical) values of this Phenomenon ('P[s-t $(i...n)$, e-m $(i...n)$]') and any hypothetical differential relativistic observer; But, note that the relativistic SROCS computational structure assumes that it is solely through the direct physical interaction between any series of (hypothetical differential) relativistic observer/s and the Phenomenon's (entire spectrum of possible spatial-temporal or energy-mass) value/s – that a particular 'Phenomenon' (spatial-temporal or energy mass) value is computed. The relativistic SROCS computational structure assumes that it is solely through the direct physical interaction between any series of (hypothetical differential) relativistic observer/s and the Phenomenon's (entire spectrum of possible spatial-temporal or energy-mass) value/s – that the particular 'Phenomenon' (spatial-temporal or energy-mass) value is computed. But, a thorough analysis of this SROCS computational structure indicates that in the specific case in which the direct physical interaction between any hypothetical differential relativistic observer/s and the *Phenomenon's* whole spectrum of potential values – leads to the *"non-existence"* of all of the other 'space-time' or 'energy-mass' values that were not measured by a particular relativistic observer ('O-i') (at the same 'di1' computational level):

SRONCS: PR\{O-*diff(i...n)*, P[s-t $(i...n)$, e-m $(i...n)$] \} $\rightarrow$ '*not* P[s-t *(not i)*, e-m *(not i)*]O-i' /di1.

However, this SRONCS computational structure inevitably leads to the (above mentioned) 'logical inconsistency' and '*computational indeterminacy*':

This is because according to this SRONCS computational structure all of the other 'Phenomenon' values (e.g., 'space-time' or 'energy-mass values) – which do not correspond to the specifically measured 'space-time' or 'energy-mass' {i} values (i.e., that are measured by a particular corresponding 'O-*diff-i* relativistic observer): P[{s-t $\neq i$} or {e-m $\neq i$}] are necessarily computed by the SRONCS paradigmatic structure to both *"exist"* AND *"not exist"* – at the same 'di1' computational level: according to this SRONCS computational structure all of the other 'Phenomenon' values (e.g., 'space-time' or 'energy-mass values) – which do not correspond to the specifically measured 'space-time' or 'energy-mass' {i} values (i.e., that are measured by a particular corresponding 'O-*diff-i* relativistic observer): P[{s-t ≠ i} or {e-m ≠ i}] are necessarily computed by the SRONCS paradigmatic structure to both *"exist"* AND *"not exist"* – at the same 'di1' computational level:

SRONCS: PR{O-*diff(i...n)*, P[s-t *(i...n)*, e-m *(i...n)*] } → '*not* P[s-t *(not i)*, e-m *(not i)*]O-*i*'/di1.

But, given the SROCS/SRONCS strong 'materialistic-reductionistic' working assumption – i.e., that the computation of the "existence" or "non-existence" of the particular P[{s-t ≠ i}, {e-m ≠ i}] values solely depends on its direct physical interaction with the series of (potential) differential observers at the 'di1' computational level, then the above SRONCS computational assertion that the particular P[{s-t ≠ i}, {e-m ≠ i}] values both "exist" and "don't exist" at the same 'di1' computational level inevitably also leads to both 'logical inconsistency' and a closely linked 'computational indeterminacy' – e.g., conceptual computational inability of such 'di1' computational level to determine whether the P[{s-t ≠ i}, {e-m ≠ i}] values "exist" or "doesn't exist"...

But, since there exists ample relativistic empirical evidence pointing at the capacity of any relativistic observer to determine whether or not a particular 'P[{s-t ≠ i}, {e-m ≠ i}]' "exists" or "doesn't exist", then a novel (hypothetical) computational *'Duality Principle'* asserts that the determination of the "existence" or "non-existence" of any given P[s-t *(i...n)*, e-m *(i...n)*] can only be computed at a conceptually higher-ordered 'D2' computational level e.g., that is in principle irreducible to any direct or even indirect physical interactions between the full range of possible 'Phenomenon' values 'P[s-t *(i...n)*, e-m *(i...n)*]' and any one of the potential range of (differential) relativistic observers:

'D2': P[{s-t *(i...n)* e-m *(i...n)*, O-*r(st-i)*}, {P[s-t *(i+n)* e-m *(i+n)*, O-*r(st-i+n))*]
≠ PR{O-*diff*, P[s-t *(i...n)* e-m *(i...n)*] }/di1

Note that the computational constraint imposed by the Duality Principle is *conceptual* in nature – i.e., as it asserts the conceptual computational inability to determine the "existence" or "non-existence" of any (hypothetical) 'Phenomenon' (e.g., 'space-time' event/s or 'energy-mass' object value/s) from within its direct physical interaction with any (hypothetical) differential relativistic observer; Indeed, a closer examination of the abovementioned SROCS/SRONCS relativistic computational structure may indicate that the computational constraint imposed by the Duality Principle is not limited to only *direct physical interaction* between any 'Phenomenon' (e.g., space-time or energy-mass value/s) and any (hypothetical) differential relativistic observer/s, but rather extends to any direct *or indirect physical interaction* (between any such 'Phenomenon' and any potential differential relativistic observer/s); In order to prove this broader applicability of the computational Duality Principle – as negating the possibility of determining the "existence" or "non-existence" of any such 'Phenomenon' from within its direct *or indirect* physical interaction/s with any (hypothetical) differential relativistic observer/s (e.g., but only from a conceptually higher-ordered hypothetical computational level 'D2') let us assume that it *is possible* to determine the precise value/s of any given 'Phenomenon' based on its *indirect interaction* with another intervening variable (or computational level) 'd2' (which may receive any information or input/s or effect/s etc. from any direct physical interaction/s between the given 'Phenomenon' and any hypothetical differential relativistic observer at the 'di1' level);

$$\text{SROCS: PR}\{O\text{-}diff(i\ldots n), P[s\text{-}t\ (i\ldots n), \text{e-m}\ (i\ldots n)]\ \}/\text{di1}$$
$$\rightarrow \{'P[s\text{-}t\ (i), \text{e-m}\ (i)]'\ \text{or}\ 'not\ P[s\text{-}t\ (i), \text{e-m}\ (i)]O\text{-}i'\}/\text{di2}.$$

But, a closer analysis of the this hypothetical 'di2' (second) intervening computational level (or factor/s) as possibly being able to determine whether any particular space-time event or energy-mass object "exists" or "doesn't exist" may indicate that it precisely replicates the same SROCS/SRONCS ('problematic') computational structure which has been shown to be constrained by the (novel) computational 'Duality Principle'. This is because despite the (new) assumption whereby the computation of the "existence" or "non-existence" of any particular 'Phenomenon' (e.g., space-time or energy-mass) value is computed at a different 'di2' computational level (or factor/s etc.), the SROCS/SRONCS *intrinsic 'materialistic-reductionistic'* computational structure is such that it assumes that the determination of the "existence"/"non-existence" of any particular Phenomenon value/s is *'solely caused'* (or *'determined'*) *by the direct physical interaction* between that 'Phenomenon' and any hypothetical (differential) relativistic observer/s, which is represented by the *causal arrow* "$\rightarrow$" embedded within the relativistic SROCS/SRONCS computational structure :

$$\text{SROCS: PR}\{O\text{-}diff(i\ldots n), P[s\text{-}t\ (i\ldots n), \text{e-m}\ (i\ldots n)]\ \}/\text{di1}$$
$$\rightarrow \{'P[s\text{-}t\ (i), \text{e-m}\ (i)]'\ \text{or}\ 'not\ P[s\text{-}t\ (i), \text{e-m}\ (i)]O\text{-}i'\}/\text{di2}.$$

Thus, even though the direct physical interaction between the 'Phenomenon' and the differential relativistic observer seem to take place at the 'di1' computational level whereas the determination of the "existence"/"non-existence" of a particular Phenomenon value appears to be carried out at a different 'di2' computational level, the actual (embedded) computational structure still represents a SROCS/SRONCS paradigm. This is because even this new SROCS/SRONCS computational structure still maintains the strict 'materialistic-reductionistic' working assumption whereby it is solely the direct physical interaction between the Phenomenon and the differential relativistic observer that determines the "existence"/"non-existence" of a particular Phenomenon value; An alternate way of proving that the SROCS/SRONCS computational structure remains unaltered (e.g., even when we assume that the computation of the "existence" or "non-existence" of the particular 'Phenomenon' value may take place at another 'di2' computational level) is based on the fact that due to its (above mentioned) 'materialistic-reductionistic' working hypothesis – the determination of the "existence" or "non-existence" of the particular 'Phenomenon' value is solely computed based on the information obtained from the direct physical interaction between the 'Phenomenon' and the series of potential differential observers (at the 'di1' computational level); Hence, in effect there is a total contingency of the determination of the "existence"/"non-existence" of the particular 'Phenomenon' value (at the hypothetical 'di2' computational level) upon the direct physical interaction between this 'Phenomenon' and any differential relativistic observer (at the 'di1' level) which therefore does not alter the 'di1' SROCS computational structure, and may be expressed thus:

$$\text{SRONCS: PR}\{O\text{-}diff(i\ldots n), P[s\text{-}t\ (i\ldots n), \text{e-m}\ (i\ldots n)]\ \}\rightarrow 'not\ P[\{s\text{-}t \neq i\}, \{\text{e-m} \neq i\}]O\text{-}i'\ /\text{di1 or di2}.$$

(Note: precisely due to the above mentioned total "existence"/"non-existence" of the particular Phenomenon value (*i*) at 'di2' upon input from the Phenomenon's direct physical interaction with any differential relativistic observer at 'di1' it may be more convenient to formally represent this SROCS computational structure as occurring altogether – either at the 'di1' or 'di2' computational level, as presented above);

However, as proven by the Duality Principle (above), given the fact that there exists ample empirical evidence indicating the capacity of relativistic (computational) systems to determine whether a particular 'Phenomenon' (space-time or energy-mass) value "exists" or "doesn't exist", then the broader extension of the Duality Principle evinces that it is not possible (e.g., in principle) to determine such "existence" or "non-existence" of any particular 'Phenomenon' (e.g., space-time or energy-mass) value from *within* any direct or indirect physical interaction/s between any such 'Phenomenon' and any (hypothetical) series of differential relativistic observer/s; Instead, the 'Duality Principle' postulates that the determination of any 'Phenomenon' (e.g., 'space-time' or energy-mass) values can only be determined by a conceptually higher ordered 'D2' computational level which is capable of determining the '*co-occurrence/s*' of specific Phenomenon values and corresponding differential relativistic observers' measurements (e.g., and which is irreducible to any direct or indirect physical interactions between such differential relativistic observer/s and any Phenomenon value/s):

$$\text{'D2': (P\{s-t }(i)\text{ e-m }(i),\text{ O-}r(st\text{-}i)\};\text{ ...P\{s-t }(i+n)\text{ e-m }(i+n\text{),}\text{ O-}r(st\text{-}i+n)\})$$
$$\neq \text{PR\{O-}diff(i...n),\text{ P[s-t }(i...n)\text{ e-m }(i...n)]\text{ \}/di1}$$

Hence, a thorough reexamination of Relativity's SROCS computational structure (e.g., which assumes that the determination of any 'space-time' or 'energy-mass' Phenomenon value is solely determined based on that particular event's or object's direct or indirect physical interaction/s with any one of a series of potential relativistic observers) has led to the recognition of a (novel) computational 'Duality Principle'; This 'Duality Principle' proves that it is not possible (in principle) to determine any such space-time or energy-mass 'Phenomenon' values based on any hypothetical direct or indirect physical interaction between such 'Phenomenon' and any hypothetical series of (differential) relativistic observer/s; Rather, according to this novel computational Duality Principle the determination of any space-time or energy-mass relativistic value can only be computed based on a conceptually higher-ordered 'D2' computational level (e.g., which is again in principle irreducible to any hypothetical direct or indirect physical interaction between any differential relativistic observer/s and any space-time or energy-mass Phenomenon); Such conceptually higher-ordered 'D2' computational level is also postulated to compute the 'co-occurrences' of any' differential relativistic observer/s' and corresponding 'Phenomenon' (e.g., space-time or energy-mass value/s)...

Intriguingly, it also hypothesized that the same precise SROCS/SRONCS computational structure may underlie the quantum probabilistic interpretation of the 'probability wave function' and 'uncertainty principle'; Indeed, it is hereby hypothesized that precisely the same SROCS/SRONCS computational structure may pertain to the quantum mechanical computation of the physical properties of any given subatomic 'target' ('*t*') (e.g., assumed to be dispersed all along a probability wave function) which is hypothesized to be determined solely through its direct physical interaction with another subatomic complimentary 'probe' (P('e/s' or '*t*/m')) entity, thus:

$$\text{SROCS: PR\{P('e/s' or 't/m'), }t\text{ [e/s }(i...n),\text{ t/m }(i...n)]\text{ \}}$$
$$\rightarrow \text{['}t\text{ [e/s }(i),\text{ t/m }(i)]' or 'not t\text{ [s/e }(i),\text{ t/m }(i)]' /di1}$$

In a nutshell, it is suggested that this SROCS/SRONCS computational structure accurately represents the (current) probabilistic interpretation of quantum mechanics in that it describes the basic working hypothesis of quantum mechanics wherein it is assumed that the determination of the particular (complimentary) 'spatial-energetic' or 'temporal-mass' values of any given subatomic 'target' particle – i.e., which is assumed to be dispersed probabilistically all along the probability wave function's (complimentary) spatial-energetic and temporal-mass values, occurs through the direct physical interaction of such probability wave function dispersed 'target' entity with another subatomic measuring 'probe' element; Moreover, it is assumed that this direct physical interaction between the probability wave function dispersed 'target' element and the subatomic probe element constitutes the sole (computational) means for the "collapse" of the target's probability wave function to a singular complimentary target value: This inevitably produces a SROCS computational structure which possesses the potential of expressing a SRONCS condition, thus:

SRONCS: $PR\{P('e/s' \text{ or } 't/m'), t [e/s (i...n), t/m (i...n)] \} \rightarrow$ 'not $t [e/s (\neq i), t/m (\neq i)]'$ /di1

wherein the probabilistically distributed 'target' element (e.g., all along the complimentary 'spatial-energetic' or 'temporal-mass' probability wave function) which possesses all the possible spectrum of such 'spatial-energetic' or 'temporal-mass' values: $t [s/e (i...n), t/m (i...n)]$ "collapses" – solely as a result of its direct physical interaction with another subatomic 'probe' element (which also possesses complimentary 'spatial-energetic' and 'temporal-mass' properties); Indeed, it is this assumed direct physical interaction between the subatomic 'probe' and probabilistically distributed 'target' wave function which "collapses" the target's (complimentary) wave function, i.e., to produce only a *single* (complimentary) spatial-energetic or temporal-mass *measured* value (e.g., t [e-s (i), t-m (i)])– which therefore *negates all of the other* "*non-collapsed*" spatial-energetic or temporal-mass complimentary *values* (e.g., 'not t [e-s ($\neq i$), t-m ($\neq i$)]') of the target's ('pre-collapsed') wave function!

But, as we've seen earlier (in the case of the relativistic SRONCS), such SRONCS computational structure invariably leads to both 'logical inconsistency' and subsequent 'computational indeterminacy': This is because the above mentioned SRONCS condition essentially advocates that all of the "non-collapsed" complimentary 'target' values (i.e., t [s-e $\neq i$ or t-m $\neq i$] seem to both "exist" AND "not exist" at the same 'di1' computational level – thereby constituting a '*logical inconsistency*'!? But, since the basic 'materialistic-reductionistic' working hypothesis underlying the SROCS/SRONCS computational structure also assumes that the determination of any particular target complimentary (spatial-energetic or temporal-mass) value can *only* be determined based on the direct physical interaction between the target probability wave function's distribution and a subatomic 'probe' element – e.g., *at the same 'di1' computational level*, then the above mentioned 'logical inconsistency' invariably also leads to '*computational indeterminacy*', e.g., a principle inability to determine whether any such "non-collapsed" complimentary 'target' values (i.e., t [s-e $\neq i$ or t-m $\neq i$]) "exists" or "doesn't exist"... However, as noted above, since there exists ample empirical evidence indicating the *capacity* of quantum (computational) systems to determine whether any such t [e-s (i...n), t-m (i...n)] quantum target value "exists" or "doesn't exist" the Duality Principle once again asserts the need to place the computation regarding the determination of any pairs of subatomic complimentary 'probe' and 'target' values at a conceptually higher-ordered 'D2' level (e.g., that is in principle *irreducible to any direct physical interactions between them*).

Finally, as shown in the case of the relativistic SROCS/SRONCS paradigm, the conceptual computational constraint imposed by the Duality Principle further expands to include not only strictly *'direct'* physical interaction/s between the subatomic 'probe' and 'target' elements but also any other hypothetical *'indirect'* interaction/s, elements, effects, or even light-signals, information, etc. – that may mediate between these subatomic 'probe' and 'target' elements;

This is because even if we assume that the determination of the "existence" or "non-existence" of any particular subatomic 'target' (spatial-energetic or temporal-mass) value can occur through a (second intervening or mediating) 'di2' computational interaction, entity, process or signal/s transfer we still obtain the same SROCS/SRONCS computational structure which has been shown to be constrained by the computational Duality Principle:

$$\text{SROCS: PR}\{P('e/s' \text{ or } 't/m'), t\,[e/s\,(i...n), t/m\,(i...n)]\}/\text{di1}$$
$$\rightarrow t\,[(e/s\,(\neq i), t/m\,(\neq i)) \text{ or } ('\text{not } t\,[e/s\,(\neq i), t/m\,(\neq i)])/\text{di2}.$$

The rational for asserting that this (novel) computational instant precisely replicates the same SROCS/SRONCS computational structure (e.g., noted above) arises (once again) from the recognition of the strict *'materialistic-reductionistic'* "*causal*" connection that is assumed to exist between the direct physical interaction between the subatomic 'probe' and target' elements (e.g., taking place at the 'di1' level) and the hypothetical 'di2' computational level – and which is assumed to *solely determine* whether a particular target value 't [s/e (i), t/m (i)]' "exists" or "doesn't exist"; This is because since the (abovementioned) basic materialistic-reductionistic causal assumption whereby the 'di2' determination of the "existence"/"non-existence" of any specific (spatial-energetic or temporal-mass) 'target' value is *solely determined by the direct 'probe-target' physical interaction* at the 'di1' level value, therefore the logical or computational structure of the (abovementioned) SROCS/SRONCS is replicated; Specifically, the case of the SRONCS postulates the "existence" of the entire spectrum of possible target values t [e-s $(i...n)$, t-m $(i...n)$] at the 'di1' direct physical interaction between the 'probe' and 'target' entities – but also asserts the "non-existence" of all the "non-collapsed" target values at the 'di2' computational level (e.g., 'not t [e/s $(\neq i)$, t/m $(\neq i)$]); This intrinsic contradiction obviously constitutes the abovementioned 'logical inconsistency' and ensuing 'computational indeterminacy' (that are contradicted by known empirical findings).

Indeed, this SRONCS structure is computationally equivalent to the abovementioned SRONCS: PR$\{P('e\text{-}s' \text{ or } 't\text{-}m'), t\,[e\text{-}s\,(i...n), t\text{-}m\,(i...n)]\} \rightarrow '\text{not } t$ [e-s (i), t-m (i)]'']/di1, since the determination of the ['t [e-s (i), t-m (i)]' or 'not t [e-s (i), t-m (i)]'] is solely determined based on the direct physical interaction at 'di1'. Therefore, also the 'logical inconsistency' and 'computational indeterminacy' (mentioned above) ensues which is contradicted by robust empirical evidence that inevitably leads to the Duality Principle's assertion regarding the determination of any (hypothetical) 'probe-target' pair/s at a conceptually higher-ordered 'D2' computational level:

$$D2: \{[P('e/s' \text{ or } 't/m')i, t\,(e/s(i), t/m(i))];$$
$$...[P('e/s' \text{ or } 't/m')n, t\,(e/s(n), t/m(n))]\}$$
$$\neq \text{PR}\{P('e/s' \text{ or } 't/m'), t\,[e/s\,(i...n), t/m\,(i...n)]\}/\text{di1}$$

Therefore, an analysis of the basic SROCS/SRONCS computational structure underlying both relativistic as well as quantum's computational paradigms has led to the identification of a novel computational 'Duality Principle' which constrains each of these quantum and relativistic SROCS/SRONCS computational paradigms and ultimately points at the inevitable existence of a conceptually higher-ordered 'D2' computational level; Based upon the Duality Principle's identification of such a conceptually higher-ordered 'D2' computational level (which alone can determine any relativistic 'Phenomenon' or any quantum spatial-energetic or temporal-mass target value, it also postulates the computational products of this conceptually higher-ordered 'D2' computational level – as the determination of the "co-occurrence" of any relativistic Phenomenon-relativistic observer pair/s or of any quantum 'probe-target' (complimentary) pair/s; Thus, the first step towards the hypothetical unification of quantum and relativistic theoretical frameworks within a singular (conceptually higher-ordered) model is the identification of a singular computational 'Duality Principle' constraining both quantum and relativistic (underlying) SROCS paradigms and its emerging conceptually higher-ordered singular 'D2' computational level (which produces 'co-occurring' quantum 'probe-target' or relativistic 'observer-Phenomenon' pairs) – as the only feasible computational level (or means) capable of determining any quantum (space-energy or temporal-mass) 'probe-target' relationship or any 'observer-Phenomenon' relativistic relationship/s.

3. 'D2': A singular 'a-causal' computational framework

There are two (key) questions that arise in connection with the discovery of the Duality Principle's conceptually higher-ordered (novel) 'D2' computational framework:

a. Is there a *singular* (mutual) 'D2' computational level that underlies *both* quantum and relativistic (basic) SROCS paradigms?

b. What may be the D2 '*a-causal*' *computational framework* – which transcends the SROCS' computational constraints imposed by the Duality Principle?

In order to answer the first question, lets apply once again the conceptual proof of the 'Duality Principle' regarding the untenable computational structure of the SROCS – which (it is suggested) is applicable (once again) when we try to determine the physical relationship/s between these two potential quantum and relativistic 'D2' computational frameworks; Specifically, the Duality Principle proves that it is not possible (e.g., in *principle*) to maintain two such "independent" (conceptually higher-ordered) 'D2' computational frameworks; Rather, that there can only exist a *singular conceptually higher-ordered 'D2' computational framework* which coalesces the above mentioned quantum and relativistic 'D2' computational levels; Let's suppose there exist two "*separate*" such conceptually higher-ordered computational frameworks: 'D21' and 'D22' as underlying and constraining quantum and relativistic modeling (e.g., as proven above through the application of the Duality Principle to the two principle SROCS/SRONCS computational paradigms underlying current quantum and relativistic modeling). Then, according to the Duality Principle this would imply that in order to be able to determine any hypothetical physical relationship between quantum ['qi{1}'] and relativistic ['ri{2}'] entities or processes (i.e., that exist at the above mentioned hypothetical corresponding D21 quantum and D22 relativistic computational levels) – we would necessarily need a conceptually higher-ordered 'D3' that is (again in principle) irreducible to the lower-ordered D21('qi{1}') and D22('ri{2}') physical interactions at the D2 computational level. This is because otherwise, the determination of the "existence" or "non-existence" of any such hypothetical quantum or relativistic phenomena would be carried out at the same computational level ('D2'} as the direct physical interaction between these (hypothetical) quantum and relativistic entities (or processes), thereby precisely replicating the SROCS structure (that was shown constrained by the Duality Principle), thus:

$$\text{SROCS/D2: PR}['qi\{D21\}', ri\{D22\}']$$
$$\rightarrow ['(qi\{D21\} \text{ or } ri\{D22\}') \text{ or } ('\text{not } qi\{D21\}' \text{ or } '\text{not } ri\{D22\}') / D2.$$

But, since we already know that the Duality Principle proves the conceptual computational inability to carry out the conceptually higher-ordered computation at the same computational level (e.g., in this case termed: 'D2') as the direct physical interaction between any two given elements, then we are forced (once again) to conclude that there must be only *one singular conceptually higher-ordered D2* computational level underlying both quantum and relativistic SROCS models. Therefore, we are led to the (inevitable) conclusion whereby there may only exist *one* conceptually higher-ordered 'D2' computational framework which underlies (and constrains) both quantum and relativistic relationships.

A critical element arising from the computational Duality Principle is therefore the recognition that it is not possible (in principle) to determine (or compute) any quantum or relativistic relationships based on any 'direct' physical relationship, (at 'di2' or indirect physical relationship/s ('di3') , (as shown above) that may exist between any hypothetical differential relativistic observer and any hypothetical 'Phenomenon' or between any complimentary subatomic 'probe' measurement and the target's (assumed) probability 'wave-function'; Hence, the untenable SROCS/SRONCS computational structure evident in the case of attempting to determine the (direct or indirect) physical relationship/s between the conceptually higher-ordered 'D21' quantum and 'D22' relativistic computational frameworks once again points at the Duality Principle's conceptual computational constraint which can only allow for only a *singular conceptually higher-ordered 'D2' computational framework* – as underlying *both* quantum and relativistic phenomena (which constitutes the answer to the first theoretical question, above).

Next, we consider the second (above mentioned) theoretical question – i.e., provided that (according to the Duality Principle) there can only be a *singular* conceptually higher-ordered 'D2' computational framework as underlying both quantum and relativistic phenomena, what may be its computational characteristics? It is suggested that based on the recognition of the Duality Principle's singular conceptually higher-ordered 'D2' computational framework – which necessarily underlies both quantum and relativistic phenomena, it is also possible to answer the second (above mentioned) question regarding the computational characteristics of such higher-ordered (singular) 'D2' framework; Specifically, the Duality Principle's (above) proof indicates that rather than the existence of any direct (or indirect) 'materialistic-reductionistic' physical interaction between any hypothetical differential relativistic 'observer' and any corresponding 'Phenomena', or between any complimentary subatomic 'probe' element and probability wave function 'target' there exists a singular conceptually higher-ordered 'D2' computational framework which simply computes the "co-occurrences" of any of these quantum or relativistic (differential) 'observer/s' and corresponding 'Phenomenon' value/s or between any quantum subatomic 'probe' and 'target' elements...

Therefore, the singular conceptually higher-ordered 'D2' computational framework produces an *"a-causal"* computation which computes the 'co-occurrences' of any range of quantum 'probe-target' or relativistic 'observer-Phenomenon' pairs thus:

1. D2: {P('e-s' or 't-m'), T [e-s (i), t-m (i)]; ... P('e+n/s+n' or 't+n/m+n'), T [(e+n) (s+n), (t+n) (m+n)]} ≠ PR{[P('e-s' or 't-m'), T (e-s (i)), t-m (i))]; ... P('e+n/s+n' or 't+n/m+n'), T (e+n) (s+n), (t+n) (m+n)]}/di1

2. 'D2': {P [s-t $(i...n)$ e-m $(i...n)$, O-$r(st-i)$]; ... P[s-t $(i+n)$ e-m $(i+n$), O-$r(st-i+n)$]} ≠ PR{O-$diff$, P[s-t $(i...n)$, e-m $(i...n)$] }/di1.

The key point to be noted (within this context) is that such 'a-causal' computation negates or precludes the possibility of any "real" 'material-causal' interaction taking place at either quantum or relativistic levels! In other words, the Duality Principle's negation of the fundamental quantum or relativistic SROCS/SRONCS computational structure (e.g., as invariably leading to both 'logical inconsistency' and 'computational indeterminacy' that are contradicted by robust quantum and relativistic empirical data) also necessarily negates the existence of any (real) *causal-material* interaction between or within any quantum or relativistic phenomena – e.g., at the conceptually higher-ordered 'D2' computational level. In order to prove that the Duality Principle constraining the basic (materialistic-reductionistic) SROCS/SRONCS computational structure also necessarily points at the conceptual computational inability of such SROCS/SRONCS paradigms to determine the existence of any (real) 'causal-material' interactions (e.g., between any exhaustive series of x and y factors, interactions etc.) let us reexamine (once again) the SROCS/SRONCS working hypothesis wherein it *is* possible to determine whether a certain 'x' factor 'causes' the 'existence' or 'non-existence' of the particular 'y' factor:

Let's suppose it is possible for the SROCS/SRONCS direct physical (quantum or relativistic) interaction between the 'x' and 'y' (exhaustive series') factors to causally determine the 'existence; or 'non-existence' of the 'y' factor. In its most general formulation this would imply that:

$$\text{SROCS: PR}\{x,y\}/\text{di1} \rightarrow ['y' \text{ or 'not } y']/\text{di1}$$

But, as we've already seen (earlier), such SROCS computational structure invariably also contains the special case of a SRONCS of the form:

$$\text{SRONCS: PR}\{x,y\} \rightarrow \text{'not } y'/\text{di1}$$

However, this SRONCS structure inevitably leads to both 'logical inconsistency' and 'computational indeterminacy' which are contradicted by empirical findings (e.g., in the case of quantum and relativistic phenomena).

Therefore, the Duality Principle inconvertibly proves that the basic materialistic-reductionistic SROCS/SRONCS paradigmatic structure underlying the current quantum and relativistic theoretical models must be replaced by a conceptually higher-ordered (singular) 'D2' computation which cannot (in principle) contain any SROCS/SRONCS *causal-material* relationships – e.g., wherein any hypothetical 'y' element is "caused" by its direct (or indirect) physical interaction with another (exhaustive) $X\{1...n\}$ series. As pointed out (above), the only such possible conceptually higher-ordered 'D2' computation consists of an 'a-causal association' between pairs of D2: {('xi', yi)... ('xn', 'yn')}.

The essential point to be noted is that the Duality Principle thereby proves the conceptual computational unfeasibility of the currently assumed 'materialistic-reductionistic' SROCS/SRONCS structure – including the existence of any hypothetical 'causal-material' interaction between any exhaustive 'x' and 'y' series! This means that in both quantum and relativistic domains the determination of any hypothetical (exhaustive) spatial-temporal event or energy-mass object, or of any complimentary spatial-energetic or temporal subatomic target – there cannot (in principle) exist any 'causal-material' interaction between the relativistic event and any differential relativistic observer or between the subatomic probe and target elements... Instead, the Duality Principle proves that the only viable means for determining any such exhaustive hypothetical relativistic or quantum relationship is through the conceptually higher-ordered singular 'a-causal' D2 association of certain pairs of spatial-temporal or energy-mass values and corresponding relativistic observer frameworks or between pairs of subatomic probe and corresponding complimentary pairs of spatial-energetic or temporal-mass target values...

However, if indeed, the entire range of quantum and relativistic phenomena must necessarily be based upon a singular conceptually higher-ordered 'D2' computational level – which can only compute the "co-occurrences" of quantum 'probe-target' or relativistic 'observer-Phenomenon' pairs, but which precludes the possibility of any "real" 'material-causal' relationship/s existing between any such quantum ('probe-target') or relativistic ('observer-Phenomenon') pairs, then this necessitates a potential significant reformulation of both quantum and relativistic theoretical models based on the Duality Principle's asserted conceptually higher-ordered singular 'D2' 'a-causal' computational framework; This is because the current formulation of both quantum and relativistic theoretical frameworks is deeply anchored in- and dependent upon- precisely such direct (or indirect) physical interactions between a differential relativistic observer and any hypothetical (range of) 'Phenomenon' (e.g., as defined earlier), or between any subatomic (complimentary) 'probe' element and a probabilistically dispersed 'target' wave function. Thus, for instance, the entire theoretical structure of Relativity Theory rests upon the assumption that the differential physical measurements of different observers travelling at different speeds relative to any given object (or event) arises from a direct physical interaction between a (constant velocity) speed of light signal and the differentially mobilized observer/s... In contrast, the (novel) Duality Principle proves the conceptual computational inability to determine any such relativistic differential Phenomenon values – based on any direct or indirect physical interaction between any (hypothetical) differential relativistic observer and any given 'Phenomenon' (at their 'di1' or even 'di2' computational levels), but only from the conceptually higher-ordered 'D2' computational level through an 'a-causal' computation of the "co-occurrences" of any (differential) relativistic observer and (corresponding) Phenomenon! Hence, to the extent that we accept the Duality Principle's conceptual computational proof for the existence of a singular higher-ordered 'a-causal D2' computational framework – as underlying both quantum and relativistic theoretical models, then Relativity's well-validated empirical findings must be reformulated based on such higher-ordered 'D2 a-causal computation' framework...

Likewise, in the case of Quantum Mechanical theory it is suggested that the current formalization critically depends on the 'collapse' of the target 'wave-function' – upon its direct physical interaction with the (complimentary) probe element, which is contradicted by the (earlier demonstrated) Duality Principle's proof for the conceptual computational inability to determine any (complimentary) 'target' values based on its direct (or even indirect) physical interactions with another subatomic (complimentary) 'probe' element. Instead, the Duality Principle asserts that all quantum (complimentary) 'probe-target' values may only be computed 'a-causally' based on the conceptually higher-ordered 'D2' computation of the "co-occurrences" of any hypothetical 'probe-target' complimentary elements... Therefore, it becomes clear that both Quantum and Relativistic theoretical models have to be reformulated based on the Duality Principle's (proven) singular conceptually higher-ordered 'a-causal D2' computational framework. A key possible guiding principle in searching for such an alternative singular conceptually higher-ordered 'D2 a-causal' computational framework formulation of both quantum and relativistic (well-validated) empirical findings is Einstein's dictum regarding the fate of a "good theory" (Einstein, 1916) – which can become a special case in a broader more comprehensive framework. More specifically, based on the Duality Principle's (abovementioned) negation of the current existing quantum or relativistic theoretical interpretations of these well-validated empirical findings including: the quantum – 'probabilistic interpretation of the uncertainty principle' (and its corresponding probabilistic 'wave function'), 'particle-wave duality' 'quantum entanglement', and relativistic constancy of the speed of light (and corresponding speed limit on transfer of any object or signal), there seems to arise a growing need for an alternative reformulation of each and every one of these physical phenomena (e.g., separately and conjointly) – which may "fit in" within this singular (conceptually higher-ordered) 'D2 a-causal' computational framework; Indeed, what follows is a 'garland' of those quantum or relativistic empirical findings – reformulated based upon the Duality Principle – as fitting within a singular 'a-causal D2' computational mechanism; In fact, it is this assembly of Duality Principle's (motivated) theoretical reformulations of the (above) well-validated empirical dictums which will invariably lay down the foundations for the hypothetical 'Computational Unified Field Theory'. Fortunately (as we shall witness), this piecemeal work of the assembly of all quantum and relativistic Duality Principle's theoretically refomalized 'garlands' may not only lead to the discovery of such singular conceptually higher-ordered 'D2' Computational Unified Field Theory' (CUFT) , but may also resolve all known (apparent) theoretical contradictions between quantum and relativistic models (as well as predict yet unknown empirical phenomena, and possibly open new theoretical frontiers in Physics and beyond)...

3.1 Single- multiple- and exhaustive- spatial-temporal measurements

Perhaps a direct ramification of the above mentioned critical difference between empirical facts and theoretical interpretation which may have a direct impact on the current (apparent) schism between Relativity Theory and Quantum Mechanics is the distinction between *single- vs. multiple- spatial-temporal* empirical measurements and its corresponding "particle" vs. "wave" theoretical constructs; It is hypothesized that if we put aside (for the time being) the 'positivistic' vs. 'probabilistic' characteristics of Relativity theory and Quantum Mechanics then we may be able to characterize *both* relativistic and quantum empirical data as representing 'single'- vs. 'multiple'- spatial-temporal measurements; Thus, for instance, it is suggested that a (subatomic) "*particle*" or (indeed) any well-localized relativistic object (or event) can be characterized as indicating a '*single*' (localized) *spatial-temporal measurement* such that the given object or event is measured at a particular (single) spatial point $\{si\}$ at any given temporal point $\{ti\}$. In contrast, the "*wave*" characteristics of quantum mechanics represent a *multi spatial-temporal measurement* wherein there are at least *two* separate spatial-temporal measurements for each temporal point $\{si\ ti\ ,\ s(i+n)\ t(i+n)\}$.

Indeed, I hypothesize that precisely such a distinction between single- and multiple-spatial-temporal measurement (and conceptualization) may stand at the basis of some of the (apparent) quantum 'conundrums' such as the 'particle-wave duality', the 'double-slot experiment', and 'quantum entanglement'; Specifically, I suggest that if (indeed) the primary difference between the 'particle' and 'wave' characterization is *single-* vs. *multiple-* spatial-temporal measurements, then this can account for instance for the (apparently) "strange" empirical phenomena observed in the 'double-slot' experiment. This is because it may be the case wherein the opening of a *single* slot only allows for the measurement of a single spatial-temporal measurement at the interference detector surface (e.g., due to the fact that a single slot opening only allows for the measurement of the change in a single photon's impact on the screen). In contrast, opening two slots allows the interference detector surface to measure *two* spatial-temporal points simultaneously thereby revealing the 'wave' (interference) pattern. Moreover, I hypothesize that if indeed the key difference between the 'particle' and 'wave' characteristics is their respective single- vs. multiple- spatial-temporal measurements, then it may also be the case wherein any *"particle"* measurement (e.g., or for that matter also any single spatial-temporal *relativistic* measurements) is *embedded* within the broader *multi- spatial-temporal 'wave'* measurement... In this case, the current probabilistic interpretation of quantum mechanics (which has been challenged earlier by the Duality Principle) may give way to a *hierarchical-dualistic* computational interpretation which regards any 'particle' measurement as merely a localized (e.g., single spatial-temporal) segment of a broader multi spatial-temporal 'wave' measurement.

One further potentially significant computational step – e.g., beyond the 'single' spatial-temporal "particle" (or object) as potentially *embedded* within the 'multiple spatial-temporal "wave" measurement – may be to ask: is it possible for both the single spatial-temporal "particle" and the multi- spatial-temporal "wave" measurements to be embedded within a conceptually higher-ordered 'D2' computational framework?

This hypothetical question may be important as it may point the way towards a formal physical representation of the Duality Principle's asserted singular conceptually higher-ordered 'D2 a-causal computational framework': This is because the Duality Principle's assertion regarding the existence of a singular higher-ordered D2 'a-causal' computation can consist of *all single- multiple- or even the entire range of spatial pixels'*{si....sn} that exist at any point/s in time {ti ...ti} which are *computed as "co-occurring" pairs* of 'relativistic observer – Phenomenon' or pairs of subatomic 'probe – target' elements (e.g., as computed at this singular conceptually higher-ordered 'D2' computational level); This implies that since there cannot be any "real" 'material-causal' interactions between any of these relativistic 'observer-Phenomenon' or quantum 'probe-target' pairs, then all such hypothetical 'spatial pixels'{si....sn} occurring at any hypothetical temporal point/s {ti ...ti} must necessarily form an exhaustive 'pool' of the entire corpus of spatial-temporal points, which according to the Duality Principle must only exist as the above mentioned quantum (subatomic) 'probe-target' or relativistic (differential) 'observer-Phenomenon' computational pairs at the singular conceptually higher-ordered 'D2 A-Causal Computational Framework'.

3.2 The 'Universal Simultaneous Computational Frames' (USCF's)

Indeed, an additional empirical support for the existence of such (hypothetical) singular conceptually higher-ordered 'D2' exhaustive pool of all "co-occurring" quantum or relativistic pairs may be given by the well validated empirical phenomenon of 'quantum entanglement'; In a nutshell, 'quantum entanglement' refers to the finding whereby a subatomic measurement of one of two formerly connected "particles" – which may be separated (e.g. at the time of measurement) by a distance greater than a lights signal can travel can 'instantaneously' affect the measure outcome of the other (once interrelated) 'entangled' particle...

The reason that 'quantum entanglement' may further constrain the operation of higher-ordered hypothetical 'D2 A-Causal' computational framework is that it points at the existence of an empirical dictum which asserts that even in those computational instances in which two spatial-temporal events seem to be physically "separated" (e.g., by a distance greater than possibly travelled by Relativity's speed of light limit) the higher-ordered 'D2 A-Causal Computation' occurs 'instantaneously'! Therefore, this 'quantum entanglement' empirical dictum indicates that the 'D2 a-causal' computation of all spatial pixels in the universe – be carried out "at the same time", i.e., "simultaneously" at the D2 computational mechanism; In other words, the above mentioned 'D2 a-causal computation' mechanism must consist of the entirety of all possible quantum 'probe-target' or relativistic 'observer-Phenomenon' pairs occupying an exhaustive three-dimensional 'picture' of the entire corpus of all spatial pixels in the universe – for any given (minimal) 'time-point';

Therefore, if (indeed) due to the empirical-computational constraint imposed by 'quantum entanglement' we reach the conclusion wherein all spatial-pixels in the (subatomic as well relativistic) universe must necessarily exist "simultaneously" (e.g., for any minimal 'temporal point') at the 'D2 a-causal computation' level"; And based on the Duality Principle's earlier proven conceptual computational irreducibility of the determination of any quantum or relativistic relationship to within any direct or indirect *physical* interaction between any hypothetical subatomic 'probe' and 'target' elements or between any relativistic differential 'observer' and any 'Phenomenon' – but only from this singular higher-ordered 'D2 A-Causal Computational Framework';

It is hereby hypothesized that the 'D2 A-Causal Computational' processing consists of a series of *'Universal Simultaneous Computational Frames'* (USCF's) which comprise the entirety of the (quantum and relativistic) 'spatial-pixels' in the physical universe (i.e., at any given "minimal time-point")... Moreover, it is hypothesized that in order for this singular conceptually higher-ordered 'A-Causal Computational Framework' to produce all known quantum and relativistic physical phenomena there must necessarily exist a series of (extremely rapid) such 'Universal Simultaneous Computational Frames' (USCF's) that give rise to three distinct *'Computational Dimensions'* – which include: 'Computational *Framework*', 'Computational *Consistency*' and 'Computational *Locus*';

3.3 Computational- framework, consistency and locus

Based on the Duality Principle's asserted singular conceptually higher-ordered 'D2' computational framework comprising of an 'A-Causal-Computation' of a rapid series of 'Universal Simultaneous Computational Frames' (USCF's) it is hypothesized that three interrelated computational dimensions arise as different computational measures relating to – the *'Framework'* of computation (e.g., relating to the entire USCF/s *'frame/s'* or to a particular *'object'* within the USCF/s), the degree of *'Consistency'* across a series of USCF's (e.g., 'consistent' vs. 'inconsistent'), and the *'Locus'* of computational measure/s (e.g., whether the computation is carried out 'locally'- from within any particular 'reference system', or 'globally'- that is, externally to a particular reference system). It is further suggested that the combination of these three independent computational factors gives rise – not only to all relativistic and quantum basic physical features of 'space', 'time', 'energy', 'mass' etc. but may in fact exhaustively replicate, coalesce and harmonize all apparently existing theoretical contradictions between quantum and relativistic theories of physical reality...

First, the (four) basic physical features of physical reality are defined as the product of the interaction between the two Computational Dimensions of 'Framework' ('frame' vs. 'object') and 'Consistency' ('consistent' vs. 'inconsistent'): thus, for instance, it is hypothesized that a computational index of the degree of 'frame-consistent' presentations across a series of USCF's gives us a measure of the "spatial" value of any given object; In contrast, the computation of the degree of 'frame-inconsistent' measure/s of any given object – gives rise to the '"energy" value (of any measured object or event). Conversely, the computational measure of the degree of 'object-consistent' presentations (e.g., across a series of USCF's) produces the object's "mass" value. In contrast, the measure of an object's (or event's) 'object-inconsistent' presentations computes that object's/event's temporal value... A (partial) rational for these hypothetical computational measures may be derived from glancing at their computational "equivalences" – within the context of an analysis of the apparent physical features arising from the dynamics of a *cinematic* (two dimensional) film;

A quick review of the analogous cinematic measure of (any given object's) "spatial" or "energetic" value/s indicates that whereas a (stationary) object's 'spatial' measure or a measure of the 'spatial' distance a moving object traverses (e.g., across a certain number of cinematic film frames) depends on the number of pixels that object occupies "*consistently*", or the number of pixels that object travelled which remained constant (e.g., consistent) – across a given number of cinematic frames. Thus, the cinematic computation of 'spatial' distance/s is given through an analysis of the number of pixels (e.g., relative to the entire frame's reference system) that were either traversed by an object or which that object occupies (e.g., its "spatial" dimensions); In either case, the 'spatial value' (e.g., of the object's consistent dimensions or of its travelled distance) is computed based on the number of consistent pixels that object has travelled through or has occupied (across a series of cinematic frames); In contrast, an object's "energetic" value is computed through a measure of the number of pixels that object has 'displaced' across a series of frames – such that its "energy" value is measured (or computed) based on the number of pixels that object has displaced (e.g., across a certain number of series of cinematic frames). Thus, an object's 'energy' value can be computed as the number of 'inconsistent' pixels that object has displaced (across a series of frames)... Note that in both the cases of the 'spatial' value of an object or of its 'energetic' value, the computation can only be carried out with reference to the (entire- or certain segments of-) 'frame/s', since we have to ascertain the number of 'consistent' or 'inconsistent' pixels (e.g., relative to the reference system of the entire- or segments of- frame/s);

In contrast, it is suggested that the analogous cinematic measures of "mass" and "time" – involve a computation of the number of "*object*-related" (i.e., in contrast to the abovementioned "frame-related") "consistent" vs. "inconsistent" presentations; Thus, for instance, a special cinematic condition can be created in which any given object can be presented at- or below- or above- a certain 'psychophysical threshold' – i.e., such that the "appearance" or "disappearance" of any given object critically depends on the number of times that object is presented 'consistently' (across a certain series of cinematic frames); Such psychophysical-object cinematic condition necessarily produces a situation in which the number of consistent-object presentations (across a series of frames) determines whether or not that object will be perceived to "exist" or "not exist"; Indeed, a further extension of the same precise psychophysical cinematic scenario can produce a condition in which there is a direct correlation between the number of times an object is presented 'consistently' (across a given series of cinematic frames) and its perceived "mass": Thus, whereas the given object – would seem to "not exist" below a certain number of presentation (out of a given number of frames), and would begin to "exist" once its number of presentations exceeds the particular psychophysical threshold, then it follows that a further increase in the number of presentations (e.g., out of a given number of frames) will increase that object's perceived "mass"... Perhaps somewhat less 'intuitive' is the cinematic computational equivalence of "time" – which is computed as the number of 'object-related' "inconsistent" presentations; It's a well-known fact that when viewing a cinematic film, if the rate of projection is slowed down ("slow-motion") the sense of time is significantly 'slowed-down'... This is due to the fact that there is much less changes taking place relative to the object/s of interest that we are focusing on... Indeed, it is a scientifically validated fact that our perception of time depends (among other factors) on the number of stimuli being presented to us (within a given time-interval, called the 'filled-duration' illusion). Therefore, it is suggested that the cinematic computation (equivalence) of "time" is derived from the number of 'object-related *inconsistent*' presentations (across a given number of cinematic frames); the greater the number of object-related inconsistent presentations the more time has elapsed – i.e., as becomes apparent in the case of 'slow-motion' (e.g., in which the number of object-related inconsistent presentations are small and in which very little 'time' seems to elapse) as opposed to 'fast-motion' (e.g., in which the number of object-related inconsistent presentations is larger and subsequently a significant 'time' period seems to pass)...

Obviously, there are significant differences between the two dimensional cinematic metaphor and the hypothetical Computational Unified Field Theory's postulated rapid series of three-dimensional 'Universal Simultaneous Computational Frames' (USCF's); Thus, for instance, apart from the existence of *two-dimensional* vs. *three dimensional* frames, various factors such as: the (differing) rate of projection, the universal simultaneous computation (e.g., across the entire scope of the physical universe) and other factors (which will be delineated below). Nevertheless, utilizing at least certain (relevant) aspects of the cinematic film metaphor may still assist us in better understanding some the potential dynamics of the USCF's rapid series; Hence, it is suggested that we can perhaps learn from the (above mentioned) 'object' vs. 'frame' and 'consistent' vs. 'inconsistent' computational features characterizing the cinematic equivalents of "space" ('frame-consistent'), "energy" ('frame-inconsistent'), "mass" ('object-consistent') and "time" ('object-inconsistent') – with reference to the CUFT's hypothesized two (abovementioned) Computational Dimensions of 'Computational Framework' (e.g., 'frame' vs. 'object') and 'Computational Consistency' (e.g., 'consistent' vs. 'inconsistent'). The third (and final) hypothesized computational dimension of 'Computational Locus' does not correlate with the cinematic metaphor but can be understood when taking certain aspects of the cinematic metaphor and combining them with certain known features of Relativity theory; As outlined (earlier), this third 'Computational Locus' dimension refers to the particular frame of reference from which any of the two other Computational Dimensions (e.g., 'Framework' or 'Consistency') are being measured: Thus, for instance, it is suggested that the measurement of any of the abovementioned (four) basic physical features of 'space' ('frame-consistent'), 'energy' (frame-inconsistent'), 'mass' ('object-consistent') and 'time' ('object-inconsistent') – can be computed from *within* the 'local' frame of reference of the particular object (or observer) being measured, or from an external 'global' frame of reference (e.g., which is different than that of the particular object or observer).

4. The 'Computational Unified Field Theory' (CUFT)

Based on the abovementioned three basic postulates of the 'Duality Principle' (e.g., including the existence of a conceptually higher-ordered 'D2 A-Causal' Computational framework), the existence of a rapid series of 'Universal Simultaneous Computational Frames' (USCF's) and their accompanying three Computational Dimensions of – 'Framework' ('frame' vs. 'object'), 'Consistency' ('consistent' vs. 'inconsistent') and 'Locus' ('global' vs. 'local') a (novel) 'Computational Unified Field Theory' (CUFT) is hypothesized (and delineated);

First, in order to fully outline the theoretical framework of this (new) hypothetical CUFT let us try to closely follow the (abovementioned) 'cinematic-film' metaphor (e.g., while keeping in mind the earlier mentioned limitations of such an analogy in the more complicated three dimensional universal case of the CUFT): It is hypothesized that in the same manner that a cinematic film consists of a series of (rapid) 'still-frames' which produce an 'illusion' of objects (and phenomena) being displaced in 'space', 'time', possessing an apparent 'mass' and 'energy' values – the CUFT's hypothesized rapid series of 'Universal Simultaneous Computational Frames' (USCF's) gives rise to the apparent 'physical' features of 'space', 'time', 'energy' and 'mass'... It is further hypothesized that (following the cinematic-film analogy) the minimal (possible) degree of 'change' across any two (subsequent) 'Universal Simultaneous Computational Frames' (USCF's) is given by Planck's 'h' constant (e.g., for the various physical features of 'space', 'time', 'energy' or 'mass')... Likewise the maximal degree of (possible) change across two such (subsequent) USCF's may be given by: 'c^2'; Note that both of these (quantum and relativistic) computational constraints – arising from the 'mechanics' of the rapid (hypothetical) series of USCF's – exist as basic computational characteristics of the conceptually higher-ordered 'D2' (a-causal) computational framework, rather than exist as part of the 'di1' physical interaction (apparently) taking place within any (single or multiple) USCF's... Indeed, it is further hypothesized that a measure of the actual rate of presentation (or computation) of the series of USCF's may be given precisely through the product of these ('D2') computational constraints of the maximal degree of (inter-frame) change/s ('$c2$') divided by the minimal degree of (inter-frame) change/s ('h'): '$c2$'/'h'!

Specifically, the CUFT hypothesizes that the computational measures of 'space', 'energy', 'mass' and 'time' (and "causation") are derived based on an 'object' vs. 'frame' and 'consistent' vs. 'inconsistent' computational combinations;

Thus, it is hypothesized that the 'space' measure of an object (e.g., whether it is the spatial dimensions of an object or event of whether it relates to the spatial location of an object) is computed based on the number of '*frame-consistent*' (i.e., cross-USCF's constant points or "universal-pixels") which that object possesses across subsequent USCF's, divided by Planck's constant 'h' which is multiplied by the number of USCF's across which the object's spatial values have been measured.

$$S: (fi\{x,y,z\}[USCF(i)] + ... fj\{x,y,z\}[USCF(n)]) / h \times n\{USCF's\}$$

such that:

$$fj\{x,y,z\}[USCF(i)]) \leq fi\{x+(hxn),y+(hxn),z+(hxn)\}[USCF(i...n)]$$

where the 'space' measure of a given object (or event) is computed based on a *frame consistent* computation that adds the specific USCF's (x,y,z) localization across a series of USCF's [1...n] – which nevertheless do *not exceed* the threshold of Planck's constant per each ('n') number of frames (e.g., thereby providing the CUFT's definition of "space" as 'frame-consistent' USCF's measure).

Conversely, the 'energy' of an object (e.g., whether it is the spatial dimensions of an object or event or whether it relates to the spatial location of an object) is computed based on the *frame's differences* of a given object's location/s or size/s across a series of USCF's, divided by the speed of light 'c' multiplied by the number of USCF's across which the object's energy value has been measured:

$$E: (fj\{x,y,z\}[USCF(n)]) - (fi\{(x+n),(y+n),(z+n)\}[USCF(i...n)]) / c \times n\{USCF's\}$$

such that:

$$fj\{x,y,z\}[USCF(n)]) > (fi\{x+(hxn),y+(hxn),z+(hxn)\}[USCF(i...n)])$$

wherein the energetic value of a given object, event etc. is computed based on the subtraction of that object's "universal pixels" location/s across a series of USCF's, divided by the speed of light multiplied by the number of USCF's.

In contrast, the of 'mass' of an object is computed based on a measure of the number of times an *'object'* is presented *'consistently'* across a series of USCF's, divided by Planck's constant (e.g., representing the minimal degree of inter-frame's changes):

$$M: \sum[oj\{x,y,z\}[USCF(n)] = o(i...j\text{-}1)\{(x),(y),(z)\}\{USCF(i...n)\} / h \times n\{USCF's\}$$
$$\{USCF(1...n)\} / h \times n\{USCF's\}$$

where the measure of *'mass'* is computed based on a comparison of the number of instances in which an object's (or event's) 'universal-pixels' measures (e.g., along the three axes 'x', y' and 'z') is identical across a series of USCF's (e.g., $\sum oi\{x,y,z\}[USCF(n)] = oj\{(x+m),(y+m),(z+m)\}[USCF(1...n)])$, divided by Planck's constant. Again, the measure of 'mass' represents an *object-consistent* computational measure – e.g., regardless of any changes in that object's spatial (frame) position across these frames.

Finally, the 'time' measure is computed based on an *'object-inconsistent'* computation of the number of instances in which an 'object' (i.e., corresponding to only a particular segment of the entire USCF) changes across two subsequent USCF's (e.g., $\sum oi\{x,y,z\}[USCF(n)] \neq oj\{(x+m),(y+m),(z+m)\}[USCF(1...n)])$, divided by 'c':

$$T: \sum oj\{x,y,z\}[USCF(n)] \neq o(i...j\text{-}1)\{(x),(y),(z)\}[USCF(1...n)] / c \times n\{USCF's\}$$

such that:

$$T: \sum oi\{x,y,z\}[USCF(n)] - oj\{(x+m),(y+m),(z+m)\}[USCF(1...n)] \leq c \times n\{USCF's\}$$

Hence, the measure of 'time' represents a computational measure of the number of *'object-inconsistent'* presentations any given object (or event) possesses across subsequent USCF' (e.g., once again- regardless of any changes in that object's 'frame's' spatial position across this series of USCF's).

Interestingly, the concept of "causality" – when viewed from the perspective of the (above mentioned) 'D2 A-Causal Computation' (rapid) series of USCF's replaces the (apparent) 'di1' "material-causal" physical relationship/s between any given 'x' and 'y' objects, factors, or phenomenon – through the existence of *apparent* (quantum or relativistic) spatial-temporal or energy-mass relationships across a series of USCF's; Thus, for instance, according to the CUFT's higher-ordered 'D2 A-Causal Computation' theoretical interpretation (e.g., as well as based on the earlier outlined 'Duality Principle' proof) in both the relativistic (assumed SROCS) direct physical interaction ('di1') between any hypothetical (differential) relativistic observer and any (corresponding) spatial-temporal or energy-mass 'Phenomenon', and in the quantum (assumed SROCS) direct physical interaction ('di1') between any subatomic 'probe' particle and any possible 'target' element –there *does not exist* any 'direct' ('di1') *material-causal* relationship/s between the relativistic observer and (measured) Phenomenon, or between the quantum subatomic 'probe' and 'target' entities which results in the determination of the particular spatial-temporal value of any given Phenomenon (e.g., for a particular differential observer) or the 'collapse' of the (assumed) probability wave function which results in the selection of only one (complimentary) spatial-energetic or temporal-mass target value... Instead, according to the CUFT's stipulated conceptually higher-ordered singular (quantum and relativistic) D2 A-Causal Computational Framework these apparently 'material-causal' subatomic probe-target or relativistic differential observer-Phenomenon pair/s are in fact replaced by A hypothetical *'Universal Computational Principle'* ("ˈ") D2 A-Causal Computation of the 'co-occurrence' of a particular set of such relativistic 'observer-Phenomenon' or quantum subatomic 'probe-target' pairs (e.g., appearing across a series of USCF's!) Indeed, a thorough understanding of the CUFT's replacement of any (hypothetical quantum or relativistic) 'material-causal' relationship/s with the conceptually higher-ordered (singular) 'D2 A-Causal Computation ('ˈ'), which simply co-presents a series of particular relativistic 'observer-Phenomenon' or subatomic 'probe-target' pairs across the series of given USCF's may also open the door for a fuller appreciation of the lack of any (continuous) "physical" or "material" relativistic or quantum object's, event/s or phenomena etc. "in-between" USCF's frames – except for the (above mentioned) 'Universal Computational Principle' ('ˈ' - at 'D2'). In other words, when viewed from the perspective of the CUFT's conceptually higher-ordered (singular) 'ˈ' computational stance of the series of (rapid) USCF's all of the known quantum and relativistic phenomena (and laws) of 'space', 'time', energy', 'mass' and 'causality', 'space-time', 'energy-mass' equivalence, 'quantum entanglement', 'particle-wave duality', "collapse" of the 'probability wave function' etc. phenomena – are replaced by an 'a-causal' (D2) computational account (which will be explicated below);

4.1 The CUFT's replication of quantum & relativistic findings
As sown above, the Computational Unified Field Theory postulates that the various combinations of the 'Framework' and 'Consistency' computational dimensions produce the known 'physical' features of: 'space' ('frame-consistent'), 'energy' ('frame-inconsistent'), 'mass' ('object-consistent') and 'time' ('object-inconsistent'). The next step is to explicate the various possible relationships that exists between each of these four basic 'physical' features and the two levels of the third Computational Dimension of 'Locus' – e.g., 'global' vs. 'local': It is suggested that each of these four basic physical features can be measured either from

the computational framework of the entire USCF's perspective (e.g., a 'global' framework) or from the computational perspective of a particular segment of those USCF's (e.g., 'local' framework). Thus, for instance, the spatial features of any given object can be measured from the computational perspective of the (series of the) entire USCF's, or it can be measured from the computational perspective of only a segment of those USCF's – i.e., such as from the perspective of that object itself (or from the perspective of another object travelling alongside- or in some other specific relationship- to that object). In much the same manner all other (three) physical features of 'energy', 'mass' and 'time' (e.g., of any given object) can be measured from the 'global' computational perspective of the entire (series of) USCF's or from a 'local' computational perspective of only a particular USCF's segment (e.g., of that object's perspective or of another travelling frame of reference perspective).

One possible way of formalizing these two different 'global' vs. 'local' computational perspectives (e.g., for each of the four abovementioned basic physical features) is through attaching a 'global' {'g'} vs. 'local' {'l'} subscript to each of the two possible (e.g., 'global' vs. 'local') measurements of the four physical features. Thus, for instance, in the case of 'mass' the 'global' (computational) perspective measures the number of times that a given object has been presented consistently (i.e., unchanged)– when measured across the (entire) USCF's pixels (e.g., across a series of USCF's) ; In contrast, the 'local' computational perspective of 'mass' measures the number of times that a given object has been presented consistently (e.g., unchanged) when measured from within that object's frame of reference;

$$M(g): \sum[oj\{x,y,z\}(g) [USCF(n)] = o(i...j\text{-}1)\{(x),(y),(z)\} (g) \{USCF(i...n)\} \;/\; h \times n\{USCF's\}$$

such that

$$[oi\{x,y,z\}USCF(n)] - [oi\{(x+j),(y+j),(z+j)\}USCF(1...n)] \le n \times h[USCF(1...n)].$$

$$M(l): \sum[oj\{x,y,z\}(l) [USCF(n)] = o(i...j\text{-}1)\{(x),(y),(z)\} (l) \{USCF(i...n)\} \;/\; h \times n\{USCF's\}$$

such that

$$[oi\{x,y,z\}USCF(n)] - [oi\{(x+j),(y+j),(z+j)\}USCF(1...n)] \le n \times h[USCF(1...n)].$$

What is to be noted is that these (hypothesized) different measurements of the 'global' vs. local' computational perspectives – i.e., as measured externally to a particular object's pixels ('global') as opposed to only the pixels constituting the particular segment of the USCFs which comprises the given object (or frame of reference) may in fact replicate Relativity's known phenomenon of the increase in an object's mass associated with a (relativistic) increase in its velocity (e.g., as well as all other relativistic phenomena of the dilation in time, shrinkage of length etc.); This is due to the fact that the 'global' measurement of an object's mass critically depends on the *number of times* that object has been presented (consistently) across a series of USCF's: e.g., the greater the number of (consistent) presentations the higher its mass. However, since the computational measure of 'mass' is computed relative to Planck's ('h') constant (e.g., computed as a given object's number of consistent presentations across a specific number of USCF's frames); and since the spatial measure of any such object is contingent upon that object's consistent presentations (across the series of USCF's) such that the object does not differ ('spatially') across frames by more than the number of USCF's multiplied by Planck's constant; then it follows that the higher an object's energy – i.e., displacement of pixels across a series of USCF's, the greater number of pixels that object has travelled and also the greater number of times that object has been presented across the series of USCF's – which constitutes that object's 'global' mass measure! In other words, when an object's mass is measured from the 'global' perspective we obtain a measure of that object's (number of external) global pixels (reference) which increases as its relativistic velocity increases, thereby also increasing the number of times that object is presented (e.g., from the global perspective) hence increasing its globally measured 'mass' value. In contrast, when that object's mass is measured from the 'local' computational perspective – such 'local mass' measurement only takes into account the number of times that object has been presented (across a given series of USCF's) as measured from within that object's frame of reference; Therefore, even when an object increases its velocity – if we set to measure its mass from within its own frame of reference we will not be able to measure any increase in its measured 'mass' (e.g., since when measured from within its local frame of reference there is no change in the number of times that object has been presented across the series of USCF's)...

Likewise, it is hypothesized that if we apply the 'global' vs. 'local' computational measures to the physical features of 'space', 'energy' and 'time' we will also replicate the well-known relativistic findings of the shortening of an object's length (in the direction of its travelling), and the dilation of time (as measured by a 'global' observer): Thus, for instance, it is suggested that an application of the same 'global' computational perspective to the physical feature of 'space' brings about an inevitable shortening of its spatial length (e.g., in the direction of its travelling):

$$S(g): (fi\{x,y,z\}(g) \ [USCF(i)] + \dots fj\{x,y,z\}(g) \ [USCF(n)]) \ / \ h \times n\{USCF's\}$$

such that:

$$fj\{x,y,z\}(g) \ [USCF(i)]) \le fi\{x+(hxn),y+(hxn),z+(hxn)\}(g)[USCF(i\dots n)]$$

It is hypothesized that this is due to the global computational definition of an object's spatial dimensions which computes a given object's spatial (length) based on its consistent 'spatial' pixels (across a series of USCF's) – such that any changes in that object's spatial dimensions must not exceed Planck's ('h') spatial constant multiplied by the number of USCF's; This is because given such Planck's minimal 'spatial threshold' computational constraint – the faster a given relativistic object travels (e.g., from a global computational perspective) the less 'consistent' spatial 'pixels' that object possesses across frames which implies the shorter its spatial dimensions become (i.e., in the direction of its travelling); in contrast, measured from a 'local' computational perspective there is obviously no such "shrinkage" in an object's spatial dimensions – since based on such a 'local' perspective all of the spatial 'pixels' comprising a given object remain unchanged across the series of USCF's.

$$S \ \{'l'\}: (fi\{x,y,z\}\{'l'\} \ [USCF(i)] + \dots fj\{x,y,z\}\{'l'\} \ [USCF(n)]) \ / \ h \times n\{USCF's\}$$

such that:

$$fj\{x,y,z\}\{'l'\} \ [USCF(i)]) \le fi\{x+(hxn),y+(hxn),z+(hxn)\} \ \{'l'\} \ [USCF(i\dots n)]$$

Somewhat similar is the case of the 'global' computation of the physical feature of 'time' which is computed based on the number of measured changes in the object's spatial 'pixels' constitution (across frames):

$$Tg : \sum oi\{x,y,z\}[USCF(n)] \neq oj\{(x+m),(y+m),(z+m)\} [USCF(1...n)] /c \times n\{USCF's\},$$

such that:

$$T: \sum oi\{x,y,z\}[USCF(n)] - oj\{(x+m),(y+m),(z+m)\} [USCF(1...n)] \leq c \times n\{USCF's\}$$

The temporal value of an event (or object) is computed based on the number of times that a given object or event has changed – relative to the speed of light (e.g., across a certain number of USCF's); However, the *measurement* of temporal changes (e.g., taking place at an object or event) differ significantly – when computed from the 'global' or 'local' perspectives: This is because from a *'global'* perspective, the faster an object travels (e.g., relative to the speed of light) the less potential changes are exhibited in that object's or event's presentations (across the relevant series of USCF's). In contrast, from a 'local' perspective, there is no change in the number of measured changes in the given object (e.g., as its velocity increases relative to the speed of light) – since the local (computational) perspective does not encompass globally measured changes in the object's displacement (relative to the speed of light)…

Note also that we can begin appreciating the fact that from the CUFT's (D2 USCF's) computational perspective there seems to be inexorable (computational) interrelationships that exist between the eight computational products of the three postulated Computational Dimensions of 'Framework', 'Consistency' and 'Locus'; Thus, for instance, we find that an acceleration in an object's velocity increases the number of times that object is presented (e.g., 'globally' across a given number of USCF frames) – which in turn also increases it 'mass' (e.g., from the 'global Locus' computational perspective), and (inevitably) also decreases its (global) 'temporal' value (due to the decreased number of instances that that object changes across those given number of frames (e.g., globally- relative to the speed of light maximal change computational constraint)… Indeed, over and beyond the hypothesized capacity of the CUFT to replicate and account for all known relativistic and quantum empirical findings, its conceptually higher-ordered 'D2' USCF's emerging computational framework may point at the unification of all apparently "distinct" physical features of 'space', 'time', 'energy' and 'mass' (and 'causality') as well as a complete harmonization between the (apparently disparate) quantum (microscopic) and relativistic (macroscopic) phenomena and laws; the apparent disparity between quantum (microscopic) and relativistic (macroscopic) phenomena and laws;

Towards that end, we next consider the applicability of the CUFT to known quantum empirical findings: Specifically, we consider the CUFT's account of the quantum (computational) complimentary properties of 'space' and 'energy' or 'time' and 'mass'; of an alternative CUFT's account of the "collapse" of the probability wave function; and of the 'quantum entanglement' and 'particle-wave duality' subatomic phenomena; It is also hypothesized that these alternative CUFT's theoretical accounts may also pave the way for the (long-sought for) unification of quantum and relativistic models of physical reality. First, it is suggested that the quantum complimentary 'physical' features of 'space' and 'energy', 'time' and 'mass' – may be due to a *'computational exhaustiveness'* (or 'complimentarity') of each of the (two) levels of the Computational Dimension of 'Framework'. It is hypothesized that both the *'frame'* and *'object'* ('D2-USCF's') computational perspectives are *exhaustively* comprised of their 'consistent' (e.g., 'space' and 'energy', or 'mass' and 'time' physical features, respectively): Thus, whether we chose to examine the USCF's (D2) computation of a 'frame' – which is exhaustively comprised of its 'space' ('consistent') and 'energy' ('inconsistent') computational perspectives or if we chose to examine the 'object' perspective of the USCF's (D2) series – which is exhaustively comprised of its 'mass' ('consistent') and 'time' (inconsistent) computational aspects: in both cases the (D2) USCF's series is exhaustively comprised of these 'consistent' and 'inconsistent' computational aspects (e.g., of the 'frame' or 'object' perspectives)...

This means that the computational definitions of each of these pairs of 'frame': 'space' (consistent) and 'energy' (inconsistent) or 'object': 'mass' (consistent) or 'time' (inconsistent) is 'exhaustive' in its comprising of the USCF's Framework (i.e., 'frame' or 'object') Dimension:

Indeed, note that the computational definitions of 'space' and 'energy' exhaustively define the USCF's (D2) Framework computational perspective of a *'frame'*:

$$S: [fi\{x,y,z\}[USCF(n)] + fj\{x,y,z\}[USCF(1...n)])] / h \times n\{USCF's\},$$

such that:

$$fi\{x,y,z\}[USCF(n)]) \leq fj\{x+(hxn),y+(hxn),z+(hxn)\}[USCF(1...n)]);$$

and

$$E: (fi\{x,y,z\}[USCF(n)]) - (fj\{(x+m),(y+m),(z+m)\}[USCF(1...n)])/c \times n\{USCF's\}$$

such that:

$$fi\{x,y,z\}[USCF(n)]) > (fj\{x+(hxn),y+(hxn),z+(hxn)[USCF(n)])$$

Likewise, note that the computational definitions of 'mass' and 'time' exhaustively define the USCF's (D2) Framework computational perspective of an *'object'*:

$$M: \sum [oi\{x,y,z\}USCF(n)] = [oi\{(x+j),(y+j),(z+j)\} USCF(1...n)] / h \times n\{USCF's\}$$

such that

$$[oi\{x,y,z\}USCF(n)] - [oi\{(x+j),(y+j),(z+j)\}USCF(1...n)] \leq n \times h[USCF(1...n)].$$

and

$$T: \sum oi\{x,y,z\}[USCF(n)] \neq oj\{(x+m),(y+m),(z+m)\} [USCF(1...n)] /c \times n\{USCF's\}$$

such that:

$$T: \sum oi\{x,y,z\}[USCF(n)] - oj\{(x+m),(y+m),(z+m)\} [USCF(1...n)] \leq c \times n\{USCF's\}$$

Thus, it is hypothesized that it is the *computational exhaustiveness* of the Framework Computational Dimension's (two) levels (e.g., of 'frame' or 'object' perspectives) which gives rise to the known quantum complimentary 'physical' features of 'space' and 'energy' (e.g., the *frame's* 'consistent' and 'inconsistent' perspectives) or of 'mass' and 'time' (e.g., the *object's* 'consistent' and 'inconsistent' perspectives). However, since this hypothetical 'computational exhaustiveness' of the Framework Dimension's (two) levels arises as an integral part of the USCF's (D2) Universal Computational Principle's operation – it manifests through both the (above mentioned) computational definitions of 'space' and 'energy, 'mass' and 'time', as well as through a singular 'Universal Computational Formula', postulated below:

4.2 The 'Universal Computational Formula'
Based on the abovementioned three basic postulates of the 'Duality Principle' (e.g., including the existence of a conceptually higher-ordered 'D2 A-Causal' Computational framework), the existence of a rapid series of 'Universal Simultaneous Computational Frames' (USCF's - e.g., which are postulated to be computed at an incredible rate of 'c^2'/ 'h') and their accompanying three Computational Dimensions of - 'Framework' ('frame' vs. 'object'), 'Consistency' ('consistent' vs. 'inconsistent') and 'Locus' ('global' vs. 'local') a singular 'Universal Computational Formula' is postulated which may underlie all (known) quantum and relativistic phenomena:

$$\frac{c^2 x'}{h} = \frac{s}{t} \times \frac{e}{m}$$

wherein the left side of this singular hypothetical Universal Computational Formula represents the (abovementioned) universal rate of computation by the hypothetical Universal Computational Principle, whereas the right side of this Universal Computational Formula represents the 'integrative-complimentary' relationships between the four basic physical features of 'space' (s), 'time' (t), 'energy' (e) and 'mass' (m), e.g., as comprising different computational combinations of the three (abovementioned) Computational Dimensions of 'Framework', 'Consistency' and 'Locus';

Note that on both sides of this Universal Computational Formula there is a coalescing of the basic quantum and relativistic computational elements – such that the rate of Universal Computation is given by the product of the maximal degree of (inter-USCF's relativistic) change 'c^2' divided by the minimal degree of (inter-USCF's quantum) change 'h'; Likewise, the right side of this Universal Computational Formula meshes together both quantum and relativistic computational relationships – such that it combines between the relativistic products of space and time (s/t) and energy-mass (e/m) together with the quantum (computational) complimentary relationship between 'space' and 'energy', and 'time' and 'mass';

More specifically, this hypothetical Universal Computational Formula fully integrates between two sets of (quantum and relativistic) computations which can be expressed through two of its derivations:

$$\frac{s}{t} = \frac{m}{e} \times \left(\frac{c^2 x'}{h} \right)$$

$$t \times m \times \left(\frac{c^2 x'}{h} \right) = s \times e$$

The first amongst these equations indicates that there is a computational equivalence between the (relativistic) relationships of 'space and time' and 'energy and mass'; specifically, that the computational ratio of 'space' (e.g., which according to the CUFT is a measure of the 'frame-consistent' feature) and 'time' (e.g., which is a measure of the 'object-inconsistent' feature) is equivalent to the computational ratio of 'mass' (e.g., a measure of the 'object-consistent' feature) and 'energy' (e.g., 'frame-inconsistent' feature)... Interestingly, this (first) derivation of the CUFT's Universal Computational Formula incorporates (and broadens) key (known) relativistic laws – such as (for instance) the 'E=Mc²' equation, as well as the basic concepts of 'space-time' and its curvature by the 'mass' of an object (which in turn also affects that object's movement – i.e. 'energy').

The second equation explicates the (above mentioned) quantum 'computational exhaustiveness' (or 'complimentary') of the Computational Framework Dimension's two levels of *'frame'*: 'space' ('consistent') and 'energy' ('inconsistent') and of *'object'*: 'mass' ('consistent') and 'time' ('inconsistent') 'physical' features, as part of the singular integrated (quantum and relativistic) Universal Computational Formula...

5. Unification of quantum and relativistic models of physical reality

Thus, the three (abovementioned) postulates of the 'Duality Principle', the existence of a rapid series of 'Universal Simultaneous Computational Frames' (USCF's – computed by the 'Universal Computational Principle' {'γ'} at the incredible hypothetical rate of 'c²/h'), and the three Computational Dimensions of 'Framework', 'Consistency' and 'Locus' have resulted in the formulation of the (hypothetical) new 'Universal Computational Formula':

$$\frac{c^2 x'}{h} = \frac{s \times e}{t \, m}$$

It is (finally) suggested that this (novel) CUFT and (embedded) Universal Computational Formula can offer a satisfactory harmonization of the existing quantum and relativistic models of physical reality, e.g., precisely through their integration within the (above) broader higher-ordered singular 'D2' Universal Computational Formula;

In a nutshell, it is suggested that this Universal Computational Formula embodies the singular higher-ordered 'D2' series of (rapid) USCF's, thereby integrating quantum and relativistic effects (laws and phenomena) and resolving any apparent 'discrepancies' or 'incongruities' between these two apparently distinct theoretical models of physical reality:

Therefore, it is suggested that the three (above mentioned apparent) principle differences between quantum and relativistic theories, namely: *'probabilistic' vs. 'positivistic'* models of physical reality, *'simultaneous-entanglement' vs. 'non-simultaneous causality'* and *'single-' vs. 'multiple-' spatial-temporal modeling* can be explained (in a satisfactory manner) based on the new (hypothetical) CUFT model (represented by the Universal Computational Formula);

As suggested earlier, the apparent 'probabilistic' characteristics of quantum mechanics, e.g., wherein an (apparent) multi spatial-temporal "probability wave function" 'collapses' upon its assumed 'SROCS' direct ('di1') physical interaction with another 'probe' element is replaced by the CUFT's hypothesized (singular) conceptually higher-ordered 'D2's' rapid series of USCF's (e.g., governed by the above Universal Computational Formula); Specifically, the Duality Principle's conceptual proof for the principle inability of the SROCS computational structure to compute the "collapse" of (an assumed) "probability wave function" ('target' element) based on its direct physical interaction (at 'di1') with another 'probe' measuring element has led to a reformalization of the various subatomic quantum effects, including: the "collapse" of the "probability wave function", the "particle-wave duality", the "Uncertainty Principle's" computational complimentary features, and "quantum entanglement" as arising from the (singular higher-ordered 'D2') rapid USCF's series:

Thus, instead of Quantum theory's (currently assumed) "collapse" of the 'probability wave function', the CUFT posits that there exists a rapid series of 'Universal Simultaneous Computational Frames' (USCF's) that can be looked at from a 'single' spatial-temporal perspective (e.g., subatomic 'particle' or relativistic well localized 'object' or 'event') or from a 'multiple' spatial-temporal perspective (e.g., subatomic 'wave' measurement or conceptualization). Moreover, the CUFT hypothesizes that both the subatomic 'single spatial-temporal' "particle" and 'multiple spatial-temporal' "wave" measurements are embedded within an exhaustive series of 'Universal Computational Simultaneous Frames' (USCF's) (e.g., that are governed by the above mentioned Universal Computational Formula). In this way, it is suggested that the CUFT is able to resolve all three abovementioned (apparent) conceptual differences between quantum and relativistic models of the physical reality: This is because instead of the 'collapse' of the assumed 'quantum probability wave function' through its (SROCS based) direct physical interaction with another subatomic probe element, the CUFT posits the existence of the rapid series of USCF's that can give rise to 'single-spatial temporal' (subatomic "particle" or relativistic 'object' or 'event') or to 'multiple spatial-temporal' (subatomic or relativistic) "wave" phenomenon; Hence, instead of the current "probabilistic-quantum" vs. "positivistic-relativistic" (apparently disparate) theoretical models, the CUFT coalesces both quantum and relativistic theoretical models as constituting integral elements within a singular rapid series of USCF's. Thereby, the CUFT can explain all of the (apparently incongruent) quantum and relativistic phenomena (and laws) such as for instance, the (abovementioned) 'particle' vs. 'wave' and 'quantum entanglement' phenomena – e.g., which is essentially a representation of the fact that all single- multiple- (or exhaustive) measurements are embedded within the series of 'Universal *Simultaneous* Computational Frames' (USCF's) and therefore that two apparently "distinct" 'single spatial-temporal' measured "particles" that are embedded within the 'multiple spatial-temporal' "wave" measurement necessarily constitute integral parts of the same singular simultaneous USCF's (which therefore give rise to the apparent 'quantum entanglement' phenomenon). Nevertheless, due to the above mentioned 'computational exhaustiveness' (or 'complimentarity') the computation of such apparently 'distinct' "particles" embedded within the same "wave" and USCF's (series) leads to the known quantum ('uncertainty principle's') complimentary computational (e.g., simultaneous) constraints applying to the measurement of 'space' and 'energy' (e.g., 'frame': consistent vs. inconsistent features), or of 'mass' ad 'time' (e.g., 'object': consistent vs. inconsistent features). Such USCF's based theoretical account for the empirically validated "quantum entanglement" natural phenomena is also capable of resolving the apparent contradictions that seems to exist between such "simultaneous action at a distance" (to quote Einstein's famous objection) and Relativity's constraint set upon the transmission of any signal at a velocity that exceeds the speed of light: this is due to the fact that while the CUFT postulates that the "entangled particles" are computed simultaneously (along with the entire physical universe) as part of the same USCF/s (e.g., and more specifically of the same multi spatial-temporal "wave" pattern).

Another important aspect of this (hypothetical) Universal Computational Formula's representation of the CUFT may be its capacity to replicate Relativity's curvature of 'space-time' based on the existence of certain massive objects (which in turn also affects their own space-time pathway etc.): Interestingly, the CUFT points at the existence of USCF's regions that may constitute: "high-space, high-time; high-mass, low-energy" vs. other regions which may be characterized as: "low-space, low-time; low-mass, high-energy" based on the computational features embedded within the CUFT (and its representation by the above Universal Computational Formula). This is based on the Universal Computational Formula's (integrated) representation of the CUFT's basic computational definitions 'space', 'time', 'energy' and 'mass' as:

$$\frac{s}{t} = \frac{m}{e} \times \left(\frac{c^2 x'}{h} \right)$$

which represents: 'space' – as the number of (accumulated) USCF's 'consistent-frame' pixels that any given object occupies and its (converse) computational definition of 'time' as the number of 'inconsistent-object' pixels; and likewise the computational definition of 'mass' – as the number of 'consistent-object' USCF's pixels and of 'energy' – as the (computational) definition of 'mass' as the number of 'inconsistent-frame' USCF's pixels.

Hence, General Relativity may represent a 'special case' embedded within the CUFT's Universal Computational Formula integrated relationships between 'space', 'time', 'energy' and 'mass' (computational definitions): This is because General Relativity describes the specific dynamics between the "mass" of relativistic objects (e.g., a 'global-object-consistent' computational measure), their curvature of "space-time" (i.e., based on an 'frame-consistent' vs. 'object-inconsistent' computational measures) and its relationship to the 'energy-mass' equivalence (e.g., reflecting a 'frame-inconsistent' – 'object-consistent' computational measures); This is because from the (above mentioned) 'global' computational measurement perspective there seems to exist those USCF's regions which are displaced significantly across frames (e.g., possess a high 'global-inconsistent-frame' energy value) – and therefore also exhibit increased 'global-object-consistent' mass value, and moreover are necessarily characterized by their (apparent) curvature of 'space-time' (i.e., alteration of the 'global-frame-consistent' space values and associated 'global-object-inconsistent' time values)…

Therefore, in the special CUFT's case described by General Relativity we obtain those "massive" objects, i.e., which arise from high 'global-frame-inconsistent' energy values (e.g., which are therefore presented many times consistently across frames – yielding a high 'global-object-consistent' mass value); These objects also produce low (dilated) global temporal values since the high 'global-object-consistent' (mass) value is inevitably linked with a low 'global-object-inconsistent' (time) value; Finally, such a high 'global-frame-inconsistent' (energy) object also invariably produces low 'global-frame-consistent' spatial measures (e.g., in the vicinity of such 'high-energy-high-mass' object). Thus, it may be the case that General Relativity's described mechanical dynamics between the mass of objects and their curvature of 'space-time' (which interacts with these objects' charted space-time pathway) represents a particular instance embedded within the more comprehensive (CUFT) Universal Computational Formula's outline of a (singular) USCF's-series based D2 computation (e.g., comprising the three above mentioned 'Framework', 'Consistency' and 'Locus' Computational Dimensions) of the four basic 'physical' features of 'space', 'time', 'energy' and 'mass' interrelationships (e.g., as 'secondary' emerging computational products of this singular Universal Computational Formula driven process)…

Indeed, the CUFT's hypothesized rapid series of USCF's (governed by the above mentioned 'Universal Computational Formula') integrates (perfectly) between the essential quantum complimentary features of 'space and energy' or 'time and mass' (e.g., which arises as a result of the abovementioned 'computational exhaustiveness' of each of the Computational Framework Dimension's 'frame' and 'object' levels, which was represented earlier by one of the derivations of the Universal Computational Formula); "quantum entanglement", the "uncertainty principle" and the "particle-wave duality" (e.g., which arises from the existence of the postulated 'Universal Simultaneous Computational Frames' [USCF's] that compute the entire spectrum of the physical universe simultaneously per each given USCF and which embed within each of these USCF's any 'single- spatial-temporal' measurements of "entangled particles" as constituting integral parts of a 'multiple spatial-temporal' "wave" patterns); Quantum mechanics' minimal degree of physical change represented by Planck's 'h' constant (e.g., which signifies the CUFT's 'minimal degree of inter-USCF's change' for all four 'physical' features of 'space', 'time', 'energy' and 'mass'); As well as the relativistic well validated physical laws and phenomena of the "equivalence of energy and mass" (e.g., the famous "E= Mc^2" which arises as a result of the transformation of any given object's or event's 'frame-inconsistent' to 'object-consistent' computational measures based on the maximal degree of change, but which also involves the more comprehensive and integrated Universal Computational Formula derivation: t x m x (c^2/h x ') = s x e .); Relativity's 'space-time' and 'energy-mass' relationships expressed in terms of their constitution of an integrated singular USCF's series which is given through an alternate derivation of the same Universal Computational Formula:

$$\frac{s}{t} = \frac{m \; x}{e} \left(\frac{c^2 x'}{h} \right)$$

Indeed, this last derivation of the Universal Computational Formula seems to encapsulate General Relativity's proven dynamic relationships that exist between the curvature of space-time by mass and its effect on the space-time pathways of any such (massive) object/s – through the complete integration of all four physical features within a singular (conceptually higher-ordered 'D2') USCF's series... Specifically, this (last) derivation of the (abovementioned) Universal Computational Formula seems to integrate between 'space-time' – i.e., as a ratio of a 'frame-consistent' computational measure divided by 'object-inconsistent' computational measure – as equal to the computational ratio that exists between 'mass' (e.g., 'object-consistent') divided by 'energy' (e.g., 'frame-inconsistent') multiplied by the Rate of Universal Computation (R = c^2/h) and multiplied by the Universal Computational Principle's operation (''); Thus, the CUFT's (represented by the above Universal Computational Formula) may supply us with an elegant, comprehensive and fully integrated account of the four basic 'physical' features constituting the physical universe (e.g., or indeed any set of computational object/s, event/s or phenomena etc.):

Therefore, also the Universal Computational Formula's full integration of Relativity's maximal degree of inter-USCF's change (e.g., represented as: 'c^2') together with Quantum's minimal degree of inter-USCF's change (e.g., represented by: Planck's constant 'h') produces the 'Rate' {R} of such rapid series of USCF's as: R = c^2/h, which is computed by the Universal Computational Principle '' and gives rise to all four 'physical' features of 'space', 'time', 'energy' and 'mass' as integral aspects of the same rapid USCF's universal computational process.

Thus, we can see that the discovery of the hypothetical Computational Unified Field Theory's (CUFT's) rapid series of USCF's fully integrates between hitherto validated quantum and relativistic empirical phenomena and natural laws, while resolving all of their apparent contradictions.

6. CUFT: Theoretical ramifications

Several important theoretical ramifications may follow from the CUFT; First, the CUFT's (novel) definition of 'space', 'time', 'energy' and 'mass' – as emerging computational properties which arise as a result of different combinations of the three Computational Dimensions (e.g., of 'Framework', 'Consistency' and 'Locus') transform these apparently "physical" properties into (secondary) 'computational properties' of a D2 series of USCF's... This means that instead of 'space', 'energy', 'mass' and 'time' existing as "independent - physical" properties in the universe they arise as *secondary integrated computational properties* (e.g., 'object'/'frame' x 'consistent'/'inconsistent' x 'global'/'local') of a singular conceptually higher-ordered 'D2' computed USCF's series...

Second, such CUFT's delineation of the USCF's arising (secondary) computational features of 'space', 'time', 'energy' and 'mass' is also based on one of the (three) postulates of the CUFT, namely: the 'Duality Principle', i.e., recognizing the computational constraint set upon the determination of any "causal-physical" relationship between any two (hypothetical) interacting 'x' and 'y' elements (at any direct 'di1' or indirect '...din' computational level/s), but instead asserting only the existence of a conceptually higher-ordered 'D2' computational level which can compute only the "co-occurrences" of any two or more hypothetical spatial-temporal events or phenomena etc. This means that the CUFT's hypothesized 'D2' computation of a series of (extremely rapid) USCF's does not leave any room for the existence of any (direct or indirect) "causal-physical" 'x→y' relationship/s. Instead, the hypothesized D2 A-Causal Computation calls for the computation of the co-occurrences of certain related phenomena, factors or events – but which lack any "real" 'causal-physical' relationship/s (phenomena or laws)...

Third, the Duality Principle's above mentioned necessity to replace any (direct or indirect) causal-physical relationship (or scientific paradigm), e.g., "x→y" by the CUFT's hypothesized D2 A-Causal Computation of the "co-occurrence" of particular spatial-temporal factors, events, phenomena etc. that constitute certain 'spatial-pixels' within a series of USCF's may have significant theoretical ramifications for several other key scientific paradigms (across the different scientific disciplines); Specifically, it is suggested that perhaps an application of the Duality Principle's identified- and constrained- SROCS computational structure (e.g., of the general form: PR{x,y}/di1→['y' or 'not y']/di1) towards key existing scientific paradigms such as: 'Darwin's Natural Selection Principle', 'Gödel's Incompleteness Theorem' (e.g., and Hilbert's failed 'Mathematical Program'), Neuroscience's (currently assumed) 'materialistic-reductionistic' working hypothesis etc. may open the door for a potential reformalization of these scientific paradigms in a way that is compatible with the novel computational Duality Principle and its ensued CUFT.

Hence, to the extent that the hypothesized CUFT may replicate (adequately) all known quantum and relativistic empirical phenomena and moreover offer a satisfactory (conceptually higher-ordered 'D2') USCF's series based computational framework that may harmonize- and bridge the gap- between quantum and relativistic models of physical reality, the CUFT may constitute a potential candidate to integrate (and replace) both quantum and relativistic theoretical models; However, in order for such (potentially) serious theoretical consideration to occur, the next required step will be to identify those particular (empirical) instances in which the CUFT's predictions may differ (significantly) from those of quantum mechanics or Relativity theory.

7. Conclusion

In order to address the principle contradiction that exists between quantum mechanics and Relativity Theory (e.g., comprising of: Probabilistic vs. deterministic models of physical reality, *"Simultaneous-entanglement"* vs. *"non-simultaneous-causality"* features and Single vs. multiple spatial-temporal modeling) a computational-empirical analysis of a fundamental 'Self-Referential Ontological Computational System' (SROCS) structure underlying both theories was undertaken; It was suggested that underlying both quantum and relativistic modeling of physical reality there is a mutual 'SROCS' which assumes that it is possible to determine the 'existence' or 'non-existence' of a certain 'y' factor solely based on its *direct physical interaction* (PR{x,y}/di1) with another 'x' factor (e.g., at the same 'di1' computational level), thus:

$$\text{SROCS: PR}\{x,y\}/\text{di1} \rightarrow [\text{'}y\text{' or 'not y'}]\text{'}/\text{di1}.$$

In the case of Relativity theory, such basic SROCS computational structure pertains to the computation of any spatial-temporal or energy-mass value/s of any given event (or object) – based (solely) on its *direct physical interaction* with any hypothetical (differential) relativistic observer:

$$\text{SROCS: PR}\{O\text{-}diff, \text{P[s-t } (i...n), \text{ e-m } (i...n)] \}/\text{di1}$$
$$\rightarrow \{\text{'P[s-t } (i), \text{ e-m } (i)]\text{' or 'not P[s-t } (i), \text{ e-m } (i)]'\}/\text{di1}$$

In the case of quantum mechanics, it is hypothesized that precisely the same SROCS/SRONCS computational structure may pertain to the quantum mechanical computation of the physical properties of any given subatomic 'target' (e.g., assumed to be dispersed all along a probability wave function) that is hypothesized to be determined solely through its direct physical interaction with another subatomic complimentary 'probe' entity, thus:

$$\text{SROCS: PR}\{P(\text{'s-e' or 't-m'}), t \text{ [s-e } (i...n), \text{ t-m } (i...n)] \}/\text{di1}$$
$$\rightarrow [\text{'}t \text{ [s-e } (i), \text{ t-m } (i)]\text{' or 'not } t \text{ [s-e } (i), \text{ t-m } (i)]'}/\text{di1}$$

However, the computational-empirical analysis indicated that such SROCS computational structure (which underlies both quantum and relativistic paradigms) inevitably leads to both *'logical inconsistency'* and ensuing *'computational indeterminacy'* (i.e., an apparent inability of these quantum or relativistic SROCS systems to determine weather a particular spatial-temporal or energy-mass 'Phenomenon' or a particular spatial-energetic or temporal-mass target value "exists" or "doesn't exist"). But, since there exists ample empirical evidence indicating the capacity of these quantum or relativistic computational systems to determine the "existence" or "non-existence" of any particular relativistic 'Phenomenon' or quantum complimentary target value, then a novel computational 'Duality Principle' asserts that the currently assumed SROCS computational structure is invalid; Instead, the Duality Principle points at the existence of a conceptually higher-ordered ('D2') *"a-causal"* computational framework which computes the "co-occurrences" of any range of quantum 'probe-target' or relativistic 'observer-Phenomenon' pairs thus:

1.   D2: {P('e-s' or 't-m'), T [e-s *(i)*, t-m *(i)*]; ... P('e+n/s+n' or 't+n/m+n'), T [(e+n) (s+n), (t+n) (m+n)]} ≠ PR{[P('e-s' or 't-m'), T (e-s *(i)*), t-m *(i)*)]; ... P('e+n/s+n' or 't+n/m+n'), T (e+n) (s+n), (t+n) (m+n)]}/di1

2.   'D2': {P [s-t *(i...n)* e-m *(i...n)*, O-r(st-i)]; ... P[s-t *(i+n)* e-m *(i+n)*, O-r(st-i+n)]} ≠ PR{O-*diff*, P[s-t *(i...n)*, e-m *(i...n)*] }/di1.

Indeed, a further application of this (new) hypothetical computational Duality Principle indicated that there cannot exist "multiple D2" computational levels but rather only one singular 'conceptually higher-ordered 'D2' computational framework as underlying both quantum and relativistic (abovementioned) 'co-occurring' phenomena.

Next, an examination of the potential characteristics of such conceptually higher-ordered (singular) 'a-causal D2' computational framework indicated that it may embody 'single'- 'multiple'- and 'exhaustive' spatial-temporal measurements as embedding all hypothetical 'probe-target' subatomic pairs or all hypothetical (differential) observer/s – 'Phenomenon' pairs; It was suggested that such D2 (singular 'a-causal') arrangement of all hypothetical quantum 'probe-target' or relativistic 'observer-Phenomenon' pairs may give rise to all known *single* spatial-temporal (quantum) *"particle"* or (relativistic) *"object"* or *"event"* measurements or all *multiple* spatial-temporal *"wave"* measurements. Moreover, when we broaden our computational analysis beyond the scope of such 'single-' or 'multiple' spatial-temporal measurements (or conceptualizations) to the entire corpus of all hypothetical possible spatial-temporal points- e.g., as 'co-occurring' at the Duality Principle's asserted conceptually higher-ordered 'D2' computational framework, then this may point at the existence of a series of *'Universal Simultaneous Computational Frames'* (USCF's). The existence of such (a series of) hypothetical conceptually higher-ordered 'D2' series of USCF's which constitute the entirety of all hypothetical (relativistic) spatial-temporal or energy-mass phenomena and all hypothetical (quantum complimentary) spatial-energetic or temporal-mass "pixels" was suggested by the well-validated empirical phenomenon of 'quantum entanglement' (e.g., relating to a 'computational linkage' between 'greater than light-speed travelling distance' of two spatial-temporal "entangled particles"); This is because based on the fact that two such disparate 'entangled' quantum "particles" (e.g., which could hypothetically comprise a probability wave function that can span tremendous cosmic distances) we may infer that the entirety of all (hypothetical) cosmic quantum (complimentary) 'probe-target' pairs or all (hypothetical) relativistic 'observer-Phenomenon' pairs may be computed as "co-occurring" simultaneously as part of such (hypothetical) 'D2' 'Universal Simultaneous Computational Frames' (USCF's).

This hypothetical (rapid series of) 'Universal Simultaneous Computational Frames' (USCF'S) was further stipulated to possess three basic (interrelated) 'Computational Dimensions' which include: Computational *'Framework'* (e.g., relating to the entire USCF/s *'frame/s'* or to a particular *'object'* within the USCF/s), Computational *'Consistency'* (which refers to the degree of 'consistency' of an object or of segments of the frame across a series of USCF's (e.g., 'consistent' vs. 'inconsistent'), and Computational *'Locus'* of (e.g., whether the computation is carried out 'locally'- from within any particular object or 'reference system', or 'globally'-

i.e., externally to a particular reference system from the perspective of the entire frame or segments of the frame). Interestingly (partially) by using a 'cinematic film metaphor' it was possible to derive and formalize each of the four basic physical features of 'space', 'time', 'energy' and 'mass' as emerging (secondary) computational properties arising from the singular 'D2' computation of a series of USCF's – through a combination of the two Computational Dimensions of 'Framework' and 'Consistency': Thus, a combination of the 'object' level (e.g., within the 'Framework' Dimension) with the 'consistent' vs. 'inconsistent' levels (of the 'Consistency' Dimension) produced the physical properties of 'mass' and 'time' (correspondingly); On the other hand, a combination of the 'frame' level (within the Framework Dimension) and the 'consistent' vs. 'inconsistent' ('Consistency' Dimension) yielded the two other basic physical features of 'space' and 'energy'. It was further hypothesized that (following the cinematic-film analogy) the minimal (possible) degree of 'change' across any two (subsequent) 'Universal Simultaneous Computational Frames' (USCF's) is given by Planck's 'h' constant (e.g., for the various physical features of 'space', 'time', 'energy' or 'mass'), whereas the maximal (possible) degree of change across two such (subsequent) USCF's is be given by: 'c^2'; Finally, the 'rate' at which the series of USCF's may be computed (or presented) was hypothesized to be given by: c^2/h!

Hence, based on the above mentioned three basic theoretical postulates of the 'Duality Principle' (e.g., including the existence of a conceptually higher-ordered 'D2 A-Causal' Computational framework), the existence of a rapid series of 'Universal Simultaneous Computational Frames' (USCF's) and their accompanying three Computational Dimensions of – 'Framework' ('frame' vs. 'object'), 'Consistency' ('consistent' vs. 'inconsistent') and 'Locus' ('global' vs. 'local') a (novel) 'Computational Unified Field Theory' (CUFT) was hypothesized; Based on a computational formalization of each of the four basic physical features of 'space' and 'energy', 'mass' and 'time' (e.g., which arise as secondary computational measures of the singular D2 rapid series of USCF's Computational Dimensions combination of 'frame': 'consistent' vs. 'inconsistent' and 'object': 'consistent' vs. 'inconsistent', correspondingly), the hypothesized 'Computational Unified Field Theory' (CUFT) can account for all known quantum and relativistic empirical findings, as well as seem to 'bridge the gap' between quantum and relativistic modeling of physical reality: Specifically, the various relativistic phenomena were shown to arise based on the interaction between the two ('global' vs. 'local') 'Framework' and (consistent vs. inconsistent) 'Consistency' computational dimensions. Conversely, a key quantum complimentary feature that characterizes the probabilistic interpretation of the 'uncertainty principle (e.g., as well as the currently assumed "collapse" of the probability wave function) was explained based on the 'computational exhaustiveness' arising from the computation of both the 'consistent' and 'inconsistent' aspects (or levels) of the Computational Dimensions' levels of 'frame' or 'object'; Thus for instance, both the 'consistent' and 'inconsistent' aspects (or levels) of the (Framework dimension's) 'frame' level (e.g., which comprise the quantum measurements of 'space' and 'energy', respectively) exhaustively describe the entire spectrum of this 'frame' computation. Thus, for instance, if we opt to increase the accuracy of the subatomic 'spatial' ('frame-consistent') measurement, then we also necessarily decrease the computational accuracy of its converse (exhaustive) 'energy' (e.g., 'frame-inconsistent') measure etc.

Indeed, such CUFT's reformalization of the key quantum and relativistic laws and empirical phenomena as arising from the singular (rapid series of) USCF's interrelated (secondary) computational measures (e.g., of the four basic quantum and relativistic physical features of 'space', 'time', 'energy' and 'mass' has led to the formulation of a singular '*Universal Computational Formula*' which was hypothesized to underlie- harmonize- and broaden- the current quantum and relativistic models of physical reality:

$$\frac{c^2 \, x'}{h} = \frac{s \times e}{t \, m}$$

wherein the left side of this singular hypothetical Universal Computational Formula represents the (abovementioned) universal rate of computation by the hypothetical Universal Computational Principle, whereas the right side of this Universal Computational Formula represents the 'integrative-complimentary' relationships between the four basic physical features of 'space' (s), 'time' (t), 'energy' (e) and 'mass' (m), (e.g., as comprising different computational combinations of the three (abovementioned) Computational Dimensions of 'Framework', 'Consistency' and 'Locus';

Note that on both sides of this Universal Computational Formula there is a coalescing of the basic quantum and relativistic computational elements – such that the rate of Universal Computation is given by the product of the maximal degree of (inter-USCF's relativistic) change 'c^2' divided by the minimal degree of (inter-USCF's quantum) change 'h'; Likewise, the right side of this Universal Computational Formula meshes together both quantum and relativistic computational relationships – such that it combines between the relativistic products of space and time (s/t) and energy-mass (e/m) together with the quantum (computational) complimentary relationship between 'space' and 'energy', and 'time' and 'mass';Significantly, it was suggested that the three (above mentioned apparent) principle differences between quantum and relativistic theories, namely: *'probabilistic'* vs. *'positivistic'* models of physical reality, *'simultaneous-entanglement'* vs. *'non-simultaneous causality'* and *'single-'* vs. *'multiple-'* spatial-temporal modeling can be explained (in a satisfactory manner) based on the new (hypothetical) CUFT model (represented by the Universal Computational Formula);

Finally, there may be important theoretical implications to this (new) hypothetical CUFT;

First, instead of 'space', 'energy', 'mass' and 'time' existing as *"independent-physical"* properties in the universe they may arise as '*secondary integrated computational properties*' (e.g., 'object'/'frame' x 'consistent'/'inconsistent' x 'global'/'local') of a singular conceptually higher-ordered 'D2' computed USCF's series...

Second, based on the 'Duality Principle' postulate underlying the CUFT which proves the conceptual computational constraint set upon the determination of any "causal-physical" relationship between any two (hypothetical) 'x' and 'y' elements (at the 'di1' computational level), we are forced to recognize the existence of a conceptually higher-ordered'D2' computational level which can compute only the "co-occurrences" of any two or more hypothetical spatial-temporal events or phenomena etc. This means that the CUFT's hypothesized 'D2' computation of a series of (extremely rapid) USCF's does not leave any room for the existence of any (direct or indirect) "causal-physical" 'x→y' relationship/s, but instead points at the . singular conceptually higher-ordered D2 A-Causal Computation which computes the co-occurrences of certain related phenomena, factors or events...

Third, an application of one of the three theoretical postulates underlying this novel CUFT, namely: the 'Duality Principle' to other potential 'Self-Referential Ontological Computational Systems' (SROCS) including: 'Darwin's Natural Selection Principle', 'Gödel's

Incompleteness Theorem' (e.g., and Hilbert's "failed" 'Mathematical Program'), Neuroscience's (currently assumed) 'materialistic-reductionistic' working hypothesis etc. (Bentwich, 2003a, 2003b, 2003c, 2004, 2006a, 2006b) (may open the door for a potential reformalization of these scientific paradigms in a way that is compatible with the novel computational Duality Principle and its ensued CUFT.

8. Acknowledgment

I would like to greatly thank Mr. Brian Fisher for his continuous input, support and encouragement of the development of the ideas that have led to the CUFT – without whom this work would not have been possible; I would like to thank my wife, Dr. Talyah Unger-Bentwich for her unfailing support and dialogue for the past fourteen years which have allowed me to work productively on this theory. I'd like to thank my dear mother, Dr. Tirza Bentwich for her lifelong nurturance of my thinking which have formed the basis for the development of this theory. Finally, I'd like to thank Dr. Boaz Tamir for a few discussions prior to the publication of this chapter.

9. References

Bagger, J. & Lambert, N. (2007). Modeling multiple M2's. *Phys. Rev. D*, Vol. 75, No. 4, 045020

Bentwich, J. (2003a). From Cartesian Logical-Empiricism to the 'Cognitive Reality': A Paradigmatic Shift, *Proceedings of Inscriptions in the Sand, Sixth International Literature and Humanities Conference*, Cyprus

Bentwich, J. (2003b). The Duality Principle's resolution of the Evolutionary Natural Selection Principle; The Cognitive 'Duality Principle': A resolution of the 'Liar Paradox' and 'Goedel's Incompleteness Theorem' conundrums; From Cartesian Logical-Empiricism to the 'Cognitive Reality: A paradigmatic shift, *Proceedings of 12th International Congress of Logic, Methodology and Philosophy of Science*, August Oviedo, Spain

Bentwich, J. (2003c). The cognitive 'Duality Principle': a resolution of the 'Liar Paradox' and 'Gödel's Incompleteness Theorem' conundrums, *Proceedings of Logic Colloquium*, Helsinki, Finland, August 2003

Bentwich, J. (2004). The Cognitive Duality Principle: A resolution of major scientific conundrums, *Proceedings of The international Interdisciplinary Conference*, Calcutta, January

Bentwich, J. (2006a). The 'Duality Principle': Irreducibility of sub-threshold psychophysical computation to neuronal brain activation. *Synthese*, Vol. 153, No. 3, pp. (451-455)

Bentwich, J. (2006b). *Universal Consciousness: From Materialistic Science to the Mental Projection Unified Theory*, iUniverse Publication

Born, M. (1954). The statistical interpretation of quantum mechanics, *Nobel Lecture, December 11, 1954*

Brumfiel, G. (2006). Our Universe: Outrageous fortune. *Nature*, Vol. 439, pp. (10-12)

Ellis, J. (1986). The Superstring: Theory of Everything, or of Nothing? *Nature*, Vol. 323, No. 6089, pp. (595–598)

Greene, B. (2003). *The Elegant Universe*, Vintage Books, New York

Heisenberg, W. (1927). Über den anschaulichen Inhalt der quantentheoretischen Kinematik und Mechanik. *Zeitschrift für Physik*, Vol. 43 No. 3–4, pp. (172–198)

Horodecki, R.; Horodecki, P.; Horodecki, M. & Horodecki, K. (2007). Quantum entanglement. *Rev. Mod. Phys*, Vol. 81, No. 2, pp. (865–942)

Polchinski, J. (2007) All Strung Out? *American Scientist*, January-February 2007 Volume 95, Number 1

Stephen W. Hawking Godel and the end of Physics. (Public lecture on March 8 at Texas A&M University)

On the right track. Interview with Professor Edward Witten, *Frontline*, Vol. 18, No. 3, February 2001

Witten, E. (1998). Magic, Mystery and Matrix. *Notices of the AMS*, Vol. 45, No. 9, pp. (1124–1129)

8

Correction, Alignment, Restoration and Re-Composition of Quantum Mechanical Fields of Particles by Path Integrals and Their Applications

Francisco Bulnes

Technological Institute of High Studies of Chalco
Mexico

1. Introduction

In the universe all the phenomena of physical, energetic and mental nature coexist of functional and harmonic management, since they are interdependent ones; for example to quantum level a particle have a harmonic relation with other or others, that is to say, each particle has a correlation energy defined by their energy density $K_{\alpha\beta}$, which relates the transition states ϕ_α, and ϕ_β, of a particle to along of the time.

There is an infinite number of paths of this kind Γ, in the space-time of the phenomena to quantum scale, that permits the transition or impermanence of the particles, that is to say, these can change from wave to particle and vice versa, or suffer energetic transmutations due to the existing relation between matter and energy, and of themselves in their infinity of the states of energy. Of this form we can realize calculations, which take us to the determination of amplitudes of transition inside a range of temporary equilibrium of the particles, that is to say, under the constant action of a field, which in this regime, remains invariant under proper movements in the space-time. Then exist a *Feynman integral* that extends on the space of continuous paths or re-walked Γ, that joins both correlated transition states. Likewise, if $\mathbb{R}^d \times I_t$, is the space-time where happens these transitions, and u, v, are elements of this space, the integral of all the continuous possible paths in $\mathbb{R}^d \times I_t$, is

$$\mathfrak{I}(L, x(t)) = \int_{C^{u,v}[0,t]} \exp\left\{\frac{i}{\hbar}(Action)(x)\right\}Dx,\tag{1}$$

where h, is the constant of Max Planck, and the (*Action*), is the realized by their *Lagrangian L*. Nevertheless, this temporal equilibrium due to the space-time between particles can turn aside, and even get lost (suffer scattering) in the expansion of the space-time when the trajectory that joins the transition states gets lost, this due to the absence of correlation of the particles, or of an adequate correlation, whose transition states do not turn out to be related, or turn out to be related in incorrect form. From a deep point of view of the knowledge of the matter, this succeeds when the chemical links between the atoms weaken and break, or get lost for lack of an electronic exchange adapted between these (*process electrons emission-*

reception). All this brings with it a disharmony matter-energy producing *collateral damage* between the immensity of interacting particles whose effects are a distortion of the field created by them. Finally these effects become visible in the matter under structural deformations or production of defective matter.

To eliminate this distortion of the field is necessary to remember the paths and to continue them of a systematised form through of *certain path or route integrals* (that belongs to a class of integral of the type (1)), that re-establishes normal course of the particles, re-integrating their field (realising the sum of all the trajectories that conform the movement of the particle), eliminating the deviations (that can derive in knots or ruptures in the space-time) that previously provoked this disequilibrium. These knots or ruptures in the space-time will be called singularities of the field.

In quantum mechanics, the spectral and vibration knowledge of the field of particles in the space, facilitates the application of corrective and restorer actions on the same field using their space of energy states through of the meaning of their electromagnetic potentials studied in quantum theory (*Aharonov-Bohm effect*). Thus these electromagnetic potentials can be re-interpreted in a spectral and vibration space that can be formulated in a set of continuous paths or re-walked, with the goal of realising corrective and restorers applications of the field, stretch to stretch, section for section, and that is inherent in this combined effect of all the possible trajectories to carry a particle from a point to other. By gauge theory is licit and consistent to manipulate the actions of correction and restoration of the field through of electromagnetic field, which ones are gauge fields of several types of interactions both strong and weak. In this last point is necessary to mention that in the class of equivalence of the electrodynamics potentials can be precisely re-interpreted like a connection on a trivial bundle of lines of $SU(2)$ (*non-Abelian part of the gauge theory using electromagnetic fields*), and admitting non-trivial bundles of lines with connection provided with more general fields (as for example, the of curvature, or the corresponding to $SU(3)$ (*the strong interactions*)). In both cases they are considered to be the phases of the corresponding functions of wave local variables and constant actions can be established across of their correspondents Lagrangians. *The path integrals* to these cases are of the same form that (1), except from the consideration of the potential states in each case. Into of this electromagnetic context and from the point of view of the solution of the wave equation through the alignments of lines of field, we can use the corresponding homogeneous bundle of lines that are used to give adequate potentials (potential module gauge, for example, those who come from the *cohomology* of $O(n-2)$, $n \geq 1$ (Bulnes & Shapiro, 2007). Of this management we can establish that {*set of fields of particles*} $\cong$ $I^{\alpha\beta}(H^1(PM^{\pm}, O(-n-2)))$, where {*potentials*}/{*gauge*} $\cong$ $H^1(PM^{\pm}, O(-n-2))$, (Bulnes & Shapiro, 2007). Here $I^{\alpha\beta}(H^1(PM^{\pm}, O(-n-2)))$, is the *cohomological class of the spectral images of the integrals of line* on the corresponding homogeneous bundle $O(-n-2)$.

Generalising the path integrals (1), we can establish that an evaluation of a global action $\mathfrak{I}$, due to the law of movement established for an operator L, that act on the space of particles $x(s)$, comes given for

$$\mathfrak{I} : x(s) \mapsto \int_M L(X(s))d(x(s)), \tag{2}$$

where M ($\cong \mathbb{R}^d \times I_t$), is the space-time of the transitions of the particles $x(s)$.

In particular, if we want the evaluation of this action to along of certain elected trajectory (path), inside of the field of minimal trajectories that governs the principle of minimal action

Correction, Alignment, Restoration and Re-Composition of Quantum Mechanical Fields of Particles by
Path Integrals and Their Applications

215

established by L, in our integrals (2), we have the execution of the action in a path Γ, given, to know

$$\text{Exe}: \Im(x(s)) \mapsto \int_{\Gamma} \left\{ \int_{M} L(X(s))d(x(s)) \right\} \mu_s = \int_{\Gamma} \Im(x(s))\mu_s, \tag{3}$$

where μ_s, the corresponding measure on the path or trajectory Γ, in M, is. The study of this integrals and their applications in the re-composition, alignment, correction and restoration of fields due to their particles are the objective of this chapter. We define as correction of a field X, to a re-composition or alignment of X. Is re-composition if is a re-structure or re-definition of X, is to say, it is realize changes of their alignment and transition states. The corrective action is an alignment only, if X, presents a deviation or deformation in one of their force lines or energy channels. A restoration is a re-establishment of the field, strengthening their force lines.

If we consider to the trajectory Γ_j, in terms of their deviation $\theta x(s)$, of the classic path $x_c(s)$, of their harmonic oscillator of $L(x(s))$, we can establish that the harmonic oscillator propagator has total action accord with our quantum model of correction and restoration action of the path integrals studied in quantum mechanics, (Bulnes et al., 2010, 2011):

$$\Im(x(s)) = correction + restoration = \Im(\theta x(s)) + \Im(x_c(s)) = \Im(Id) + \int_{\Gamma} \Im(x(s))\mu_s, \tag{4}$$

Observe that the term $\Im(\theta x(t))$, corresponds to the actions that is realised using rotations. This term belongs to space $Hom_G(X(M), L(M)) = \alpha Id$, being Id, the identity and $\alpha \in R$, (Bulnes, 2005), in the dual space of a restoration action of the field.

Now well, the relation between field and matter is realised through a quantum jump and only to this level succeeds. In the quantum mechanics, all the particles like pockets of energy works like points of transformation (states defined by energy densities). The field in the matter of a space-time of particles is evident like answer between these energy states, as it is explained in the *Feynman diagrams*. Due to that exist duality between wave and particle, a duality also exists between field and matter in the nature sense. Both dualities are isomorphic in the sense of interchange answers of interaction of a field. The answers between densities are realized in accordance with the correlation densities established in certain commutative diagrams that can be shaped by spaces L^2, on the space-time of the particles (Oppenheim et al., 1983). Coding this region of transition states of the corresponding Feynman diagram on a logic algebra $\mathcal{A}(\cap, \cup, -)$ (*like full states or empty of electrons like particle/wave, is to say, $\phi(0) = 0$, (is not the particle electron, but is like wave) $\phi(1) = 1$, (is not the wave electron, but is like particle) and their complements*), where the given actions in (2), are applied and re-interpreting the region of the space-time of the particles like a electronic complex of a hypothetical logic nano- floodgate (is to say, like a space L^2, with a logic given by $\mathcal{A}(\cap, \cup, -)$, on their transition states), we can define the *Feynman-Bulnes integrals*, as those that establish the transition amplitude of our systems of particles through of a binary code that realize the action of correction and restoration of the field established in (4). Likewise a Feynman-Bulnes integral (Bulnes, 2006c; Bulnes et al., 2010), is a *path integral of digital spectra* with the composition of the fast Fourier transform of densities of states of the corresponding Feynman diagrams. Thus, if ϕ_1, ϕ_2, ϕ_3 and ϕ_4, are four transitive states corresponding to a Feynman diagram of the field $X(M)$, then the path integral of Feynman-Bulnes is:

$$I = \int_{\phi^-}^{\phi^+} \phi_{n_1} F(n_1) \phi_{n_2} F(n_2) \phi_{n_3} F(n_3) \phi_{n_4} F(n_4) = 0001101001..., \tag{5}$$

The integrals of Feynman-Bulnes, establish the amplitude of transition to that the input of a system with signal $x(t)$, can be moved through a synergic action of electronic charges $\mathfrak{I}$, doing through pre-determined waves functions by $L(x(t))$, and encoded in a binary algebra (pre-defined by states $\phi(0)$, and $\phi(1)$), (in the *kernel* the space of solutions of the wave equation $o_X \Delta_F{}^{AA'} = \nabla^{AA'} \delta(x - x')$, (Bulnes & Shapiro, 2007) of a point to other into a circuit Γ_j. Their integral is extended to all space of paths or re-walked included into the region of Lagrangian action $\Gamma (= \bigcup_j^k \left(\bigcap_j^k \Gamma_j \right))$, with a topology of signals in $L^2(\cap, \cup, \backslash)$ (Bulnes et al.,

2011). If we want corrective actions for stretch Γ_j, of a path Γ, we can realize them using diagrams of strings of corrective action using the direct codification of path integrals with states of emission-reception of electrons (*by means of one symbolic cohomology of strings*) (Bulnes et al., 2011). Then the evaluation of the Feynman-Bulnes integrals reduces to the evaluation of the integrals: $I(\Gamma, \Omega) = \int_{C_\Gamma^0(\Psi)} \omega(\Gamma)$, where Ω, is the orientation of $C_\Gamma^0(\Psi)$, Ψ, is

the corresponding model of graph used to correct after identifying the singularity of the field X, that distorts it. For example, observe that it can vanish the corrective action of erroneous encoding through a sub-graph: $\int_{\Gamma 0} \omega(\Gamma) = 1 \cup -1 \cup 0 \cup [\phi(0) \cup 0 \backslash \phi(1)] \cup 1 = 0 \backslash \phi(1) \cup 1 = -1 \cup 1 = 1 + (-1) = 0$. The corresponding equation in the cohomology of strings is (Watanabe, 2007; Bulnes et al., 2011):

$$\tag{6}$$

The total correction of a field requires the action to a deep level as the established in (2), and developing in (4), and only this action can be defined by a logic that organises and correlates all and each one of the movements of the particles $x(s)$, *through codes given by* (5), *in a beautiful symphony that orders the field*. Finally we give an application to medicine obtaining the cure and organic regeneration to nano-metric level by quantum medicine methods programming our Path Integrals. Also, we give some applications to nano-materials.

2. The classic and non-classic Feynman integrals and their fundamental properties. The synergic and holistic principles

We consider a space of quantum particles under a regime of permanent energy defined by an operator of conservation called the Lagrangian, which establishes a field action on any trajectory of constant type. A particle has energy of interrelation defined by their energy density which relates the states of energy of the particle over to along of time considering the path or trajectory that joins both states in the space - time of their trajectories. Thus an infinite number of possible paths exist in the space - time that can take the particle to define their transition or impermanence in the space - time, the above mentioned due to the constant action of the field in all the possible trajectories of their space - time. In fact, the particle transits in simultaneous form all the possible trajectories that define their movement. Likewise, if $\Omega(\Gamma) \subset \mathbb{R}^3 \times I_t$, is the set of these trajectories subject to a field X, whose action $\Im_\Gamma$, satisfies for any of their trajectories that $\delta(\Im_\Gamma) = 0$, then their Lagrangian L, acts in such a form that the particle minimizes their movement energy for any trajectory that takes in the space $\Omega(\Gamma)$, doing it in a combined effect of all the possible trajectories to go from one point to another considering a statistical weight calculated on the base of the statistical mechanics. This is the exposition of Feynman known as exposition of the added trajectories (Feynman, 1967). Come to this point, the classic conception of the movement of a particle question: How can a particle continue different trajectories simultaneously and make an infinite number of them?

In the quantum conception the perspective different from the movement of a particle in the space - time answers the previous question enunciating:

"The trajectory of movement of a particle is this that does not manage to be annulled in the combined effect of all the possible trajectories to go from one point to on other in the space - time"

2.1 Classic Feynman integrals and their properties

We consider $M \cong \mathbb{R}^3 \times I_t$, the space - time of certain particles $x(s)$, in movement, and be L, an operator that expresses certain law of movement that governs the movement of the set of particles in M, in such a way that the energy conservation law is applied for the entire action of each of his particles. The movement of all the particles of the space M, comes given geometrically by their tangent vector bundle TM. Then the action due to L, on M, comes defined as (Marsden et al., 1983):

$$\Im_L : TM \to \mathbb{R}, \tag{7}$$

with rule of correspondence

$$\Im(x(s)) = FluxL(x(s))x(s), \tag{8}$$

and whose energy due to the movement is

$$E = \Im - L, \tag{9}$$

But this energy is due to their Lagrangian L, defined as (Sokolnikoff, 1964)

$$L(x(s), \dot{x}(s), s) = T(x(s), \dot{x}(s), s) - V(x(s), \dot{x}(s), s), \tag{10}$$

If we want to calculate the action defined in (7) and (8), along a given path $\Gamma = x(s)$, we have that the action is

$$\Im_\Gamma = \int_\Gamma L(x(s), x(s), s)ds, \tag{11}$$

For a classic trajectory, it is observed that the action is an extreme (minimum), namely,

$$\delta(\Im_\Gamma) = \delta \left(\int_{s(p_0)}^{s(p_1)} L(x(s), \dot{x}(s), s)ds \right) = 0, \tag{12}$$

Thus there are obtained the famous equations of Euler-Lagrange equivalent to the movement equations of Newton,

$$\left\{ \frac{d}{dt}\left(\frac{\partial L}{\partial \overset{\bullet}{x}}\right) - \frac{\partial L}{\partial x} \right\}_{\Gamma_{classic}} = 0, \quad s_0 \leq s \leq s_f, \tag{13}$$

That is to say, we have obtained a differential equation of the second order in the time for the freedom grade $x(s)$. This generalizes for a system of N grades of freedom or particles, with N, equations eventually connected. An alternative to solve a system of differential equations as the described one is to reduce their order across the formulation of Hamilton-Jacobi that thinks about how to solve the equivalent problem of 2N equations of the first order in the time (Marsden et al., 1983). Identifying the momentum as

$$p_i = \frac{\partial L}{\partial \overset{\bullet}{x}}, \tag{14}$$

we define the Hamiltonian or energy operator to the ith-momentum p_i, as:

$$\mathcal{H} = \sum_i^n p_i \overset{\bullet}{x}i - L, \tag{15}$$

and Hamilton's equations are obtained

$$\overset{\bullet}{p}_i = \frac{\partial \mathcal{H}}{\partial x_i}, \quad \overset{\bullet}{x}i = \frac{\partial \mathcal{H}}{\partial p_i}, \tag{16}$$

Nevertheless, it is not there clear justification of this extreme principle that happens in the classic systems, since any of the infinite trajectories that fulfill the minimal variation principle, the particle can transit, investing the same energy. Nevertheless the Feynman exposition establishes that it is possible to determine the specific trajectory that the particle has elected as the most propitious for their movement to go from $s(p_0)$, to $s(p_1)$, in the space-time being this one the one that is not annulled in the combined effect of all. Thus the quantum mechanics justifies the extreme principle affirming that the trajectory of movement of a particle is the product of the minimal action of a field that involves to the whole space- time where infinite minimal trajectories, that is to say, exist where the extreme condition exists, but that statistically is the most real. Likewise, the nature saves energy in their design of the movement, since the above mentioned trajectory belongs to an infinite set of minimal trajectories that fulfill the principle of minimal action established in (12).

The concrete Feynman proposal is that the trajectory or real path of movement continued by a particle to go from one point to another in the space-time is the amplitude of interference of all the possible paths that fulfill the condition of extreme happened in (12) (to see figure 1 a)). Now then, this proposal is based on the probability amplitude that comes from a sum of all the possible actions due to the infinite possible trajectories that set off initially in x_0, to end then in x_f.

Correction, Alignment, Restoration and Re-Composition of Quantum Mechanical Fields of Particles by
Path Integrals and Their Applications

219

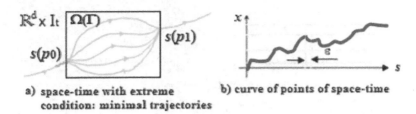

a) space-time with extreme   b) curve of points of space-time
   condition: minimal trajectories

Fig. 1. a). The extreme condition in paths of the space $\Omega(\Gamma)$. b). Curve of the space-time in the $R^3 \times I_t$.

Using the duality principle of the quantum mechanics we find that the particle as wave satisfies for this superposition

$$\psi(x,s) = A(s)\sum_{\gamma}\exp\left(\frac{i\Im}{\hbar}\right),\tag{17}$$

where the term $A(s)$, comes from the standardization condition in functional analysis (Simon & Reed 1980)

$$\int_{-\infty}^{+\infty}|\psi(x,t)|^2\,dx = 1,\tag{18}$$

where to two arbitrary points in the space-time (s_0, x_0), and (s_f, x_f), whose amplitude takes

the form $A' = \sqrt{\dfrac{m}{2\pi i\hbar(s_f - s_0)}}$. In effect, we approximate the probability amplitude taking

only the classic trajectory. Of this way a Green function is had (propagator of x_0, s_0 to x_f, s_f) of the form:

$$D_F(x_f,s_f;x_0,s_0) = A'\exp\left(\frac{i\Im_{cl}}{\hbar}\right),\tag{19}$$

Then we consider to this classic trajectory:

$$x(s) = x_0\frac{x_f - x_0}{s_f - s_0}(s - s_0),\tag{20}$$

$$\upsilon(s) = \frac{x_f - x_0}{s_f - s_0},\tag{21}$$

Of this manner, the action on this covered comes given for

$$\Im_{clásica} = \int_{s_0}^{s_f}L(s)ds = \frac{m}{2}\frac{(x_f - x_0)^2}{s_f - s_0},\tag{22}$$

It reduces us to calculate then the standardization term A', for it we must bear in mind the following limit that in our case happens in the probability amplitude for $s_f \to s_0$:

$$\delta(x_f - x_0) = \lim_{\Delta \to 0} \frac{1}{(\pi \Delta^2)^{1/2}} \exp\left[-\frac{(x-x_0)^2}{\Delta^2}\right], \tag{23}$$

Identifying then to term of normalization like $A' = \Delta$, in (23) and using (22) it is had:

$$A' = \left(\frac{m}{2\pi i\hbar(s_f - s_0)}\right)^{1/2}, \tag{24}$$

Therefore, the exact expression is had in the probability amplitude

$$D_F(x_f, s_f; x_0, s_0) = \left(\frac{m}{2\pi i(s_f - s_0)}\right)^{1/2} \exp\left(i\frac{m}{2\hbar}\frac{(x_f - x_0)^2}{s-s_0}\right), \tag{25}$$

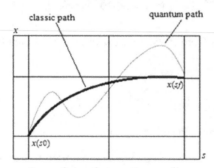

Fig. 2. Trajectories in the space-time plane, the continuous line corresponds to a classic trajectory while the pointed line corresponds to a possible quantum trajectory.

This type of exact results from the Feynman expression can also be obtained for potentials of the form:

$$V(x,\dot{x},s) = a + bx + cx^2 + d\dot{x} + ex\dot{x}, \tag{26}$$

But the condition given in (12), establishes that the paths that minimize the action are those who fulfill with the sum of paths given in terms of a functional integral, that is to say, those paths on the space $\Omega(\Gamma) \subset \mathbb{R}^3 \times I_t$, to know:

$$\sum_\gamma \exp\left(\frac{i\Im}{\hbar}\right) \longrightarrow \int_{\Omega(\Gamma)} \exp[i\Im_{x(s)} / \hbar] d(x(s)) = \int_\Omega \omega(\Gamma), \tag{27}$$

An interesting option that we can bear in mind here is to discrete the time (figure 2). Thus if the number of temporary intervals from s_0, to s_f, is N, then the temporary increase is $\delta s = (s_f - s_0)/N$, which implies that $s_n = s_0 + n\delta s$. We express for x_n, to the coordinate x, to the time s_n, that is to say $x_n = x(s_n)$. Then for the case of a free particle, it had that the action is given like:

$$\Im = \int_{s_0}^{s_f} L(s)ds = \frac{m}{2}\dot{x}^2, \tag{28}$$

Correction, Alignment, Restoration and Re-Composition of Quantum Mechanical Fields of Particles by
Path Integrals and Their Applications

221

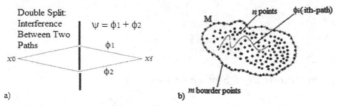

Fig. 3. a) Possible trajectories in an experiment of double split. The final amplitude result of the interference in between paths. b) The configuration space $C_{n,m}$, is the model created by the due action to each corresponding trajectory to the different splits. It is clear here that it must be had in mind all the paths in the space-time M, that contributes to the interference amplitude in this space.

Thus, on a possible path γ, it is had:

$$\Im_\gamma = \sum_{n=0}^{N-1} \frac{m}{2}\left(\frac{x_{n+1}-x_n}{\delta s}\right)^2 \delta s, \tag{29}$$

Thus it is observed that the propagator D_F, will be given for:

$$D_F(x_f,s_N,x_0,s_0) - \lim_{\substack{N\to\infty \\ \delta s \to 0}} A \int_{-\infty}^{+\infty} dx_1 \cdots \int_{-\infty}^{+\infty} dx_{N-1} \exp\left[\frac{im}{2\hbar}\sum_{n=0}^{N-1}\frac{(x_{n+1}-x_n)^2}{\delta s}\right], \tag{30}$$

Realizing the change $y_n = \left[\frac{m}{2\hbar\delta s}\right]^2 x_n$, we re-write:

$$D_F = \lim_{\substack{N\to\infty \\ \delta s \to 0}} A' \int_{-\infty}^{+\infty} dy_1 \cdots \int_{-\infty}^{+\infty} dy_{N-1} \exp\left[-\sum_{n=0}^{N-1}\frac{(y_{n+1}-y_n)^2}{i}\right], \tag{31}$$

where $y_n = \left[\frac{2\hbar\delta T}{M}\right]^{(N-1)/2}$ A. Developing the first integral of (31), we have

$$\int_{-\infty}^{+\infty} dy_1 \exp\left[-\frac{(y_2-y_1)^2-(y_1-y_0)^2}{i}\right] = \left(\frac{i\pi}{2}\right)^{1/2}\exp\left[-\frac{(y_2-y_0)^2}{2i}\right], \tag{32}$$

Then integrating for y_2, we consider the second member of (32) and the following one, $(y^3 - y^2)^2$:

$$\left(\frac{i\pi}{2}\right)^{1/2}\int_{-\infty}^{+\infty} dy_2 \exp\left[-\frac{(y_3-y_2)^2}{i}\right]\exp\left[-\frac{(y_2-y_0)^2}{2i}\right] = \left(\frac{(i\pi)^2}{3}\right)^{1/2}\exp\left[-\frac{(y_3-y_0)^2}{3i}\right], \tag{33}$$

Then a recurrence has in the integrals of such form that we can express the general term as $\frac{(i\pi)^{(N-1)/2}}{N^{1/2}}\exp\left[-\frac{(y_N-y_0)^2}{Ni}\right]$. Therefore it is had for the propagator (30), which

$$D_F(x_N,s_N,x_0,s_0) = A\left(\frac{2\pi\hbar i\delta s}{m}\right)^{(N-1)/2}\exp\left[\frac{m(x_N-x_0)^2}{2\hbar N\delta s}\right], \tag{34}$$

identifying in this case:

$$A = \frac{1}{B} = \left[\frac{2\pi\hbar i \delta s}{m}\right]^{1/2}, \tag{35}$$

it is had that the integration in paths is given for:

$$\int_{\Omega(\Gamma)} \omega(x(s)) = \lim_{\substack{N \to \infty \\ \delta s \to 0}} \frac{1}{B} \int_{-\infty}^{+\infty} \frac{dx_1}{B} \cdots \int_{-\infty}^{+\infty} \frac{dx_{N-1}}{B}, \tag{36}$$

where in the first member of (36) we have expressed the Feynman integral using the form of volume $\omega(x(s))$, of the space of all the paths that join in $\Omega(\Gamma)$, to obtain the real path of the particle (therefore we can choose also quantized trajectories (see figure 2)). Remember that the sum of all these paths is the interference amplitude between paths that happens under an action whose Lagrangian is $\omega(x(s)) = \Im_{x(s)}dx(s)$, where, if $\Omega(M)$, is a complex with M, the space-time, and C(M), is a complex or configuration space on M, (*interfered paths in the experiment given by multiple split (see the figure 3, to case of double split)*), endowed with a pairing

$$\int : C(M) \times \Omega^*(M) \to \mathbb{R}, \tag{37}$$

where $\Omega^*(M)$, is some dual complex ("forms on configuration spaces"), that is to say. such that "Stokes theorem" holds:

$$\int_{\Omega \times C} \omega = \;<\Im, d\omega>, \tag{38}$$

then the integrals given by (36) we can be write (to m-border points and n-inner points (see figure 3 b))) as:

$$\int_{\Omega(\Gamma)} \Im(x(s))dx(s) = \int_{\Gamma_{t^1} \times \ldots \times \Gamma_{t^m} \times \ldots} \Im_q dx_1{}^{m_1} \ldots dx_n{}^{m_n} = \int_{\Gamma_{t^1}} (\int_{\Gamma_{t^2}} \ldots (\int_{\Gamma_{t^m}} \Im dx_1{}^{m_n}) \ldots dx_n{}^{m_1}), \tag{39}$$

This is due to the infiltration in the space-time by the direct action $\Im$, that happens in the space $\Omega \times C$, to each component of the space $\Omega(\Gamma)$, through the expressed Lagrangian in this case by ω. In (39), the integration of the space realises with the infiltration of the time.

Two versions of (36), that use the evolution operator and their unitarity are their differential version and numerical version of Trotter-Suzuki[1] (numerical version of (36)). The first version is re-obtain the Schröndinger equation from the Feynman path integral. In this case the wave function involves the corresponding electronic propagator given in (30) with a temporal step δs, to pass from $\psi(x, 0)$, to $\psi(x, \delta s)$, having the amplitude (Holstein, 1991)

---

[1] $D_F(s, s_0) = \theta(s - s_0) \lim_{N \to \infty} \int_{-\infty}^{+\infty} dx_1 \cdots \int_{-\infty}^{+\infty} dx_{N-1} \prod_{j=0}^{N-1} \left\langle x_{j+1} \left| e^{\frac{-i\lambda T}{N} \frac{-i\lambda V}{N}} \right| x_j \right\rangle.$

Here, T, and V, are kinetic and potential energies in discrete form using their separate evolutions in slices. $\theta(s - s_0)$, is the weight of compensation in numerical compute.

$$\psi(x,\delta s) = \left[\frac{m}{2\pi i\hbar\delta s}\right] \int\limits_{-\infty}^{+\infty} \exp\left[i\frac{m\delta x^2}{2\hbar\delta s}\right]\left[1 - i\frac{\delta s}{\hbar}V(x,0) \quad +\delta x\frac{\partial}{\partial x} + \frac{\delta s^2}{2}\frac{\partial^2}{\partial x^2}\right]\psi(x,0)d\delta x, \quad (40)$$

Realising the integral we obtain the differential version of the Feynman integral (36).

Let H, be the Hopf algebra (*associative algebra used to the quantized action in the space-time*) (Kac, 1990), of a class of Feynman graphs G (Barry, 2005). If Γ, is such a graph, then configurations are attached to its vertices, while momentum are attached to edges in the two dual representations (Feynman rules in position and momentum spaces). This duality is represented by a pairing between a "configuration functor" (typically C_Γ, (configuration space of subgraphs and strings (Watanabe, 2007), and a "*Lagrangian*" (e.g. ω, determined by its value on an edge, e.g. by a propagator D_F). Together with the pairing (typically integration) representing the action, they are thought as part of the Feynman model of the state space of a quantum system.

Since it has been argued in (Ionescu, 2004), this *Feynman picture* is more general than the manifold based "Riemannian picture", since it models in a more direct way the observable aspects of quantum phenomena ("interactions" modeled by a class of graphs), without the assumption of a continuity (or even the existence) of the interaction or propagation process in an ambient "space-time", the later being clearly only an artificial model useful to relate with the classical physics, i.e. convenient for "quantization purposes".

Likewise, an action on G ("$\mathfrak{I}_{int}$"), is a character $\mathfrak{I}_{int} : H \to \mathbb{R}$, (defined similarly to the given in (7)) which is a cocycle in the associated DG-co-algebra $(T(H^*), D)$, that is to say, the action in this context is an endomorphism (matrix) of transition of the certain densities of field given by ϕ_i.

A QFT (*Quantum Field Theory*) defined via Feynman Path Integral quantization method is based on a graded class of Feynman graphs. For specific implementation purposes these can be 1-dimensional CW-complexes or combinatorial objects. For definiteness we will consider the class of Kontsevich graphs $\Gamma \in G_n$, the admissible graphs from (Kontsevich, 2003). Nevertheless we claim that the results are much more general, and suited for a generalisation suited for an axiomatic approach; a Feynman graph will be thought of both as an object in a category of Feynman graphs (categorical point of view), as well as a *co-bordism* between their boundary vertices (TQFT point of view). The main assumption the class of Feynman graphs needs to satisfy the existence of subgraphs and quotients.

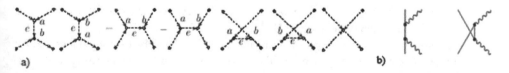

a)   b)

Fig. 4. The Kontsevich class is the quantized class used by the Feynman rules.

Example 1. In the compute of path integrals on the graph configuration space $C_\square(\Omega)$, The graphs $\Gamma \in G_n$, will be used in the string schemes given by BRST-quantization on gauge theory. For example, the BRST-quantization is always nilpotent around a vertex: $Q_{BRST} \bullet \upsilon = \oint dz j_{BRST}(z)\upsilon(0) = 0$. The Kontsevich class not has loops everywhere (figure 4 a)). The Feynman diagrams (figure 4 b)) conforms a subclass in the Kontsevich class, that is to say, restricted in the deformation quantization in respective micro-local structure of the Riemannian manifold (Kontsevich, 2003).

While the concept of subgraph of γ, is clear (will be modeled after that of a subcategory), we will define the quotient of Γ, by the subgraph γ as the graph Γ', obtained by collapsing γ, (vertices and internal edges) to a vertex of the quotient. Then it satisfies the graph class succession under *Hom*, that will define all the types of graphs with connecting arrows:

$$\gamma \to \Gamma \to \Gamma/\gamma, \tag{41}$$

We enunciate the following basic properties of the classic Feynman integrals. Let γ, $\gamma' \in \Gamma$, where $\Gamma \in G$, and $\omega(\Gamma)$, their corresponding Lagrangian with the property like in (38). We consider the path integral I_Γ, like a map given in (37). Let D_F, their corresponding propagator (the value of $\omega(\Gamma)$, in the corresponding edge γ). Then are valid the following properties:

a. $\forall\ D_F$, propagator there is an unique extension to a Feynman rule on (39), that is to say
$\omega(\Gamma) = \omega(\gamma) \wedge \omega(\gamma')$, with $\Gamma/\gamma = \gamma'$.

b.   If $\omega(\Gamma)$, is a Lagrangian on (41) with $\Gamma/\gamma = \gamma'$, then

$$\int \omega(\Gamma) = \int \omega(\gamma) \wedge \omega(\gamma'), \tag{42}$$

c.   From (38) $\int_{\partial C} \omega = \Im \circ D_F$, then $\forall$ extension (41),

$$\int_{\partial\gamma\Im(\Gamma)} \omega(\Gamma) = \Im(\gamma) \bullet \Im(\gamma'), \tag{43}$$

d. $\forall\ \Gamma \in G$ (Feynman graph),

$$\sum_{e \in \Gamma_{\text{int}}} \int_{\partial e \bar{C}_\Gamma} \omega(\Gamma)\Im(\mathrm{d}^{\text{int}}\Gamma), \tag{44}$$

where e, is a simple sub-graph of Γ, without boundary.

e.   As consequence of the integral (44), we have the composition formulas

$$\Im \circ D_{F_e} = \Im \circ \mathrm{d}^{\text{int}}, \quad \Im \circ D_{F(i\text{-}e)}, \tag{45}$$

f.   Feynman integrals over codimension one strata corresponding to non-normal subgraphs vanish. A graph $\Gamma \in G$, is normal if the corresponding quotient Γ/γ, belongs to the same class of Feynman graphs G.

g.   The remaining terms corresponding to normal proper subgraphs meeting the boundary $[m]$, of $\Gamma \in G_a$, yield a forest formula, like intM (figure 3, b)) corresponding to the co-product D_{Fb}, of G. Then for a Feynman graph $\Gamma \in G_a$:

$$\sum_{\gamma \to \Gamma \to \Gamma/\gamma^{\text{in}}}{}^G \int_{\partial_\gamma \bar{C}_\Gamma} \omega(\Gamma)\Im(D_{F_b}, \Gamma), \tag{46}$$

where the proper normal subgraph γ, meets non-trivially the boundary of Γ.

h.   If the Lagrangian $\omega(\Gamma)$, is a closed form then the corresponding Feynman integral $\Im$, is a cocycle. Then

Correction, Alignment, Restoration and Re-Composition of Quantum Mechanical Fields of Particles by
Path Integrals and Their Applications

225

$$\int_{\partial \tilde{C}_r} \omega(\Gamma) = \int_{\tilde{C}_r} d\omega(\Gamma) = 0, \tag{47}$$

2.2 Non-classic Feynman integrals and their properties
2.2.1 Twistor version (Bulnes & Shapiro, 2007)

Consider the space of hypercomplex coordinates (coordinates in the m-dimensional complex projective space $\mathbb{P}^m$) that determine through the position, quantum states of particles in free state $Z_1{}^a$, $Z_2{}^a$, ..., $Z_m{}^a$, and we define the functional space of Feynman

$$\Phi_D = \{ \, F_U \mid F_U(z) = \int d^n z \; \phi_1(z) \; \phi_2(z) \dots \; \phi_n(z) = 000 \dots \text{0-box diagram}\}, \tag{48}$$

This space is the corresponding to the group of Feynman ϕ^n-integral for the 000 ... 0-box diagram (that is to say, $C(M)$ given in the before section) with certain configuration space $C_{n, m}$, like was defined in section 1. 2, (with n-states ϕ_i, and m-edges or lines) with arrange

$$\tag{49}$$

This functional belongs to the integral operator cohomology on homogeneous bundles of lines $H^1(\mathbb{P}T, O(-2-2))$, where $\mathbb{P}T = \mathbb{P}T^+ \cup \mathbb{P}T^-$ for example, for $n = 4$, one has the diagram of Feynman for the ϕ^4-integral one that corresponds to the 0000-box diagram

$$\tag{50}$$

The elements F_U, can be expressed in a low unique way the map in the complex manifold $\mathbb{P}^m$, like

$$\Phi_D \to \mathcal{L}(\mathbb{P}^m(\mathbb{C}), \mathbb{C}), \tag{51}$$

with rule of correspondence

$$\int d^n z \; \phi_1(z) \; \phi_2(z) \dots \phi_n(z) \mapsto \int d^n Z^\alpha \; W_\alpha \; \phi_1(Z^\alpha W_\alpha) \; \phi_2(Z^\alpha W_\alpha) \dots \phi_n(Z^\alpha W_\alpha), \tag{52}$$

that allows us to identify Φ_D, with $\mathcal{L}(\mathbb{P}^m(X), X)$. Building the twistor space $T = \{[x]_U \mid [x]_U = \dots + (x)_{-2} + (x)_{-1} + (x)_0 + (x)_1 + \dots\}$, where $(x)_{-n} = (x)^{-n}/n!$ (contours with opposite $x = 0$) and $(x)_{n+1} = -n!(x)^{-n-1}/2\pi I$ (in an environment around $x = 0$). This twistor space satisfies that $T \cong \mathcal{L}(\mathbb{P}^m(\mathbb{C}), \mathbb{C})$. Then $\int d^n z \; \phi_1(z) \; \phi_2(z) \dots \phi_n(z) = \oint [x]_U \; \forall \; x \in \mathbb{C}^m$. Then these integrals have their equivalent ones as integral of contour in the cohomology of contours $H^d(\Pi - \Upsilon, \mathbb{C})$, where Π, it is the product of all the twistor spaces (and dual twistor) and it is the subspaces union on those which the factor $(Z^{\alpha}W_{\alpha})^{-1}$, are singular. To check this course with demonstrating, for the case $m = 1$, and using the integral operator of Cauchy has more than enough contours (jointly with the residue theorem) that the integral $\int d^n z\ \phi_1(z)\ \phi_2(z)\ \dots\ \phi_n(z)$ takes the form of the formalism of Sparling given by the integral one

$$\oint [x]_U = \{\oint \int_0^{\infty} + \int \oint\} \frac{e^z}{(x+z)^2} dz, \tag{53}$$

which bears to the isomorphism among the cohomological spaces

$$H^1(\mathbb{P}(\mathbb{C}), \Omega) \cong H^1(\mathbb{P}\mathcal{T}, O(-2\text{-}2)), \tag{54}$$

which would be a quaternion version of these integrals? It would be the one given for integral of type Cauchy of functions of $\mathbb{H}$-modules (Shapiro & Kravchenko, 1996), on opened D, that turn out to be Liapunov domains in R^n. Since one has the you make twistor projective bundle $S^2 \backslash \mathbb{P}^3(\mathbb{C}) \to \mathbb{P}(\mathbb{H}) \cong S^4$, and $H^1(\mathbb{P}\mathcal{T}, O(-2\text{-}2)) \cong H^1(\mathbb{P}(\mathbb{C}), \Omega)$, then the cohomology $H^1(\mathbb{P}(\mathbb{C}), \Omega) \cong H^1(\mathbb{P}(\mathbb{C})$, space in corresponding differential forms). But $S^1 \to \mathbb{P}^3(\mathbb{C}) \to \mathbb{P}(\mathbb{C})$, it is a principal bundle with $\mathbb{P}(\mathbb{C}) \to S^2$, and since S^1, and S^3, are the underlying groups in the structure of the hyper-complexes and quaternion ($\mathbb{H} \cong \mathbb{C}^2$) then $S^1 \to S^3 \to S^2$, represents in quantum mechanics a spin system $\frac{1}{2}$ which can be represented by the cohomology of a diagram formed as an alternating chain of 0-lines and 2-lines, that is; $H^1(\mathbb{P}\mathcal{T}^{\pm}, O(-n-2))$, that is to say for the system of quantum state of spin $\frac{1}{2}$ that is the corresponding to a 4-integral one given by the 0000-box diagram

$$= \iint d^4x\, d^4y\, \phi(x)\, \psi_A(x) \leq_x D_F^{AA'}(x-y)\, \psi_{A'}(y)\, \theta(y), \tag{55}$$

But this cohomology of diagrams of contour integrals is applicable to 1-functions for $\mathbb{P}(\mathbb{C})$, in $\mathbb{P}\mathcal{T}^{\pm}$, that which is not chance, since it is a consequence of the G-structure of the manifold F, (where they are defined these quantum phenomena) which is induced in the S^3-structure of the underlying spinors (Penrose & Rindler, 1986).

If we consider that the for-according complex manifolds have a pseudo-Hermitian complex structure not symmetrical and induced by the sheaf of quadratic forms $O(8^2T^*(M))$, it can expand the symmetry according classic of the diagrams of Feynman from their contour integrals to the construction of according structures that can be induced to the pseudo-Hermitian complex structure of the mentioned manifolds, giving the possibility to obtain a single integral operators cohomology of Feynman type for analytic manifolds (Huggett, 1990).

2.2.2 Instanton version

The Feynman integrals are invariants in R^3, under rotations of Wick, that is to say

Correction, Alignment, Restoration and Re-Composition of Quantum Mechanical Fields of Particles by
Path Integrals and Their Applications

227

$$\int\limits_{\Omega(\Gamma)} \exp[i\Im[\phi]]D[\phi] \mapsto \int\limits_{\Omega(\Gamma)} \exp[-\Im[\phi]]D[\phi], \tag{56}$$

to a coordinates system in E^4, given by (x_0, x_1, x_2, x_3), with $x_0 = s$, then $\forall$ coordinates transformation given by $s \to i\tau$, we have that

$$M \to E^4, \tag{57}$$

then $\Omega(\Gamma)$, represents a region $W(C)$, in E^4 (a Wick region in the space time). This action has place in S^4, to the solutions of the Yang-Mills equations on S^4. The action realised in this transformation has Euclidean action

$$\Im_E = \int\limits_{\tau_1}^{\tau_2} \left(\frac{1}{2} m\dot{x}(\tau)^2 + V(x) \right) d\tau, \tag{58}$$

where the potential energy $V(x)$, changes to $-V(x)$, with the Wick rotation.

2.2.3 Feynman-Bulnes version

Considers a microelectronic device that is fundamented in the functional space $L^2(\cup, \cap, /)$ encoding in a logic algebra $\mathcal{M}_{1,0}$. The corresponding functional equation to inputs and outputs of information signals using certain liberty based in the artificial process of thought to create "intelligent" computers needs the use of path to plantee their solution (Bulnes, 2006c). Then extrapoling the Feynman integrals to calculate the amplitud of interference of the many paths (criteria) to resolve a automation problem that designs a cybernetic complex that at least to theorical level has a quantum programming with Feynman rules and an adequate neuronal net[2].

Def. 1 (Path Integrals of Feynman-Bulnes). A integral of Feynman-Bulnes is a path integral of digital spectra with composition with Fast Transform of densities of state of Feynman diagrams.

If ϕ_1, ϕ_2, ϕ_3, and ϕ_4, are four densities of states corresponding to the Feynman diagrams to the poles of field $X(M)$, then the path integral of Feynman-Bulnes is:

$$I_{FB} = \int\limits_{Z^-}^{Z^+} \phi_{n_1} F(n_1) \phi_{n_2} F(n_2) \phi_{n_3} F(n_3) \phi_{n_4} F(n_4), \tag{59}$$

---

[2] The integrals of Feynman-Bulnes give solution to the functional equation of a automatic micro-device to control (micro-processor) $F(XZ^+, YZ^-) = 0$ (Bulnes, 2006c). The informatics theory assign a cybernetic complex to $\mathbb{C}$, (Gorbatov, 1986) and each cube in this cybernetic net establish a path on the which exist a vector of input XZ^+, and a vector of output YZ^-, signed with a time of transition τ, to carry a information given in XZ^+, on a curve γ_i, (path) to a state YZ^-, through logic certain (conscience), that include all the circuit $\mathbb{C}$ (Bulnes, 2006c). In the case of $\mathbb{C}$, the logic is the real conscience of interpretation of $\mathbb{C}$, (criteria of $\mathbb{C}$). As $\mathbb{C}$, has a real conscience of recognition; into of their corrective action and reexpert, elect the adequate path to the application of the corrective action. For it, the integrals of Feynman-Bulnes can be explained on the electable model Ψ_β, (path, see figure 1 a)) as:

$$correcci\grave{o}n = \int\limits_{Z^-}^{Z^+} D_F(z(t))\mu_{n_1} F(n_1)d(z(t)) = \Sigma_n^N \delta(\Psi_\alpha - \Psi_\beta)\phi(\Psi_\alpha).$$

3. Combination of quantum factors and programming diagrams of path integrals: The coding and encoding problems

Since a duality exists between wave and particle, a duality also exists between field and matter in the natural sense (Schwinger, 1998). Both dualities are isomorphic in the sense of the exchange of states of quantum particles and the interaction of a field. Indeed in this quantum exchange of information of the particles, that happen in the space-time $\Omega(\Gamma)$ region, the pertinent transformations are due to realise to correct, restore, align or re-compose (put together) a field X.

Elements of fiel	Nano-metric application	Effect obtained on field
0-lines	localization of anomalous points	encoding nodes to application
1-lines	application of electronic propagator	alignment of lines of field
-1-lines	inversion of actions*	reflections of restoration

* Creation of contours around of points of application

Table 1. Combination of quantum factors of the field X.

Any anomalous declaration in a quantum field shows like a distortion, deviation, non-definition or not existence of the field in the space-time where this must exist like physical declaration of the matter (existence of quantum particles in the space). The quantum particles are transition states of the material particles. We remember that from the point of mathematical view, a singularity of the space $X(M)$, is a discontinuity of the flux of energy, where $FluxD_{Fj}(z)z \neq 0$, $\forall z \in M$ (Marsden et al., 1983). This discontinuity creates a space of disconnection where the alignment atoms stay unenhanced due to not have electrons that they do unify them under the different chemical links that exist and through the ionic interchange foreseen in the space $TM^{\pm,3}$(Landau & Lifshitz, 1987), (vector bundle of the particles in M, and responsible of the geometrical configuration of the field in M, and that promote the ionic restoration in $X(M)$, (*Gauge theory*). In a topological sense of the field, the detection of these anomalies of the field X will do through anomalies in the trajectories of flux $\gamma(z)$, such that $FluxD_{Fj}(z)z \neq 0$, $\forall z \in \gamma(z) \subset M$.

Def. 2. [10] If *Flux*: $\mathbb{R} \times M \to M$, is a flux and $z \in M$, the curve $\gamma_z : \mathbb{R} \to M$, with rule of correspondence $t \mapsto \phi_t(z) = \phi(t, z)$, is a line of flux.

A anomaly in a trajectory and thus in M, will be a singular point which can be a knot (multiform points), a discontinuity (a hole (source or fall hole)) in M)) or a indeterminate point (without information of the field in whose point or region in M). But we require their electromagnetic mean into the context of X, for we obtain their corresponding diagnosis using an electromagnetic device that establish an univocal correspondence between detected anomalies and Feynman diagrams used to the spectral encoding through of the integrals of Feynman-Bulnes.

If we consider the space $C^0_\Gamma(\psi)$ (Watanabe, 2007), as space of configuration associate with sub-graph (Γ, ψ), where ψ, is the corresponding smooth embedding to n-knot that which is identified as a ∞, in an integral as the given in (6), we can define rules to sub-graphs that coincides with the rules of signs in the calculate of integrals like (36). Thus we can identify the three fundamental forms given for $\omega(\Gamma) = \Pi sgn(z)$, (figure 5).

---

[3] $X(M)$, is a section of $TM^{\pm}$, in a mathematical sense (Marsden et al., 1983).

Correction, Alignment, Restoration and Re-Composition of Quantum Mechanical Fields of Particles by
Path Integrals and Their Applications

229

In this study the path integrals and their applications in the re-composition, alignment, correction and restoration of fields due to their particles realise using certain rules of fundamental electronic state and their sub-graphs, through considering the identification. We define as correction of a field X, to a re-composition or alignment of X. Is re-composition if it is a re-structure or re-definition of X, is to say, it is realize changes of their alignment and transition states (properties of the table 1 and additional properties with the algebra $\mathcal{M}_{0,1}$). The corrective action is an alignment only, if X, present a deviation or deformation in one of their force lines or energy channels (properties of contours on particles: ⟶, ⟶, ⟶). A restoration is a re-establishment of the field, strengthening their force lines (properties of the Dirac and Heaviside function on particles: with $t \geq s$, $w(s, t)\phi(-1) \leq \phi(1)U_0(s, t)$, etc) (Fujita, 1983). Consider the following corrective action by the string diagrams to the states of emission-receptor of electrons (see figure 5 a)). The evaluating of the integrals of Feynman-Bulnes is reduced to evaluating the integrals:

$I(\Gamma;\Omega) = \int_{C^0\Gamma(\Psi)}\omega(\Gamma)$, where Ω, is the orientation on $C^0_\Gamma(\Psi)$, where Ψ, is the corresponding model of graph used to correct and identify the anomaly and Γ, the corresponding sub-graph of the transitive graph determined by a re-composition field treatment. The space $\mathcal{M}_{0,1}$, conforms a reticular sub-algebra in mathematical logic. In the figure 5 b), the corrective action in the memory of an Euclidean portion of the space time $\Omega(\Gamma)$, through a sub-graph Γ,

of strings, in the re-composition of the alignment of field comes given as: $z_j = \int_{C^0\Gamma(\Psi)}\theta_1\theta_2\theta_3\theta_4 =$ $<\{[1 \cup \phi(1)] \cup 0\} \cup \phi(1)> \cup 1 = 1$, (Bulnes, 2006c). Observe that it can vanish the corrective

action of encoding memory through another sub-graph: $\int_{\Gamma_0}1 \cup -1 \cup 0 \cup [\phi(0) \cup 0 \cup \phi(1)] \cup 1$ $= 0 \cup \phi(1) \cup 1 = -1 \cup 1 = 1 + (-1) = 0$ (see the equation (6)).

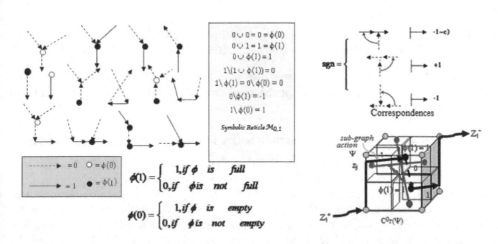

Fig. 5. a) String diagrams of corrective action using direct encoding by path integral. b) Euclidean portion of the space time $\Omega(\Gamma)$.

All anomalies in the space-time produce scattering effects that can be measured by the proper states using the following rules, considering these anomalies like a process of scattering risked by the particle with negative potential effect of energy:

	Particle $\phi(1)$	Anti-particle $\phi(-1)$
Input	$\phi(1)$ •——————▸o Positive future	$\phi(1)$ •◀——————o negative future
Output	$\phi(1)$ o——————▸• Positive past	$-\phi(1)$ o◀——————• negative past

Table 2. Past and future in the scattering effect of the field X.

The negative actions in one perturbation created by an anomaly in the quantum field X, acts deviating and decreasing the action of the "healthy" quantum energy states ϕ_i ($i = 1, 2, \ldots$), in the re-composition of field (see the example explained in the figure 6).

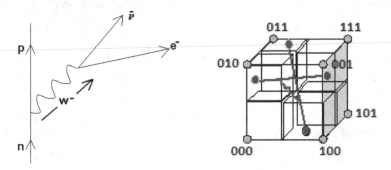

Fig. 6. a) Feynman diagram to a negative boson field. b) Cube of the net of the configuration space.

Example 2. The energy in this Feynman diagram is the given by $E_{output} = W^- = -E$, (negative boson in the field $\mathfrak{d}^a{}_{\mu\nu}$(of interactions given in $SU(3)$) (Holstein, 1991). Then their path integral to output energy is: $I^{\alpha\beta} = \int_C 0_{(\Gamma)}\phi = -1$ (see figure 6 a)). For other side, the cube of the net of the configuration space $C^0{}_\Gamma(\Psi)$ of the space-time $\Omega(\Gamma)$ is the 3-dimensional cube to arrangements in 000-box (see figure 6 b)).

4. On a fundamental theorem to correction and restoring of fields and their corollaries

One result that explains and generalises all actions of correction and restoring of a quantum field including the electromagnetic effects that observes with vector tomography is:
Theorem 1 (F. Bulnes) (Bulnes et al., 2010). Be $M = X(M)\backslash M$. Be a set of singular points of M, such that the states of $X(M)$, in these points are distorted states of the field X. An integral of line $I_{\alpha\beta} \in H^1(PM^{\pm}; O(-k))$, to k a helicity in M, determine an answer of the transformation $I_{\alpha\beta}X(M)$, that it is an appropriate width to correct the field $X(M)$, under the action of the operator $D_F(M)$, such that (10) is satisfied, then the integral of line that re-establishes the field and recomposes the part $X(M)$ comes given for

$$I^{\alpha\beta} = \int_\gamma \left[\iint_M D_F(x(s))X(Z^\alpha, W_\alpha, X^\alpha, Y)\omega \right] \mu_s,$$

Correction, Alignment, Restoration and Re-Composition of Quantum Mechanical Fields of Particles by
Path Integrals and Their Applications

231

The effect on the field is re-construct and re-establish their lines of field (channels of enery) by synergic action (see figure 8).

Proof. (Bulnes, 2006a; Huggett, 1990).

The fundamental consequences are great, and they have to do with the reinterpretation of the anomalies of the field in an electromagnetic spectra (Schwinger, 1998), (see the figure 7), which we can measure across detectors of electromagnetic radiation, detectors and meters of current, voltage or amperage calibrated in micro or nano-units (Bulnes et al., 2011).

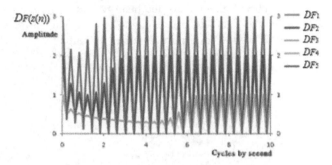

Fig. 7. Electronic propagators measuring corrective and restorer actions.

An important result (that can be a consequence in a sense of the previous one (for example in integral geometry and gauge theory)) that applies the vector tomography to electromagnetic fields used to measure fields of another nature and classify the anomalies by their electromagnetic resonance (at least in the first approach) is given by:

Theorem 2 (Bulnes, F) (Bulnes, 2006b). If the Radon transform (tomography on $X(M)$) is not defined, is infinite or has the value of zero, the corresponding pathologies are: great emission of electromagnetic radiation, current or voltage (points unless polarity *due to the atomic degradation (isotopes)*, have a node with variation not bounded of current, voltage or resistance (it is loose or is much *(ponds of energy))* *due to an existence of positron states* (like the defined in table 2)), has a peak or is a node, due to that have not unique value or this is indeterminate (not have determined direction, can have a *source of increase scattering*).

Proof. (Bulnes, 2006a).

In the demonstration of the theorem 1, the Stokes theorem guarantees the invariance of the value of the integrals of path under the application of an electromagnetic field (Landau & Lifshitz, 1987), like gauge of a quantum field, since the value of these integrals does not depend on the contour measured for the detection of a field anomaly (Bulnes et al., 2011).

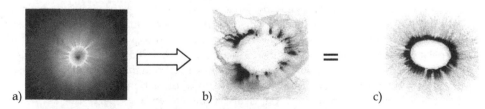

a) b) c)

Fig. 8. The field in a) is the radiation electromagnetic spectra to recompose and restore the field X, given in b). The corresponding image in c), is the field restored and corrected after the application a) in b).

5. Some applications to nano-medicine, nano-engineering and nano-materials

5.1 Application to nanomedicine

In nanomedicine the applications of the corrective actions and restorers of a field are essential and they are provided by the called integral medicine, which bases their methods on the regeneration of the codes of cellular energy across the conduits of energy of the vital field that keeps healthy the human body, the above mentioned for the duality principle of mind-body. But the transformations are realised in the quantum area of the mind of the body, that is to say the electronic memory of the healthy body. The mono-pharmacists of integral medicine contain codes of electronic memory at atomic level that return the information that the organs have lost for an atomic collateral damage.

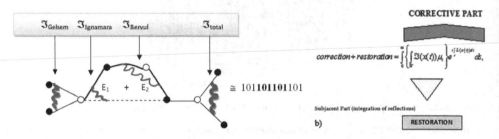

Fig. 9. Diagram of strings and path integrals of intelligence code of cure.

Diagram of strings belonging to the cohomology of strings equivalent to the code of electronic memory spilled to a patient sick with the duodenum (Bulnes et al., 2011) (see figure 9 a)). In nanomedicine, the path integrals are intelligence codes of corrective and restoration actions to cure all sicknesses. In the (see figure 9 b), W, is the topological group of the necessary reflections to the recognition of the object space of the cure (Bulnes et al., 2010). This recompose the amplitude of the wave defined in the spectra $I^{\alpha\beta}X(B)$, (with B, the human body) in the context of the space-time that to our nano-metric scale this space is constituted of pure energy. Into this space transits the geodesics or paths, to each particle where to each one of those paths exist a factor of weight given by $\exp(i\Im/\hbar)$, with $\hbar$, the constant of Max Planck and $\Im$, is the classical action associate to each path (see figure 8 b)).

5.2 Application to nanomaterials

The study of the resultant energy due to the meta-stables conditions that it is obtains in the quasi-relaxation phenomena establishes clearly their plastic nature for the suffered deformations on the specimen. Nevertheless their study can to require the evaluation of the field of plastic deformation on determined sections to a detailed study on the liberated energy in the produced dislocations when the field of plastic deformation acts. Thus, it is doing necessary the introduction of certain evaluations of the actions of the field to along of the dislocation trajectories in mono-crystals of the metals with properties of asymptotic relaxation. Then we consider like specimens, mono-crystals of *Molybdenum* (Mo), (see figure 10) subject to stress tensor that produce the plastic deformation given by the action inside of path integral

Correction, Alignment, Restoration and Re-Composition of Quantum Mechanical Fields of Particles by
Path Integrals and Their Applications

233

$$\int_M L\varepsilon(t)d(\varepsilon(t)) = \int_0^{+\infty}\left[\int_M d\varepsilon_{PT}\right]\varphi(\tau)e^{-\tau t}d\tau, \tag{60}$$

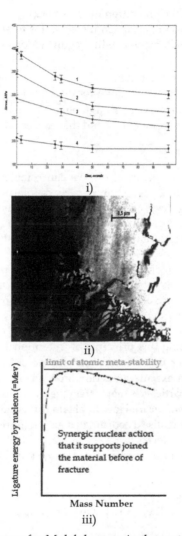

i)

ii)

iii)

Fig. 10. i). Quasi-relaxation curves for Molybdenum single crystal: 1.- σ_0 = 396 MPa, 2.- σ_0 = 346 MPa, 3.- σ_0 =292 MPa, 4.- σ_0 = 208 MPa. Mo <100> {100}, at T= 293 °C. ii). Image of the electronic microscope of high voltage, HVTEM of Molybdenum single crystal in regime of quasi-relaxation. iii). Atomic meta-stability condition.

By the theorem of Bulnes-Yermishkin (Bulnes, 2008), all functional of stress-deformation to along of the time must satisfy for hereditary integrals in the quasi-relaxation phenomena that have considered the foreseen actions inside of trajectory of quasi-relaxation like path integrals measuring field actions on crystal particles of metals:

$$\Gamma(\sigma - \varepsilon, t) = \int\limits_{0}^{+\infty}\left[\int\limits_{0}^{t}\sigma(t)d\varepsilon(t)\right]\varphi(\tau)e^{-\tau t}d\tau, \tag{61}$$

The square bracket in (60), is the one differential form $\omega(\Gamma_\varepsilon)$, using the property (42), on the space-time $\Omega(\Gamma)$. The figure 10 iii), establish the behavior to atomic level in the tendency of the mono-crystals to be joined in meta-stability regime (Alonso & Finn, 1968).

5.3 Nanoengineering and nanosciences

Since it has been mentioned previously if we consider a set of particles in the space $\mathbb{E}$, under certain law of movement defined by their Lagrangian L, we have that the action defined by a field that expires with this movement law and that causes it is defined by the map:

$$\mathfrak{I} : T\mathbb{E} \to \mathbb{R},$$

with rule of correspondence as given in (8), we can establish that the global action in a particles system with instantaneous action can be re-interpreted locally as a permanent action of the field considering the synergy of the instantaneous temporary actions under this permanent action of the field. This passes to the following principle:

Principle. The temporary or instantaneous action on a global scale can be measured like a local permanent action.

The previous proposition together with certain laws of *synchronicity of events* in the space-time will shape one of the governing principles of the nanotechnology, why? Because at microscopic level the permanence of a field is constant in proportion to the permanence and the interminable state of energy that exists in the atoms. As a result of it a nano-technological process will be directed to the manipulation of the microstructures of the components of the matter using this principle of "intentional" action. Then supposing that the field X, can control under finite actions like the described ones for $\mathfrak{I}$, and under the established principle, we can execute an action on a microstructure always and when the sum of the actions of all the particles is major than their algebraic sum (*to give an order to only one particle so that the others continue it*). How to obtain this combined effect of all the particles that move and that is wanted realise a coordinated action (of tidy effect) and simultaneously (synchronicity), with the only effect?

Inside the universe of minimal trajectories that satisfies the variation functional (12) we can choose a $\gamma_t \in \Omega(\Gamma)$, such that

$$\text{Exe}\mu_t = \int\limits_{\gamma_t}\left(\int\limits_{p_1}^{p_2}L(x(s))d(x(s))\right)\mu_{\gamma_t}, \tag{62}$$

which is not arbitrary, since we can define any action on γ_t, like

$$\mathfrak{I}_{\gamma_t} = \int\limits_{p_1}^{p_2}L(x(s),\dot{x}(s),s)ds, \tag{63}$$

that is to say, there exists an intention defined by the field action that infiltrates into the whole space of the particles influencing or "infecting" the temporary or instantaneous actions doing that the particles arrange themselves all and with added actions not in the

algebraic sense, but in the holistic sense. This action is the "conscience" that has the field to
exercise their action in "intelligent" form that is to say, in organized form across his path
integrals like the already described ones. Then extending the above mentioned integral to
the whole space $\Omega(\Gamma)$, we have the synergic principle of the whole field X,

$$\Im_{TOTAL} \geq \sum_{j} \int_{\gamma_t} \Im_j(x(s)) d(x(s)), \tag{64}$$

the length and breadth of E. The order conscience is described by the operator of execution
of a finite action of a field X, on a target (region of space that must be infiltrated by the
action of the field which is that for which we realise our re-walked $\Omega(\Gamma)$).
How to measure this transference of conscience of transformation due to the field X, on an
object defined by a portion of the space $\Omega(\Gamma)$? What is the limit of this supported action or
transference of conscience so that it supplies effect in the portion of the space $\Omega(\Gamma)$, and the
temporary or instantaneous actions for every particle x^i, are founded on only one synergic
global action on Ω?
We measure this transference of conscience (or intention) of X, on a particle $x(s)$, by means of
the value of the integral of the intelligence spilled (path integral) given like (Bulnes et al., 2008):

$$< \tau_\alpha X(x(s)), x_\gamma > = \int_{\Omega(\Gamma)} \Im(x_\delta \circ x_\sigma \circ x_\delta^{-1})(\phi_\sigma(x_\eta)\phi_\sigma(x_\gamma)\phi_\sigma^{-1}(x_\eta))\mu_\sigma, \tag{65}$$

We let at level conjecture and based on our investigations of nanotechnology and advanced
quantum mechanics, that a sensor for the quantum sensitisation of any particle that receives an
instruction given by a field X, must satisfy the inequality of Hilbert type (Bicheng, 2009), for
this transference of conscience defined in (65) on the region $\Omega(\Gamma)$, to know (see figure 11 c)):

$$< \tau_\alpha X(x(s)) t_\gamma > \leq \left\| \log \phi_\sigma(x_\eta) \right\|^a \left\| \log \Im(x_\sigma) \right\|^b, con \quad a = b = 2, \tag{66}$$

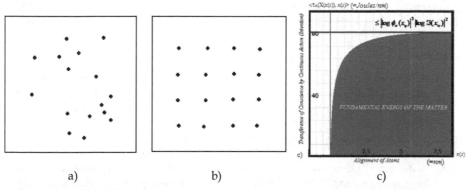

Fig. 11. a) Free particles. b) Transference of conscience in the particles. c) Transference of
conscience by continuous action.

Example 3. A force is spilled $F(x(s)^i)$, generated by a field that generates a "conscience" of
order given by their Lagrangian. For it does not have to forget the principle of energy

conservation re-interpreted in the equations of Lagrange, and given for this force

like $\dfrac{d}{dt}\left(\dfrac{\partial T}{\partial \overset{\bullet}{x}{}^{j}}\right) - \dfrac{\partial T}{\partial x^{j}} = F^{j}{}_{(x(s))}$, (also acquaintances as "living forces") transmitting their

momentum in each ith-particle of the space E, creating a region infiltrated by path integrals of trajectories $\Omega(\Gamma)$, where the actions have effect. Here T, is their kinetic energy. It was considered to be a transference of conscience (intention) given by the product $<\tau_{\alpha}\Im(x(s))$, $x(s)> = [(\log(x) + 2)^{\wedge}2]^{*}[[(\log(x) - 4.00000005)^{\wedge}2]$. Observe that the object obtains their finished transformation in an established limit. The above mentioned actions of alignment might be realised by displacements in (= nm) (see figure 11 b)).

6. Conclusions

Finally and based on the development that the quantum mechanics has had along their history, we can affirm that the classic quantum mechanics evolves to the advanced quantum mechanics (created by Feynman) and known like quantum electrodynamics reducing the uncertainty of Heissenberg of the frame of the classic quantum mechanics, on having established and having determined a path or trajectory of the region of space-time where a particle transits. Therefore the following step will demand the evolution of the quantum mechanics of Feynman to a synchronous quantum mechanics that should establish rules of path integrals that they bear to an effect of simultaneity and coordination of temporary actions on a set of particles that must behave under the same intensity that could be programmed across their "revisited" path integrals, producing a joint effect called synergy. The time and the space they are interchangeable in the quantum area as we can observe it in the integrals (61). Where a particle will be and when it will be there, are aspects that go together. This way the energy is not separated from the space-time and forms with them only one piece in the mosaic of the universe.

7. Acknowledgment

I am grateful with Carlos Sotero Zamora, Eng., and with Juan Maya Castellanos, Msc., for the help offered for the digital process of the images that were included in this chapter.

8. References

Alonso, M. & Finn, E. J. (1968) *Fundamentals University Physics Vol. 3: Quantum and statistical physics* (1st Edition) Addison Wesley Publishing Co., ISBN: 0-201-00232-9, USA.

Barry, S. (2005). *Functional Integration and Quantum Physics*. (2nd edition) AMS, 08218-3582-3, USA

Bicheng, Y. (2009). *Hilbert-Type Integral Inequalities* (Bentham eBooks) <http://benthamscience.com/ebooks/9781608050550/index.htm>

Bulnes, F., Bulnes F. Hernández, E. & Maya, J. (2011) Diagnosis and spectral encoding in integral medicine through electronic devices designed and developed by path integrals. *Journal of Nanotechnology in Engineering and Medicine*, Vol. 2, 2, 021009, (May, 2011), pp.(021009(1-10)) ISSN: 1949-2944

Bulnes, F., Bulnes, F. H., Hernández, E., Maya, J. & Monroy, F. (2010) Integral medicine: new methods of organ-regeneration by cellular encoding through path integrals applied

Correction, Alignment, Restoration and Re-Composition of Quantum Mechanical Fields of Particles by
Path Integrals and Their Applications

237

to the quantum medicine *Journal of Nanotechnology in Engineering and Medicine*, Vol. 1, Issue 3, 031009 issue (August, 2010), pp.(031009(1-7)), ISSN: 1949-2944

Bulnes, F. (2008) Analysis of Prospective and Development of Effective Technologies through Integral Synergic Operators of the Mechanics, *Proceedings of Cuban Mechanical Engineering Congress/CCIA*, ISBN: ISBN: 978-959-261-281-5, ISPJAE, Habana, Cuba, 12-08

Bulnes, F. & Shapiro, M. (2007) *On general theory of integral operators to analysis and geometry*, (1st edition) J. Cladwell, *SEPI-IPN, IMUNAM*, ISBN: 978-970-360-442-5, Mexico

Bulnes, F. (2006a). *Doctoral course of mathematical electrodynamics*, *Proceedings of Conferences and Advanced Courses of Appliedmath 2*, ISBN: 970-360-358-0, Sepi-ESIME-IPN Mexico, 10-06

Bulnes, F. (2006b). *Integral theory of the universe*, *Proceedings of Conferences and Advanced Courses of Appliedmath 2*, ISBN: 970-360-358-0, Sepi-ESIME-IPN Mexico, 10-06

Bulnes, F. (2006c). *Design of Algorithms to the Master in Applied Informatics* (1st Edition), UCI, ISBN: 978-959-08-0950-7, Cuba

Bulnes, F. (2005). *Conferences of Lie Groups* (1st Edition) Sepi-IPN, IMUNAM, ISBN: 970-36-028-43 Mexico

Feynman, R. (1967) *The Character of Physical law*, (1st Edition) MIT Press Cambridge Mass, ISBN: 026-256-00-38, Massachusetts, USA

Fujita, S. (1983). *Introduction to non-equilibrium quantum statistical mechanics* (2nd Edition), W. Krieger Pub. Co. (Malabar, Fla.), ISBN: 0898745934, USA

Gorbatov, V. A. (1986). Foundations *of the discrete mathematics* (3rd Edition), Mir Moscu, Russia

Holstein, B. R. (1991) *Topics in Advanced Quantum Mechanics* (3rd Edition), Addison-Wesley Publishing Co., (Advanced Book Program), ISBN: 0201-50820-6, USA

Huggett, S. (1990). Cohomological and Twistor Diagrams, In: *Twistors in Mathematics and Physics*, Bayley T. N. and Baston, R. J., (156), (pp218-245), Cambridge University Press, ISBN: 978-052-139-783-4, UK

Ionescu, L. M. (2004) Remarks on quantum theory and noncommutative geometry. *Int. J. Pure and Applied Math.*, Vol. 11, No.4, (10-08), pp.(363-376), ISSN:1842-6298

Kac, V. (1990) *Infinite dimensional Lie Algebras* (3rd Edition), Cambridge University Press, ISBN: 0-521-37215-1, Cambridge, USA

Kontsevich, M. (2003). Deformation quantization of Poisson manifolds I, *First European Congress of Mathematics*, *Lett.Math.Phys* Vol. 66, issue 3, pp.(157-216), ISSN: 0377-9017

Landau, L., and Lifshitz. (1987) *Electrodynamics in Continuum Media* Vol. 8 (2nd Edition) Elsevier Butterworth Heinemann, Oxford, UK

Marsden J. E., Abraham R. & Ratiu, T. (1983) *Manifolds, tensor analysis and applications* (2nd Edition), Addison-Wesley Publishing Co, ISBN: 0-201-10168-8, Massachusetts, USA

Oppenheim, A, V., Willsky, A. S. & Young, I. T. (1983). *Signals and Systems* (1st Edition), Prentice Hall, ISBN: 0-13-809731-3, USA

Penrose R. & Rindler, W. (1986). *Spinors, and twistor methods in space-time geometry* (1st Edition), Cambringe University Press, ISBN: 0-521-25267-9, CA, UK

Schwinger, J. (1998). *Particles, sources and fields Vol. 1* (3rd Edition), Perseus Books, ISBN: 0-7382-0053-0, USA

Shapiro, M. & Kravchenko, V. (1996) *Integral Representations for Spatial Models of Mathematical Physics* (1st Edition), CRC Press. Inc., ISBN: 0582297419, USA

Simon, B. & Reed, M. (1980) *Functional analysis, Vol. 1 (Methods of modern mathematical physics)* (1st Edition), Academic Press, ISBN: 0125850506, N. Y., USA

Sokolnikoff, S. I. (1964). *Tensor analysis: Theory and applications to geometry and mechanics of continua* (2nd Edition), John Wiley and Sons, ISBN: 0471810525 9780471810520, N. Y, USA

Watanabe T. (2007), Configuration space integral for long n-knots and the Alexander polynomial. *Journal of Algebraic & Geometric Topology Mathematical Sciences Publishers*, Vol. 7, February 2007, pp.(47–92) ISSN 1472-2747

9

Theoretical Validation of the Computational Unified Field Theory (CUFT)

Jonathan Bentwich
Brain Perfection LTD
Israel

"No better destiny could be allotted to any physical theory, than that it should of itself point out the way to the introduction of a more comprehensive theory, in which it lives on as a limiting case"

(Einstein, 1916)

1. Introduction

A previous article (Bentwich, 2011c) hypothesized the existence of a novel 'Computational Unified Field Theory' (CUFT) which was shown to be capable of replicating quantum and relativistic empirical phenomena and furthermore may resolve the key inconsistencies between these two theories; The CUFT (Bentwich, 2011c) is based upon three primary postulates including: the computational *'Duality Principle'* (e.g., consisting of an empirical-computational proof for the principle inability to determine the "existence" or "non-existence" of any hypothetical 'y' element based on its direct physical interaction with another exhaustive set of 'x' factors) (Bentwich, 2003 a,b,c; 2004; 2006a); the existence of an extremely rapid series (e.g., c^2/h) of *'Universal Simultaneous Computational Frames'* (USCF's) which comprise the entire corpus of the physical spatial universe at any given minimal (quantum temporal 'h') point (which is computed by a 'Universal Computational Principle', '?'); and the existence of three *'Computational Dimensions'* – e.g., of 'Framework' ('frame' vs. 'object'), 'Consistency' ('consistent' vs. 'inconsistent') and Locus ('global' vs. 'local'); Taken together these three basic theoretical postulates give rise to the CUFT's 'Universal Computational Formula':

$$\frac{c^2 x'}{h} = \frac{s \times e}{t \times m}$$

which fully integrates between the four basic physical properties of 'space', 'time', 'energy' and 'mass' and is capable of replicating all known quantum and relativistic phenomena, while resolving the apparent contradictions between Quantum Mechanics and Relativity Theory (such as for instance the existence of the relativistic speed of light limit as opposed to the quantum entanglement's instantaneous phenomenon).

Moreover, even beyond the capacity of the CUFT to replicate all known quantum and relativistic phenomena as well as resolve their key theoretical inconsistencies (and differences), the CUFT was postulated to broaden the scope of our theoretical understanding of physical reality thereby qualifying as a potential candidate for a 'Theory of Everything' (TOE) (Brumfiel, 2006; Einstein, 1929, 1931, 1951; Ellis, 1986; Greene, 2003). Specifically, based on the (abovementioned) integrated postulates of the 'Duality Principle', existence of the 'Universal Simultaneous Computational Frames' (USCF's) and three (Framework, Consistency and Locus) Computational Dimensions the CUFT describes the four basic physical properties of 'space', 'time', 'energy' and 'mass' as emerging (secondary) computational properties that arise as a result of various 'Framework x Consistency x Locus' combinations – as computed (by the Universal Computational Principle, '˥') based on the rapid series of USCF's…

However, in order to fully validate this new (hypothetical) 'Computational Unified Field Theory' (CUFT), it is necessary to further extend its theoretical framework to bear on (at least) two important aspects: i.e., to identify particular instances in which the predictions of the CUFT *critically differ* from those of both Quantum and Relativistic models, and to demonstrate the potency of the CUFT in broadening our understanding of key scientific phenomena (e.g., while demonstrating the need to perhaps reformulate these key scientific computational paradigms based on the CUFT's new broader theoretical scientific framework);

Hence, the current chapter comprises two segments : the first critically contrasts (at least) three specific instances in which the critical predictions of the CUFT significantly differs from that of both quantum mechanics and Relativity theory; and the second, which delineates the application of one of the three major theoretical postulates of the CUFT (namely: the 'Duality Principle') in the particular cases of three key scientific (computational) paradigms (including: Darwin's 'Natural Selection Principle' and associated 'Genetic Encoding Hypothesis' and Neuroscience's basic materialistic-reductionistic 'Psycho-Physical Problem');

We therefore begin by identifying (at least) three specific empirical predictions of the CUFT which may critically differ from those predicted by the existing quantum and relativistic theoretical models;

a. Contrasting between the CUFT's *Universal Computational Formula's* '1' and '2' derivatives and their corresponding relativistic and quantum empirical predictions;

b. Contrasting between the CUFT's critical prediction regarding the differential number of times that a "massive" compound (or atom/s) will be presented (consistently) across a series of subsequent USCF's relative to the number of times that a "lighter" compound (or atom/s) will be presented across the same number of (serial) USCF's, and the corresponding predictions of Quantum or Relativistic theories.

c. Critically contrasting between the CUFT's prediction of the possibility of reversing any given object's spatial-temporal sequence (e.g., based on a computation of that object's serial electromagnetic values across a series of USCF's and reversal of these recorded values based an application of the appropriate electromagnetic field to that object's recorded serial USCF's electromagnetic values) – and the *negation* of any such capacity to reverse the 'flow of time' by both Quantum and Relativistic theories

2. The CUFT's universal computational formula's relativistic & quantum derivatives

The first (of these three) differential critical predictions for which the CUFT's empirical predictions may differ significantly from those of both quantum and relativistic theoretical models is based upon the CUFT's Universal Computational Formula:

$$\frac{c^2 x'}{h} = \frac{s \times e}{t \times m}$$

Specifically, whereas Relativity theory recognizes the equivalence of mass and energy (e.g., $E = Mc^2$), the unification of 'space' and 'time' as a four-dimensional continuum, and its curvature by mass – *Relativity Theory does not allow for the complete unification* (or transformation) of all of these *four basic physical features* (e.g., within one computational formula); In contrast, the CUFT's defines each of these four basic physical features in terms of their (particular) combination of three Computational Dimensions, e.g., 'Framework' ('frame'/'object'), 'Consistency' ('consistent'/'inconsistent') and 'Locus' ('global'/'local') which are all anchored in the *same singular* (higher-ordered D2) *rapid series of 'Universal Simultaneous Computational Frames' (USCF's)*. Thus, for instance it was shown (Bentwich, 2011c) that the computational definition of 'space' and 'energy', or of 'mass' and 'time' constitute exhaustive computational pairs delineating a *frame's* – or an *'object's* - consistent vs. inconsistent computational measures (e.g., across a series of subsequent USCF's, respectively); (In fact, it was precisely this *'computational exhaustiveness'* of these frame- or object- consistency measures that was suggested to offer an alternative explanation for the currently accepted quantum's probabilistic interpretation of the "collapse" of the 'probability wave function'.)

Indeed, it is hereby hypothesized that this unique capability of the CUFT's Universal Computational Formula to comprehensively unify between all four basic physical features (e.g., 'space', 'time', 'energy' and 'mass') – not only goes beyond the capacity of the existing (relativistic or quantum) theoretical models, but also produces particular (verifiable) empirical predictions that may critically from those offered by either quantum or relativistic theoretical models:

Thus, it is suggested that the two previously outlined (Bentwich 2011c) computational derivatives of the Universal Computational Formula:

$$\frac{s}{t} = \frac{m}{e} x \frac{c^2 x'}{h} \tag{1}$$

$$t \ x \ m \ x \ (\frac{c^2 x'}{h}) = s \ x \ e \tag{2}$$

may (in fact) provide precisely such (differential) critical predictions of the CUFT as opposed to their (respective) relativistic (1) and quantum (2) predictions.

This is because according to the CUFT's Universal Computational Formula's (1) derivative a (relativistic or quantum) object's 'space' value divided by its 'time' value is equivalent to that object's 'mass' value divided by its 'energy' value – multiplied by the square of the speed of light (c^2) divided by Planck's constant ('h') (e.g., and based on the higher-ordered D2 'ı' Universal Computational Principle's computation of the given series of USCF's); In the case of the CUFT, the computational rational for this equivalence of 's'/'t' = 'm'/'e' x (c^2/h x 'ı') stems from its stipulation that the ratio between an object's 'frame-consistent' ('s') and 'object-inconsistent' ('t') values should be the same as between that object's 'object-consistent' ('m') and 'frame-inconsistent' ('e') values – multiplied by the rate of universal computation (e.g., c^2/h) (e.g., as delineated in the previous publication: Bentwich, 2011c). However, when contrasting this particular CUFT's Universal Computational Formula (I) with its counterpart in Relativity Theory we find that even though relativistic theory possesses specific equivalences between 'energy' and 'mass' (e.g., the famous $'E = Mc^2')$ and the unification of 'space-time' as a 'four-dimension' continuum, it fails to account for any such comprehensive equivalence of $'s'/'t' = 'm'/'e' \times (c^2/h \times '^\neg')$; In fact, if we focus on the (above) relativistic 'energy-mass' equivalence $('E = mc^2')$ we can notice that the CUFT's Universal Computational Formula's (I) derivative $'s'/'t' = 'm'/'e' \times (c^2/h \times '^\neg')$ in fact *contains this 'energy-mass' equivalence but goes beyond that equivalence* to incorporate also its precise (hypothetical) relationship with the (ratio between) 'space' and 'time' as well as with the (hypothetical) rate of universal computation (e.g., $c^2/h \times '^\neg'$); These broader CUFT Universal Computational Formula's (I) relationships between 'space' and 'time' and the (hypothesized) 'universal computational rate' $(c^2/h \times '^\neg')$ – and Relativity's (known) 'energy-mass equivalence' could be represented in this manner:

$$'e' \times 's'/'t' = 'm' \times c^2/h \,(x \,'^\neg')$$

In this way, we can see that the CUFT Universal Computational Formula's (1 derivative) contains (and replicates) Relativity's core 'energy-mass equivalence' $('E = mc^2')$, but also goes beyond that particular relationship as embedded within a broader more comprehensive (singular) Universal Computational Formula's unification of the four basic physical features (e.g., of 'space', 'time', 'energy' and 'mass'). As such, the above first derivative of the Universal Computational Formula (1) points at a particular empirical instant in which one of the CUFT's critical predictions *differs* from those offered by Relativity Theory.

A second instance in which the CUFT's (critical) empirical prediction may differ from that of Quantum Mechanic is in the case of the CUFT Universal Computational Formula's (2) derivative – as it relates to an extension of quantum's (current) particular complimentary relationships between a subatomic object's or event's 'spatial' and 'energetic' or 'temporal' and 'mass' values: According to the current quantum mechanical account of Heisenberg' 'Uncertainty Principle' (Heisenberg, 1927) there exist (strict) complimentary relationships between an object's (or event's) 'spatial' and 'energetic' values or between its 'temporal' and 'mass' properties – e.g., such that their simultaneous measurement accuracy level cannot (in principle) exceed Planck's minimal 'h' value… theoretically, this is due to the (currently prevailing) 'probabilistic interpretation' of quantum mechanics which posits that it is due to the direct physical interaction between the subatomic 'probe' and 'target' elements that the probability wave function "collapses" – giving rise to a particular 'complimentary' spatial-energetic' or 'temporal-mass' value, and therefore that any increase in the accuracy measurement of any of these pairs' complimentary values (e.g., 'e' vs. 's'; or 't' vs. 'm') necessarily also brings about a proportional decrease in the measurement accuracy of the other complimentary pair's dyad); Hence, the current (probabilistic interpretation of) Quantum Mechanical theory posits a strict complimentary relationship between any subatomic target's (simultaneous) measurement of its 'spatial' and 'energetic' values or between its 'temporal' and 'mass' values – as necessarily constrained by the Uncertainty Principle:

$$'e' \times 's' \leq 'h';$$

or

$$'t' \times 'm' \leq 'h'.$$

In contrast, the second (2) derivative of the CUFT's Universal Computational Formula further broadens these apparently disparate quantum complimentary relationships (e.g., of an object's 'spatial' and 'energetic', or 'temporal' and 'mass' values) – to form a direct computational equivalence, e.g., based on the hypothesized 'mechanics' of a Universal Computational Principle's higher-ordered 'D2'/'ᵓ' integrated computation of a rapid series of USCF's that singularly define each of these 'complimentary computational' pairs;

$$t \times m \times \left(\frac{c^2 x^{\,\imath}}{h}\right) = s \times e$$

Specifically, the CUFT's Universal Computational Formula's (2) derivative hypothesizes that due to the fact that each of the four basic physical properties (e.g., of 'space', 'time', 'energy' and 'mass') is defined based on the same three fundamental (hypothesized) Computational Dimensions (i.e., of 'Framework' ['frame'/'object'], 'Consistency' ['consistent'/'inconsistent'] and 'Locus' ['global'/'local']) – which are all produced by the same *singular* (higher-ordered 'D2'/'ᵓ') rapid series *USCF's series*, then each of these complimentary computational pairs (i.e., of 'space' and 'energy', or 'mass' and 'time') exhaustively defines an object's given computational (USCF's based) measurement (Bentwich, 2011c). Thus, for instance, the CUFT hypothesized (Bentwich, 2011c) that an object's (or event's) 'temporal' value (e.g., which represents an 'object-inconsistent' USCF's Index) exhaustively compliments that object's 'mass' ('object-consistent') value, and likewise that an object's ('frame-consistent') 'spatial' USCF's measure exhaustively compliments its ('frame-inconsistent') 'energetic' measurement. Moreover, based (again) on the unification of all of these four basic physical properties as (secondary) computational combinations of the same three basic (abovementioned) Computational Dimensions (e.g., of 'Framework', 'Consistency' and 'Locus'), it was hypothesized that the computational relationship (i.e., multiplication) between an 'object-inconsistent' ('t') and 'object-consistent' ('m') measures should be equivalent to the (multiplication) relationships between a 'frame-consistent' ('s') and 'frame-inconsistent' ('t') values – i.e., while taking into considerations their production by a higher-ordered (D2) Universal Computational Principle's ('ᵓ') rapid universal computational rate (e.g., $\frac{c^2 x^{\,\imath}}{h}$);

Hence, the (2) derivative of the CUFT's Universal Computational Formula is expressed by the above:

$$t \times m \times \left(\frac{c^2 x^{\,\imath}}{h}\right) = s \times e$$

But, note that this second derivative of the CUFT's Universal Computational Formula goes beyond the (abovementioned) current Quantum Mechanical Uncertainty Principle's complimentary measurement constraint stipulation:

$$\text{'e' x 's'} \leq \text{'h'};$$

or

$$\text{'t' x 'm'} \leq \text{'h'}.$$

This is because whereas the CUFT's (2) derivative explicitly stipulates that the multiplication relationship between the complimentary pair of {'t' and 'm'} is *equivalent* to that of {'e' and 'm'} (i.e., while taking into account its delineated relationship with the hypothetical rate of universal computation: $\frac{c^2 x^{\,\imath}}{h}$), current quantum mechanical (probabilistic) formulation only allows for a (partial) direct (multiplication) relationship between each of these complimentary pairs (e.g., independently).

Hence, an empirical contrast between the CUFT's Universal Computational Formula's (2) derivative and its corresponding quantum predictions also points at (potentially) significant differences between these theoretical models;

2.1 The CUFT's differential USCF's presentations of "massive" vs. "light" elements

Another interesting instance in which the predictions of the CUFT and Relativistic or Quantum models may critically differ is associated with the CUFT's computational definitions of an object's (relativistic or subatomic) "mass" value; Based on the CUFT's previous (Bentwich, 2011c) computational definition of an object's "mass" as the number of 'object-consistent' presentations across any series of USCF's – it was shown that the computation of the number of such 'object-consistent' ("mass" measure) from the 'global' ('g') framework of a relativistic object may indeed produce an increased 'mass' measure relative to its 'local' "mass" measure. This was due to the greater number of times such a relativistic object would be presented from the 'global' perspective relative to its 'local' perspective' based on the increased number of ('global') pixels such a relativistic object would have to traverse across a given series of USCF's (e.g., which would nevertheless not affect its 'locally' measured number of presentations).

However, a further extension of the CUFT's basic computational definition of an object's "mass" (e.g., as the number of 'object-consistent' presentations across a certain given number of USCF's) – when viewed from the *'local'* framework perspective and when contrasting between relatively "massive" objects and "lighter" objects may in fact point an another interesting instance in which the critical predictions of the CUFT and Relativistic or Quantum models may differ significantly; This is because since the CUFT defines the "mass" of an object as the number of 'object-consistent' presentations across a series of USCF's, then when we compute a (relatively) "massive" object as opposed to a (relatively) "light" (relativistic) object – i.e., from the 'local' framework perspective, we obtain that the more "massive" object is necessarily presented a greater number of times across the same given number of USCF's than the "lighter" object! This means that according to the CUFT's critical prediction, we should obtain a difference in the number of (consistent) presentations of any two objects that (significantly) differ in their "mass" value... Thus, to the extent that we are capable of measuring a sufficiently minute number of consecutive USCF's, the CUFT predicts that more "massive" objects should appear consistently on a larger number of such consecutive USCF's – as opposed to (relatively) "lighter" objects which should appear less frequently across such a series of consecutive USCF's; More specifically, it is predicted that if we chose to examine the number of (consecutive) USCF's in which a (relatively) more "massive" compound (or element) appears – relative to a less "massive" compound (or element), then according to the CUFT we should expect to detect the more "massive" compound (or element) on a larger number of (consecutive) USCF's relative to the less "massive" compound or element; In contrast, according to the existing quantum or relativistic models of physical reality, the difference between more "massive" objects and less "massive" objects arises from the number of atoms comprising such objects, or differences in the weight of their nucleus etc.; However, there are no such known (quantum or relativistic) differences across elements (compounds or atoms, etc.) which possess differential mass values in terms of the "frequency" of their presentations (i.e., across a series of subsequent USCF's)...

Obviously, with the discovery of the CUFT's (hypothetical) far more rapid rate of USCF's computation (e.g., c2/h) than the currently assumed quantum or relativistic (direct or indirect) relationships, and augmented by the CUFT's USCF's emerging (secondary computational) properties of "mass", "space", "energy" and "time" – such a critical contrast between the CUFT's empirically predicted *greater* number of (object-consistent)presentations of (relatively more)"*massive*" relative to a *lesser* number of (object-consistent) presentations for "*lighter*" objects, and the complete lack of such a prediction by either quantum or relativistic models may test the validity of the CUFT (as opposed to either quantum or relativistic theories).

2.2 Reversibility of USCF's spatial-temporal sequence

Another (intriguing) critical prediction of the CUFT which (significantly) differs from the current quantum or relativistic models of physical reality is regarding the potential capacity to alter the "spatial-temporal sequence" of any given (quantum or relativistic) phenomenon; The critical issue is that according to both quantum and relativistic theories the "*flow of time*" may only proceed in *one direction* (e.g., from the 'past' towards the 'future' – but not vice versa), which is often termed: the "arrow of time"; This is because from the standpoint of Relativity Theory there exists a strict 'speed of light' limit set upon the transmission of any signal or upon the speed at which any relativistic object can travel – which therefore prohibits our capacity to "catch-up" with any signal emanating from an event in the 'past'- or with any actual- event/s that has happened in the 'past'; The only tentative (hypothetical) possibility to re-encounter any such 'past' space-time events from the standpoint of Relativity is in a case in which there is an extreme curvature of space-time (due to the presence of extremely massive objects) which may create closed 'space-time loops' that may allow past signals to "turn around" and arrive back to the observer... But, even in this (rare) hypothetical instance, our hypothetical capacity would be only to *witness* a light signal that has *emanated from an event* that took place in the '*past*' – rather than any "real" capacity to "*reverse*" *the spatial-temporal sequence of events or occurrences associated with the flow of time*... Hence, from the perspective of Relativity Theory, we cannot "reverse" the flow of time – e.g., cause spatial-temporal events (or objects) to occur in the "reversed" order...

Likewise, from the perspective of Quantum Mechanics, there seems to exist a clear limit set upon our capacity to "reverse" the "flow of time" – due to the fact that our entire knowledge of any subatomic 'target' (phenomenon) is strictly dependent upon- (and is therefore also constrained by-) that 'target' element's direct (or indirect) physical interaction with another subatomic 'probe' element. Hence, according to the (current) probabilistic interpretation of Quantum Mechanics, the determination of the (complimentary) 'space-energy' (s/e) or 'temporal-mass' (t/m) values of any given subatomic 'target' phenomenon (or phenomena) can only be determined *following* its direct (or indirect) *physical interaction/s* with another subatomic 'probe' element; But, note that according to such (probabilistic interpretation of) Quantum Mechanical theory the subatomic 'target' element is dispersed 'all along' a 'probability wave function' *prior to its interaction with the probe element* – but "collapses" into a single (complimentary) 'space-energy' or 'temporal-mass' value immediately following its direct (or indirect) physical interaction with the 'probe' element. This means that according to current (probabilistic) quantum mechanical theory, there exists a clear "unidirectional" (asymmetrical) 'flow of time' – i.e., one in which the determination of any subatomic (target) phenomena can be determined only *following the collapse of the probability wave function*' which takes place as a result of the direct (or indirect) physical interaction between the (probability wave function's) 'target' and other subatomic 'probe' element (Born, 1954); Now, as shown previously (Bentwich, 2011b) the computational structure (implicitly) assumed by the (above) probabilistic interpretation of quantum mechanics produces a 'Self-Referential Ontological Computational System' (SROCS) – which was shown to inevitably lead to both 'logical inconsistency' and 'computational indeterminacy' that are contradicted by known quantum empirical findings and which therefore also pointed at the computational 'Duality Principle' (e.g., asserting the existence of a conceptually higher-ordered 'D2' computational level that is capable of computing the simultaneous "co-occurrence" of any exhaustive hypothetical 'probe-target' pairs series). But, even beyond the Duality Principle's challenging of the current (implicit SROCS computational structure underlying) the probabilistic interpretation of Quantum Mechanics, note that it is precisely this SROCS assumed computational structure – which prohibits the capacity of any "collapsed" target element (or phenomenon) to "revert back to its 'un-collapsed' probability wave function state"! Therefore, it becomes clear that from the perspective of (probabilistic) Quantum Mechanical theory we cannot reverse any spatial-temporal quantum event/s, phenomenon or phenomena...

In contrast, the CUFT postulated the existence of a (conceptually higher-ordered) rapid series (e.g., c^2/h x 'ר') of 'Universal Simultaneous Computational Frames' (USCF's) which give rise to all (secondary) computational properties of 'space', 'time' 'energy' and 'mass'; Specifically, the computational definition of "time" was given through a measure of the number of instances that an object is presented *inconsistently* across a given series of USCF's – relative to the USCF's displacement of the speed of light: $t : \sum oj\{x,y,z\}$ [USCF(n)] $\neq o(i...j-1)\{(x),(y),(z)\}$ [USCF($1...n$)] /c x n\{USCF's\}

Therefore, the less instances in which a given object is presented 'inconsistently' (across a given series of USCF's), the less 'time' passes for that object – e.g., as may be observed from a 'global' framework in the case of its measurement of a high speed relativistic observer or in the case of a 'massive' object etc.

In much the same way, an object's "spatial" or "mass" or "energy" values– are all derived based on differential (e.g., 'frame-consistent', 'object-consistent' or 'frame-inconsistent', respectively) secondary computational measures of the various combinations of the three (abovementioned) Computational Dimensions. As a matter of fact, all of these four basic physical features of 'space', 'time', 'energy' and 'mass' were entirely integrated within the singular 'Universal Computational Formula' (Bentwich, 2011c) which outlined their intricate relationships with the singular (conceptually higher-ordered 'D2/'ר') rapid series (c^2/h) of USCF's. One final key factor associated with the CUFT's conceptualization of the "flow of time" may arise from its replacement of any (hypothetical) '*causal*-material' ($x{\rightarrow}y$) relationship between any two hypothetical 'x' and 'y' factors – by a conceptually higher-ordered '*D2 a-causal*' computation which can compute the "co-occurrences" of any two such given 'x' and 'y' elements at any given spatial-temporal point/s (for any particular USCF(i) frame.

This is because when we take into consideration the CUFT's integrated postulates of the Duality Principle's (conceptual) proof for the inability to determine the "existence" or "non-existence" of any (hypothetical) 'y' element based on its direct physical interaction (e.g., at di1...din) with another (exhaustive set of) 'x' factor/s; and the existence of a (rapid series of) 'Universal Simultaneous Computational Frames' (USCF's) which are *computed simultaneously for all of the exhaustive pool of 'spatial pixels' that exist per any given (discrete) USCF(i)* (e.g., by the Universal Computational Principle, 'ר') – this leads to the CUFT assertion that there cannot exist any real "causal" relationship/s between any two hypothetical 'x' and 'y' elements (Bentwich, 2011c)... Instead, the CUFT postulated that all (exhaustive series) of 'universal spatial pixels' must be computed simultaneously as part of a particular (discrete) USCF(i) frame. Therefore, the CUFT also asserted that the appearance of any "material-causal" relationship between any two given 'x' and 'y' elements may only arises as a result of a certain (apparent) 'spatial-temporal patterns' emerging across a series of USCF's – rather than as the result of any "real" (e.g., direct or indirect) physical interaction/s between the 'x' and 'y' factors (e.g., constituting a given SROCS quantum or relativistic paradigm);

Therefore, the CUFT's standpoint (and ensuing empirical critical predictions) with regards to the issue of the "flow of time" may differ significantly from the strict 'unidirectional' and 'un-altered' "flow of time" assumed by both quantum and relativistic models; This is due to the fact that according to the CUFT, the computational "time" measure of any object – i.e., whether it relates to the 'passage of time' (e.g., including the possibility of 'time dilation') or to the 'direction or time' (e.g., including the currently assumed "arrow of time" by relativistic and quantum theories) is entirely contingent upon the number of inconsistent presentations of that object across a given series of USCF's, as well as the particular USCF's spatial-temporal "sequence" underlying the development of a given phenomenon (or particular 'sequence of events'); In order to explicate the CUFT's critical prediction regarding the possibility to "reverse the flow of time" – i.e., at least as it applies to a particular given object, let us analyze the (standard) "flow or time" as it applies say to the developmental processes taking place in a small plant (or ameba); According to the CUFT, the "*flow of time*" associated with such a small plant's growth essentially comprises a particular *sequence of spatial-temporal as well as energetic-and mass- changes* taking place in the particular plant – *across a series of USCF's*. In fact, based on the CUFT's postulated (higher-ordered 'D2') series of discrete USCF's that are comprised of an exhaustive universal pool of "spatial-pixels" (being computed for each individual USCF), a further postulate of the CUFT is that each of this exhaustive pool of 'universal spatial pixels' constitutes a particular *electromagnetic value* which is specific to a given spatial point within a particular USCF frame (e.g., as a single electromagnetic value). Thus, the "flow of time" associated with the growth of this given plant essentially comprises a particular series of specific electromagnetic value/s that are localized to particular 'universal spatial pixel/s' appearing at any particular (series of) USCF's frames...

But, if the above description of the CUFT's 'mechanics' underlying the "flow of time" is accurate, then based on the (earlier mentioned) computational definition of 'time' as the number of 'object-inconsistent' presentations (across a series of USCF's) and of the "flow of time" as the particular sequence of 'electromagnetic-spatial pixels' series underlying a given sequence of events (or phenomenon), then it should be possible (in principle, at least) to "reverse the flow of time" for a given object (e.g., such as for the abovementioned developing plant) through a manipulation of the sequential order of electromagnetic-spatial pixel values of that plant across a series of USCF's... Let there be a particular sequence of spatial-electromagnetic pixels points across a series of USCF's that exhaustively define that plant's growth curve; Now, based on the CUFT's strict definition of the "flow of time" for that given (developing) plant which comprises the particular sequence of spatial-electromagnetic pixels (series) across the given series of USCF's frames, it should be possible to exert a differential electromagnetic field manipulation of each of the given spatial-electromagnetic pixels per each of the USCF's frames so as to produce a "reversal" of the "flow of time" – i.e., the spatial-electromagnetic pixels series' values arranged in the reversed order (such that instead of a USCF's series running from '1... to n' it would run from 'n... to 1')!

The key point to be noted (here) is that whereas both relativistic and quantum theories assume a strict *"unidirectional"* and *"unaltered"* flow of time, the CUFT's computational definition of 'time' as the number of 'object-inconsistent' USCF's presentations and of the "flow of time" (direction) strictly depending on the *particular sequence of 'spatial-electromagnetic pixel' values* allows the CUFT to predict a (differential) critical prediction whereby it may be possible to "reverse the flow of time" of a given object through a manipulation of the sequence (e.g., order) of the series of the particular 'spatial-electromagnetic universal pixels' comprising the series of USCFs' object presentation... Note that according to the CUFT there does not seem to exist any "objective", "unidirectional", or "unaltered" "flow of time" underlying the (quantum or relativistic) physical phenomena, but only a particular configuration of a certain sequence of 'spatial-electromagnetic universal pixels' that is presented in a particular sequence (e.g., comprising a given series of USCF's). Therefore, to the extent that we are able to manipulate (e.g., technologically) the 'spatial-electromagnetic pixels' values of an object across a series of USCF's (such that it follows the "reversed order" of the original USCF's series) then we have in fact "reversed the flow of time" for that particular object (or event)...

Moreover, from a purely technological standpoint, the process by which such a potential reversal of the (original) sequence of 'spatial-electromagnetic universal pixels' may be achieved (i.e., through a manipulation of the electromagnetic value/s of a given object's 'spatial-electromagnetic universal pixels' in order to produce the "reversed" spatial electromagnetic universal pixels' USCFs' sequence) does not necessarily require the capacity to identify, compute and manipulate each and every individual USCF (i...n) frame, but instead necessitate the identification (computation) and manipulation of a "sufficiently large" number of USCF's from within a given pool of consecutive USCF's. (Due to the novelty of the possibility to manipulate the series of spatial-electromagnetic pixels values comprising a given object's "flow of time" the determination of the particular number or rate of such 'sampled' specific spatial-electromagnetic universal pixels (across a certain number of USCF's)that is necessary to accurately reverse that object's "time flow" sequence would have to be tested experimentally.)

Finally, it is clear that to the extent that these particular CUFT's empirical predictions regarding the possibility to "reverse the flow of time" for any given object (or event) based on the manipulation of its specific sequence of 'spatial-electromagnetic universal pixels' may be validated experimentally, this may open the door for a series of potentially far reaching scientific and technological advances in our understanding of the physical universe, as well as in some of its potential human clinical and other potential applications; Thus, for instance, if it may be possible to "reverse the flow of time" for a relatively small object it should be possible (e.g., at least in principle) to "reverse the flow of time" for an entire organism or for a particular (healthy or pathological, young, diseased or aged) cell/s, tissue/s or organ/s... Likewise, based on an extension of the identification of any given object's precise 'spatial-electromagnetic universal pixels' composition (e.g., across a certain series of USCF's) and the potential for altering that object's (single or multiple) pixels values trough an electromagnetic manipulation of its particular pixels' values – it should be possible (e.g., again at least in principle) to also "encode" comprehensively the particular spatial-electromagnetic values pixels of any object, cell/s tissue/s or even an entire organism or physical object and subsequently either alter its composition (or condition), or even "de-materialize" it based on the application of appropriate electromagnetic field (that may 'counteract' the particular 'spatial- electromagnetic pixels' values of that object or certain elements within it) and subsequently "materialize" it elsewhere based on the appropriate application of the precise electromagnetic field that can produce that object's particular spatial-electromagnetic values (e.g., at any accessible point in space) (Bentwich, 2011a)...

Thus, a critical contrasting of three particular instances in which the CUFT's empirical predictions may significantly differ from the corresponding predictions offered by contemporary Quantum or Relativistic theories may validate the Computational Unified Field Theory – as not only replicating all known quantum and relativistic empirical phenomena as well as bridging the gap between their apparent theoretical inconsistencies (Bentwich, 2011c), but in fact may demonstrate the advantage of the CUFT over existing Quantum and Relativistic theoretical models (e.g., while incorporating all known quantum and relativistic phenomena within a broader novel theoretical framework); Hence, as outlined earlier, the aim of the second half of this chapter is to broaden the validation of the Computational Unified Field Theory (CUFT) through the application of one of its key theoretical postulates, namely: the computational 'Duality Principle' to a series of key (computational) scientific paradigms. Once again, to the extent that the computational-empirical analysis of each of these key scientific paradigms may be shown (below) to be constrained by the CUFT's postulated 'Duality Principle', this would both extend the construct validity of the CUFT (to other key scientific disciplines), as well as call for these scientific paradigms' reformulation based on this (novel) more comprehensive Computational Unified Field Theory (of which the Duality Principle is one of three principle theoretical postulates). Needless to say that given the new (hypothetical) Computational Unified Field Theory's aim – to not only unify between Quantum Mechanics and Relativity Theory (e.g., as shown previously: Bentwich, 2011c) but to fulfill the requirements of a 'Theory of Everything' (TOE) (Brumfiel, 2006; Einstein, 1929, 1931, 1951; Ellis, 1986; Greene, 2003), a demonstration of the potential applicability of the CUFT to a series of key scientific paradigms may be significant as part of its theoretical validation process;

3. The 'Duality Principle': Potential resolution of key 'Self-Referential Ontological Computational Systems' (SROCS) scientific paradigms

To the extent that a series of key scientific paradigms can be shown to be constrained by the Computational Unified Field Theory's (CUFT) postulated 'Duality Principle' (Bentwich, 2011c), there emerges a need to re-formalize each of these central scientific paradigms based on the Duality Principle's higher-ordered 'D2 A-Causal' computational framework as embedded within the broader Computational Unified Field Theory.

As noted previously (Bentwich, 2011c), one of the three principle theoretical postulates underlying the CUFT is the computational 'Duality Principle' which constrains any 'Self Referential Ontological Computational System' (SROCS) of the general form:

$$PR\{x,y\} \rightarrow ['y' \text{ or 'not } y']/di1...din$$

Indeed, it was shown (there) that Quantum Mechanics' probabilistic interpretation which is based on the assumption whereby the determination of the complimentary values of any subatomic 'target' element solely depends on its direct (or indirect) physical interaction with another 'probe' element (at the 'di1' to 'din' computational levels), thus:

$$SROCS: PR\{P, t\} \rightarrow ['t' \text{ or } '\neg t']/di1...n.$$

Likewise, Relativity's computational structure was also shown to constitute precisely such a SROCS computational structure:

$$\text{SROCS: PR\{P, } O\textit{diff}\} \rightarrow [\text{P or } \neg\text{P}]/\text{di}1...\text{din}$$

wherein it is assumed that the determination of the "existence" or "non-existence" of any (specific) relativistic 'Phenomenon' (e.g., 'spatial-temporal' or 'energy-mass') is solely based on that Phenomenon's direct (or indirect) physical interaction with a differential series of relativistic observer/s.

Moreover, it was shown that both of these quantum and relativistic SROCS paradigms also necessarily contain the "negative" case of a 'Self-Referential Ontological *Negative* Computational System' (SRONCS) which inevitably leads to 'logical inconsistency' and ensuing *'computational indeterminacy'* – i.e., a principle *computational inability* of such a SROCS/SRONCS computational structure to *determine* whether any particular (quantum) 't' or (relativistic) 'P' value– *"exists" or "doesn't exist"*; However, since in both of these (quantum and relativistic) cases there is robust empirical data indicating the *capacity* of these quantum or relativistic computational systems to determine the "existence" or "non-existence" of any such particular 't' or 'P' value/s, then the (novel) computational *'Duality Principle'* asserted the *conceptual inability to compute* the "existence" or "non-existence" of any (particular) relativistic "P" or quantum "t" value *from within their direct physical interaction with another relativistic (differential) observer/s or with another subatomic 'probe' element* – but only from a conceptually higher-ordered 'D2' computational level which is irreducible to any direct (or indirect) physical interactions between any such quantum 'probe-target' or relativistic 'observer-Phenomenon' interactions;

Indeed, according to this new hypothetical computational 'Duality Principle', the only possible determination of any such quantum or relativistic relationships can be carried out based on such conceptually higher-ordered 'D2' computation which computes the "co-occurrences" of any relativistic ('spatial-temporal' or 'energy-mass') 'observer-Phenomenon' values, or the "co-occurrences" of any quantum (computational) complimentary ('spatial-energetic' or 'temporal-mass') 'probe-target' values... In fact, based on the identification of such a (singular) conceptually higher-ordered 'D2 A-Causal' computational framework which underlies both quantum and relativistic models of physical reality the (hypothetical) 'Computational Unified Field Theory' (CUFT) was hypothesized which postulated the existence of a series of extremely rapid 'Universal Simultaneous Computational Frames' (USCF's) that give rise to all quantum and relativistic physical phenomena (and may also point at new hypothetical critical physical predictions as described above which may arise from the discovery of the singular 'Universal Computational Principle' which computes this rapid series of USCF's).

More generally, the incorporation of the computational 'Duality Principle' as one of the three central postulates of the 'Computational Unified Field Theory' (CUFT) has pointed at the possibility that to the extent that other (key) scientific paradigms may also constitute such SROCS computational structures, then they should also be constrained by the 'Duality Principle' and the CUFT (e.g., of which the Duality Principle forms an integral part). Specifically, it was suggested that there may exist a series of key scientific paradigms (including: Darwin's Natural Selection Principle and associated 'Genetic Encoding' hypothesis and Neuroscience's basic 'materialistic-reductionistic' Psycho-Physical Problem) which may all comprise such basic SROCS computational structure, and therefore may be constrained by the 'Duality Principle'; Again, to the extent that each of these key scientific (computational) paradigm may be shown to constitute a SROCS structure and therefore be constrained by the Duality Principle, then these scientific paradigms will have to be reformulated based on the Duality Principle's conceptually higher-ordered 'D2 a-causal' computational framework – thereby becoming an integral part of the CUFT's delineation of the 'D2' rapid series of the USCF's...

Therefore, what follows is a delineation of the SROCS computational structure underlying each of these key scientific paradigms – which shall therefore inevitably point at the Duality Principle's assertion regarding the (conceptual) impossibility of determining the "existence" or "non-existence" of any SROCS' (particular) 'y' element from within its direct or indirect physical (or computational) interaction with any exhaustive series of 'x' factor/s (e.g., that are particular to that specific SROCS scientific paradigm); Instead, the application of the Duality Principle to each of these scientific SROCS paradigms may point at the existence of a conceptually higher-ordered 'D2' computational framework which computes an 'a-causal' "co-occurrence" of a series of 'x-y' pairs (e.g., which alone can explain the empirical capacity of these scientific paradigms to determine the "existence" or "non-existence" of any particular 'y element);

According to the hypothesized computational Duality Principle (Bentwich, 2011c), any empirical scientific paradigm that is based on such a SROCS computational structure may inevitably lead to both 'logical inconsistency' and 'computational indeterminacy' that are contradicted by that (particular) scientific paradigm's empirically proven capacity to determine whether any specific 'y' element "exists" or "doesn't exist"; This empirically proven capacity of the given scientific paradigm to compute the "existence" or "non-existence" of the 'y' element points at the Duality Principle's asserted conceptually higher-ordered 'D2' computational framework which computes the "co-occurrences" of any (hypothetical) series of ['x-y'(st-i); ... 'x-y'(st-i+n)] pairs; Indeed, this conceptually higher-ordered computation of the "co-occurrence" of any such 'x-y' pairing (proven by the Duality Principle) was termed: 'D2 **A-Causal Computation**'. This is due to the fact that according to the Duality Principle the only possible means through which these empirically validated scientific paradigms are able to compute the "existence" or "non-existence" of any given 'y' element is through a conceptually higher-ordered 'D2 a-causal' Computation which determines the "co-occurrences" of any 'x-y' pair/s (e.g., but which was principally shown to be irreducible to any hypothetical direct or even indirect 'x-y' physical interactions)...

Indeed, as shown in the previous article (Bentwich, 2011c), since the Duality Principle's constraint of the SROCS computational structure is *conceptual* in nature – e.g., in that any SROCS computational structure is bound to produce both logical inconsistency and subsequent computational indeterminacy (e.g., which are contradicted by empirical evidence indicating the capacity of their corresponding computational systems to determine whether the particular 'y' element "exists" or "doesn't exist"), then it was shown that the Duality Principle's assertion regarding the need to place the computation of the "existence" or "non-existence" of the particular 'y' element at a conceptually higher-ordered 'D2' level overrides (and transcends) any direct *or indirect* physical relationship between the 'y' and 'x' elements (e.g., occurring at any hypothetical exhaustive computational level/s, 'di1...din'). This is because even if we assume that the computation determining whether the 'y' element "exists" or "doesn't exist" takes place at an intermediary (second) 'di2' computational level (or factor/s), then this does not alter the basic computational (causal-physical) SROCS structure; This is due to the basic materialistic-reductionistic working hypothesis underlying all key scientific SROCS paradigms wherein the sole determination of the "existence" or "non-existence" of the (particular) 'y' element is determined solely based on the direct physical interaction between the 'y' element – e.g., as signified by the "causal-arrow" within the (above mentioned) SROCS computational structure: SROCS: PR{x,y}→ ['y' or '¬y']/di1. Thus, whether we attribute the computation of the "existence/non-existence" of the 'y' element as taking place at the same 'direct physical interaction' (e.g., of the 'x' and 'y' elements at the 'di1' computational level) or whether we attempt to 'rise higher' to an additional hypothetical computational level/s (or factor/s etc.) the basic 'materialiastic-reductionistic' assumption underlying the SROCS computational structure inevitably ties the direct physical 'x-y' interaction with a 'causal-material' determination of the "existence" or "non-existence" of the (ensuing) 'y' element (e.g., occurring either at the 'di1' or 'di2' computational levels). In other words, whether we assume that the determination of the 'existence/non-existence' of the 'y' element takes place at the same (di1) computational level as the direct physical 'x-y' interaction or whether we assume that the determination of the "existence"/"non-existence" of the 'y' element occurs (e.g., somehow) through one or more intermediary computational levels (di2...din) or factor/s the basic SROCS computational structure which assumes that it is this direct physical interaction between the 'x' and 'y' element which solely can determine whether the 'y' element "exists" or "doesn't exist" is therefore inevitably constrained by the computational Duality Principle:

Moreover, the Duality Principle's computational constraint asserts the conceptual inability to determine whether the (particular) 'y' element "exists" or "doesn't exist" from *within* any direct or indirect *physical interaction* between that 'y' element and any other 'x' factor (at any 'di1... 'din' computational levels), but *only from a conceptually higher-ordered 'D2' computational framework* which can only determine an 'a-causal' computational relationship/s between any hypothetical 'x' and 'y' factor/s; Indeed, as shown in the previous article (and noted above), such conceptually higher-ordered 'a-causal D2' computation cannot (in principle) be reduced to any direct or indirect physical 'x-y' interactions but instead involves an association of a series of 'x-y' pairs occurring at different 'spatial-temporal' points, thus: D2: [{x1, y1}st1; {x1, y1}st2 ... [{xn, yn}stn]. In other words, the (novel computational) Duality Principle effectively constrains- and replaces- any scientific SROCS paradigm (e.g., of the general form: SROCS: PR{x,y}→ ['y' or '¬y']/di1) – with a conceptually higher-ordered 'D2' computational framework of the form: D2: [{x1, y1}st1; {x1, y1}st2 ... [{xn, yn}stn] which is based on the (higher-ordered 'D2') computation of the co-occurrences of certain {'x-y'}*sti...n* pairs (occurring at different spatial-temporal points). Indeed, it is suggested that such higher-ordered D2 computational metamorphosis replaces (and transcends) the strict materialistic-reductionistic working hypothesis underlying the current SROCS' scientific paradigm's focus with a conceptually higher-ordered 'non-material', 'non-causal' associative computational mechanism. Therefore, based on the Duality Principle's (above) computational-empirical proof for the basic computational constraint imposed on any scientific SROCS paradigm – e.g., which must necessarily be replaced by an alternative conceptually higher-ordered 'a-causal D2' computation, then any (existing or new) scientific paradigm that can be accurately demonstrated to replicate the (above mentioned) SROCS computational structure must be replaced by the Duality Principle's asserted conceptually higher-ordered 'D2' computational framework: D2: [{x1, y1}st1; {x1, y1}st2 ... [{xn, yn}stn].

3.1 Darwin's natural selection principle & genetic encoding hypothesis

We therefore first examine the key scientific paradigms of Darwin's 'Natural Selection Principle (Darwin, 1859) (e.g., and its closely related 'Genetic Encoding' and 'Protein

Synthesis - Genetic Expression' hypotheses) in order to show that they all (in fact) constitute such 'Self-Referential Ontological Computational System' (SROCS) computational paradigms which are necessarily constrained by the computational 'Duality Principle'; In a nutshell, it is hypothesized that Darwin's evolutionary theory comprises three (intimately linked) scientific SROCS paradigms which are: the (primary) 'Natural Selection' SROCS, the (secondary) 'Genetic Encoding' (plus associated random mutations assumption) SROCS, and (tertiary) Protein Synthesis (phenotype) – Genetic Expression SROCS computational paradigms;

i. **Natural Selection SROCS:** Darwin's 'Natural Selection' principle comprises a SROCS paradigm since it asserts that the "existence" or "non-existence" of any given organism (e.g., 'o' – and by extension, also all of its potential descendent organisms constituting a single specie) is solely dependent upon its direct (or indirect) physical interaction with an exhaustive series of 'Environmental Factors' ('E{1...n}'):

$$\textit{SROCS I [Natural Selection]: } PR\{ E\{1...n\}, o\} \rightarrow [\text{'o' or '}\neg o\text{'}]/di1.$$

But, as we've seen (above), such SROCS computational structure inevitably leads to both 'logical inconsistency' and 'computational indeterminacy' – in the case of the SRONCS: $PR\{ E\{1...n\}, o\} \rightarrow \text{'}\neg o\text{'}/di$.
This is because such a SRONCS asserts that the direct physical interaction between a given organism and an (exhaustive series of) Environmental Factors gives rise to the "non-existence" of that organism, which essentially implies that that particular organism both "exists" and "doesn't exist" at the same 'di1...din' computational level – which obviously constitutes a 'logical inconsistency'. But, due to the SROCS/SRONCS computational structure (e.g., which assumes that the only means of determining whether the organism "exists" or "doesn't exist" is through the direct physical interaction 'di1' between the organism and its exhaustive Environmental Factors), then such 'logical inconsistency' inevitably also leads to 'computational indeterminacy' – i.e., a principle inability of the SROCS/SRONCS scientific paradigm to determine whether that organism ('o') "exists" or "doesn't exist"! However, since there exists ample empirical evidence indicating the *capacity* of evolutionary biological systems to determine whether any given organism ('o') "exists" (e.g., survives) or "doesn't exist" (e.g., is extinct), then the Duality Principle asserts the conceptual computational inability of Darwin's Natural Selection principle to determine whether any given organism "exists" or "doesn't exist" based on its (strictly) assumed materialistic-reductionistic SROCS/SRONCS computational structure (e.g., direct physical interaction between the organism and an exhaustive set of Environmental Factors); Instead, the computational Duality Principle asserts that the only means for determining the evolution of any given biological species is based on a conceptually higher-ordered 'D2' computational framework which computes the "co-occurrences" of a series of any (hypothetical) organism/s and corresponding Environmental Factors, thus:

$$D2: [\{E_{\{1...n\}}, o\}st_1; \{E_{\{1...n\}}, o\}st_2 ... [\{E_{\{1...n\}}, o\}st_n].$$

Note that as in the above mentioned generalized format of the SROCS computational structure (e.g., $PR\{x,y\} \rightarrow [\text{'y' or '}\neg y\text{'}]/di1$), the computational constraint imposed by the Duality Principle is *conceptual* – i.e., it applies regardless of whether we're dealing with any 'direct' or 'indirect' physical relationships between the 'x' and 'y' factor/s; In the same manner, we can see that even if we assume that the interaction between any given organism ('o') and any exhaustive hypothetical Environmental Factors ('E{1...n}') comprises more than one Environmental Factor/s ('E{1...n}') or more than one (intermediary) computational level/s, the computational *structure* of Darwin's 'Natural Selection' SROCS paradigm inevitably leads to both 'logical inconsistency' and 'computational indeterminacy'; This is due to the fact that the fundamental 'materialistic-reductionistic' working hypothesis underlying the Natural Selection SROCS paradigm unequivocally stipulates that the determination of the "existence" or "non-existence" of any given organism ('o') is solely (and strictly) computed based on the direct (or even indirect) physical interaction/s between the organism and any exhaustive hypothetical series of Environmental Factors. Therefore, even if we assumed that Darwin's Natural Selection principle involves multiple Environmental Factors ('E{1...n}') and/or multiple computational levels ('di1'... 'di$z$'), thus:

$$PR\{ E_{\{1...n\}}, o\}/di1...din \rightarrow ['o' \text{ or } '\neg o']/diz$$

then it still (inevitably) replicates the same SROCS computational structure that invariably produces the above mentioned 'logical inconsistency' and 'computational indeterminacy' (which give rise to the Duality Principle's above mentioned computational constraint). This is due to the fact that regardless of the number of (hypothetical) intervening (or mediating) Environmental Factors or computational level/s ('di1'...'di$z$'), the SROCS strict 'materialistic-reductionistic' computational structure assumes that the determination of the "existence" or "non-existence" of the organism is solely determined based on the direct physical interaction between the organism and its Environmental Factors.

Likewise, even if we assume that Darwin's Natural Selection process operates via innumerable organism-environment interactions taking place at different 'spatial-temporal' points {st1...stn}, then due to the (abovementioned) 'materialistic-reductionistic' implicit assumption embedded within the SROCS/SRONCS computational structure (i.e., which assumes that the "existence" or "non-existence" of the organism 'o' is *solely determined* based on any direct or indirect physical interactions between that organism and an exhaustive set of Environmental Factors), this does not alter the basic SROCS/SRONCS computational structure which was shown (above) to be constrained by the Duality Principle:

$$PR\{ E_{\{1...n\}}, o\}/sti1...sti \rightarrow ['o' \text{ or } '\neg o']stn /di1... diz$$

Essentially, this Natural Selection (primary) SROCS computational structure asserts that the determination of the "existence" or "non-existence" of any particular organism ('o') is solely determined based on its single- or multiple- spatial-temporal interactions with an exhaustive set of Environmental Factors (and even that hypothetically the actual computation or determination of the "existence" or "non-existence" of the particular organism {'o'} may take place at a later spatial-temporal point than the actual direct or indirect physical interaction between the organism and the exhaustive set of Environmental Factors);

Note, however, that this basic SROCS/SRONCS computational structure embeds within it the fundamental 'materialistic-reductionistic' implicit assumption wherein there cannot be any other factor/s outside the direct (or indirect) physical interaction/s between the organism and the (exhaustive set of) Environmental Factors which determines or computes the "existence" (e.g., survival) or "non-existence" (e.g., extinction) of that organism; This strong (implicit) 'materialistic-reductionistic' assumption underlying the SROCS/SRONCS computational structure is represented by the **causal** '$\rightarrow$' connecting between the direct (or indirect) physical interaction between the organism and the Environmental factors and the determination of the "existence"/"non-existence" of the particular organism... Therefore, based on this strict 'materialistic-reductionistic' assumption underlying Darwin's Natural Selection SROCS paradigm –the direct (or indirect) physical relationship between the organism and its Environmental Factors and its (strict) *causal effect* in determining the "existence" or "non-existence" of that organism must necessarily constitute a *singular computational level* (e.g., di1...dix), regardless of the number of (hypothetical) spatial-temporal points that occupy either the direct or indirect 'organism-Environmental Factors' interaction/s or the specific spatial-temporal point/s at which the determination (or computation) of the "existence" or "non-existence" of the organism take place!

Therefore, from a purely computational standpoint, both the direct physical interaction between the organism and its Environmental Factors (at 'di1') and the determination of the ensuing "existence" or "non-existence" of that organism (e.g., assumed to take place at any hypothetical level 'di1'...'diz') – must be considered to occur at the *same computational level* (e.g., at either 'di1'...'diz')! Indeed, it is precisely this materialistic-reductionistic SROCS/SRONCS paradigmatic structure which assumes that the determination of the "existence" or "non-existence" of the particular organism occurs at the *same computational level* (e.g., at either 'di1'...'diz') as the direct physical interaction between that organism and an exhaustive set of Environmental Factors which was shown (above) to inevitably lead to both 'logical inconsistency' and 'computational indeterminacy', which were shown to be contradicted by robust empirical findings – thereby pointing at the Duality Principle's assertion regarding the need for a conceptually higher-ordered 'D2' computational level that can compute the "co-occurrences" of any spatial-temporal pairing of any given organism and its corresponding Environmental Factors:

$$D2: [\{E\{1...n\}, o\}st1; \{E\{1...n\}, o\}st2 ... \{E\{1...n\}, o\}stn].$$

Hence, based on the Duality Principle's logical-empirical analysis of the SROCS/SRONCS computational structure underlying Darwin's Natural Selection scientific paradigm, the Duality Principle has proven the conceptual computational inability to determine the "existence" or "non-existence" of any (hypothetical) organism based on any direct or indirect physical interaction/s between that organism and any (hypothetical) exhaustive set of Environmental Factor/s (at the same di1...dix computational level)- but only from a conceptually higher-ordered 'D2' computational level which simply computes the "co-occurrences" of any hypothetical series of spatial-temporal 'organism-Environmental Factors' pairing...

It is also worthwhile to note that the Duality Principle's proof for the conceptual computational inability of Darwin's Natural Selection Principle's SROCS computational structure to determine the "existence" or "non-existence" of any (hypothetical) organism ('o') from within its direct or indirect physical interaction within any (hypothetical) exhaustive series of Environmental Factors (e.g., at any 'di1... dix' computational level) also *negates* the existence of any "*causal-material*" relationship between the particular organism and any (exhaustive set of) Environmental Factors; This was previously shown (Bentwich, 2011c) through a thorough analysis of the Duality Principle's proof for the existence of a D2 'A-Causal' computational characteristics – which replaces the SROCS (implicit) 'material-causal' relationship between any two (hypothetical) 'x' and 'y' elements (e.g., at any di1...din computational level) with the D2's computation of the "co-occurrences" of any (hypothetical) series of 'x-y' pairs. The Duality Principle's conceptual proof for the principle inability of any (exhaustive series of) Environmental Factors to *causally* determine the "existence" or "non-existence" of any particular organism based on any hypothetical single- or multiple- level/s of computation or single- or multiple- spatial-temporal points was also shown (above) based on the Duality Principle's proof for the basic (implicit) material-causal assumption underlying the SROCS computational structure which inevitably leads to both 'logical-inconsistency' and 'computational indeterminacy' which are contradicted by robust empirical evidence (e.g., indicating the *capacity* of evolutionary-biological systems to determine the "existence" or "non-existence" of any particular organism). Hence, a key emerging property of the Duality Principle (e.g., in this case as it applies to Darwin's Natural Selection SROCS paradigm) is that it replaces the basic (implicit) **material-causal** assumption embedded within the SROCS computational structure with a *conceptually higher-ordered* **'D2 A-Causal' computational framework** which merely computes the *"co-occurrences"* of any (hypothetical) series of *'organism-Environmental Factors'* pairs – i.e., but which cannot (in principle) possess any 'material-causal' relationship between them...

Interestingly though (as noted above), despite the Duality Principle's conceptual computational proof that Darwin's Natural Selection Principle (computational structure) constitutes a SROCS and is therefore constrained by the Duality Principle, i.e., indicating the conceptual computational inability to determine the "existence" or "non-existence" of any (hypothetical) organism ('o') based on any of its direct or indirect material-causal interaction/s with any exhaustive set of Environmental Factors E{1...n} (but only from a conceptually higher-ordered 'D2 a-causal' computational framework) – it seems that Darwin's evolutionary theory further contingents Darwin's Natural Selection Principle's SROCS paradigm upon two other (hierarchical-dualistic) SROCS paradigms, i.e., the (abovementioned) 'Genetic Encoding' hypothesis and 'Protein Synthesis' SROCS paradigms;

ii. **Organism Phenotype - Genetic Encoding SROCS**: It is hypothesized that Darwin's (above mentioned) Natural Selection SROCS paradigm is anchored in- and based upon an additional (secondary) *'Organism Phenotype - Genetic Encoding' SROCS paradigm}*: PR{G{1...n},o-*phi*}→ ['o-*phi*' or '¬o-*phi*']/di1...din.

wherein the "existence" or "non-existence" of any particular phenotypic property of any given organism ('o-*ph*') (e.g., appearing in Darwin's Natural Selection primary SROCS paradigm) is assumed to be solely determined based on its direct (or indirect) physical interaction/s with any exhaustive set of Genetic factors (e.g., at the 'di1...din' computational levels). Note that from the (entire) dualistic relationship existing between the 'organism' and the Environmental Factors in Darwin's Natural Selection Principle SROCS paradigm – only the 'organism ('o') element is utilized within the secondary *'Organism Phenotype - Genetic Encoding' SROCS* paradigm:

$$PR\{G_{(1...n)}, o\text{-}phi\} \rightarrow [\text{'}o\text{-}phi\text{'} \text{ or '}\neg o\text{-}phi\text{'}]/di1...din.$$

In other words, the "existence" or "non-existence" of any particular organism possessing a specific phenotypic property is totally contingent upon its direct (or indirect) physical interaction with an exhaustive series of relevant Generic factors; It is to be noted that the implicit assumption underlying this 'hierarchical-dualistic' computational structure is the (tacit) contingency that exists between Darwin's (primary) Natural Selection Principle's *organism's particular phenotypic property* (e.g., which interacts directly or indirectly with the exhaustive set of Environmental Factors, thereby solely determining the "existence" or "non-existence" of that particular organism) – and the exhaustive set of relevant Genetic Factors which together (solely) determine the "existence" or "non-existence" of that particular phenotypic property! Thus, it may be said that there exists a dual 'hierarchical-dualistic' computational structure which constitutes Darwin's entire evolutionary theory that can be broken down to two interrelated SROCS computational structures, thus:

SROCS I {Natural Selection}: $PR\{E_{\{1...n\}}, \text{'o-}phi'\} \rightarrow [\text{'o-}phi'' \text{ or '}\neg\text{o-}phi']/\text{di1}.$

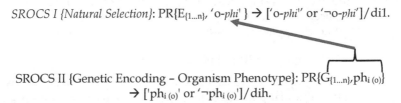

$$SROCS \text{ II \{Genetic Encoding – Organism Phenotype\}}: PR\{G_{\{1...n\}}, phi_{(o)}\}$$
$$\rightarrow [\text{'ph}i_{(o)}\text{' or '}\neg\text{ph}i_{(o)}\text{'}]/\text{di}h.$$

However, to the extent that it can be proven that this (secondary) 'Genetic Encoding – Organism Phenotype' computational structure replicates and constitutes a SROCS computational structure, then it automatically follows that both the primary and secondary SROCS paradigms comprising Darwin's (currently accepted) evolutionary theory must be replaced (and transcended) by a conceptually higher-ordered 'D2' computational framework; Thus, we now set to evince that Darwin's (secondary) Genetic Encoding- Organism's Phenotype computational structure constitutes a SROCS paradigm and is therefore also necessarily constrained by the (same) computational Duality Principle:

$$PR\{G_{\{1...n\}}, phi_{(o)}\} \rightarrow [\text{'ph}i_{(o)}\text{' or '}\neg\text{ph}i_{(o)}\text{'}]/\text{di}h.$$

As shown above, this computational structure precisely replicates the generalized SROCS structure of the form: $PR\{x,y\} \rightarrow [\text{'y' or '}\neg\text{y'}]$, which was shown to inevitably lead to both 'logical inconsistency' and ensuing 'computational indeterminacy' in the case of the 'Self-Referential Ontological Computational System' (SRONCS).

This is simply because if it is assumed that the "existence" or "non-existence" of any particular phenotypic property ('ph$i_{(o)}$') is solely dependent upon its direct physical interaction with any exhaustive series of 'Genetic Factors' ('G$\{1...n\}$'), then in the case of the (abovementioned) SRONCS paradigm the specific phenotypic property 'ph$i_{(o)}$' appears to both "exist" and "not exist" at the same 'di$h$' computational level: $PR\{G\{1...n\}, phi_{(o)} \rightarrow \neg phi_{(o)}\}/\text{di}h$, which obviously produces a 'logical inconsistency' – which also inevitably leads to an (apparent) 'computational indeterminacy', e.g., an apparent inability of the computational system to determine whether that particular phenotypic property "exists" or "doesn't exist"... But, since there exist ample empirical evidence indicating that genetic (computational) system do in fact possess the capacity to determine whether any given phenotypic property 'ph$i_{(o)}$' "exists" or "doesn't exist" within a given organism, then we must conclude that the (currently assumed) SROCS computational structure is invalid!

As shown previously, it is important to note that the computational constraint imposed by the Duality Principle is *conceptual* in nature – i.e., it applies to any single- or multiple-hypothetical computational levels that may be involved in any direct or even indirect (e.g., di$h$1...di$h$n) physical interactions between the particular phenotypic property and any exhaustive series of 'Genetic Factors' (G$\{1...n\}$);

As shown (above), the reason for this *conceptual* computational constraint imposed on the 'Genetic Encoding' SROCS by the Duality Principle stems from the existence of an implicit '*materialistic-reductionistic*' *assumption* embedded within the SROCS computational structure which is represented by the 'causal arrow' $\rightarrow$ which connects between any direct physical interaction between the exhaustive set of 'Genetic Factors' and the particular phenotypic property (at 'di$h1$') and any single- or multiple- direct or indirect physical interactions or computational levels that may mediate between this direct 'Genetic Factors – phenotype' physical interaction (at dihn) and between the determination of the "existence" or "non-existence" of the particular (relevant) phenotypic property; Therefore it may be appropriate to represent the conceptual constraint imposed by the Duality Principle upon the (secondary) Genetic Encoding-phenotype SROCS structure in this manner:

$$PR\{G\{1...n\},phi\ (o)\}di1 \rightarrow ['phi\ (o)'\ or\ '\neg phi\ (o)']dih1...dihn.$$

wherein any (hypothetical) direct or indirect physical interaction between an exhaustive set of Genetic Factors and the particular phenotypic property 'phi (o) – which can take place either at their direct physical interaction level ('di$h1$') or at any subsequent (indirect) computational level/s (e.g.,'dihn') *causally leads* to the determination of the "existence" or "non-existence" of that particular phenotypic property 'phi (o) at a hypothetical 'dihz' computational level;

However, even for this (expanded) Genetic Factors – phenotypic property SROCS computational structure it becomes clear that the (abovementioned) 'materialistic-reductionistic' implicit assumption embedded within it – inevitably leads to both 'logical inconsistency' and subsequent 'computational indeterminacy' that are contradicted by robust empirical findings indicating the *capacity* of biological evolutionary systems to determine the "existence" or "non-existence" of any particular phenotypic property in any given organism... This is due to the fact that despite the hypothesis that the determination of the "existence" or "non-existence" of the particular phenotypic property may occur at (single- or multiple) computational level/s (di$h1$... dihz) that may be different than the direct physical interaction between the particular phenotype and the (exhaustive set of) Genetic Factors, due to the above mentioned 'materialistic-reductionistic' implicit assumption embedded within this (expanded) SROCS structure the determination of the "existence" or "non-existence" of that particular phenotypic property 'phi (o) is solely- and strictly- *caused* by the direct physical interaction between the (exhaustive set of) Genetic Factors (at di$h1$) and that phenotypic property 'phi (o); But, this implies that the determination of the "existence" or "non-existence" of the phenotypic property 'phi (o) takes place at the *same computational level*/s as the direct physical interaction level (di$h1$...dihz) between the Genetic Factors and the phenotypic property, which may be represented thus:

$$PR\{G\{1...n\},phi\ (o)\} \rightarrow ['phi\ (o)'\ or\ '\neg phi\ (o)']/\ dih1...dihz$$

which precisely replicates the above SROCS computational structure which has been shown to be constrained by the Duality Principle...

In other words, whether the interaction between the Genetic Factors and the phenotypic property takes place at the same computational level (e.g., at 'di$h1$') as the determination of the "existence" or "non-existence" of the phenotypic property, or takes place at a different (single or multiple) computational level/s (e.g., 'di$h1$... dihz') – due to the implicit materialistic-reductionsitic assumption embedded within the (expanded) SROCS computational structure this inevitably leads to both 'logical inconsistency which inevitably leads to both 'logical inconsistency' and 'computational indeterminacy' that were contradicted by empirical evidence and which therefore lead to the Duality Principle's assertion regarding the need for a conceptually higher-ordered 'D2' computational level which merely computes the "co-occurrences" of any hypothetically pairs of 'Genetic Factors – phenotypic property'.

In fact, the Duality Principle's conceptual computational proof for the principle inability to determine the "existence" or "non-existence" of any particular phenotypic property from within any direct or indirect (di1..diz) physical interaction between the Genetic Factors and the phenotypic property also includes any spatial-temporal span in which these direct or indirect physical interactions occur, or in which the determination of the "existence" or "non-existence" of the (particular) phenotypic property takes place; This can be seen if we formalize each of these direct or indirect physical interaction/s between the Genetic Factors and the particular phenotypic property- as well as to the determination of the "existence"/"non-existence" of the phenotypic property any (hypothetical) spatial-temporal value/s, thus:

$$PR\{G_{\{1...n\}}, phi\ (o)\}_{st1..stj} \rightarrow ['phi\ (o)'\ or\ '\neg phi\ (o)']_{stn}/\ dih1...dihz$$

Wherein the direct physical interaction between the (exhaustive set of) Genetic Factors and the particular phenotypic property takes place at either single- or multiple- time points (st1...stj) that may be different than the spatial-temporal point/s at which the determination of the "existence" or "non-existence" of the (particular) phenotypic property takes place (e.g., 'dih1...dihz'), This is because even if we assume that the spatial temporal points at which the direct physical interaction between these Genetic Factors and the particular phenotypic property ($PR\{G\{1...n\}, phi\ (o)\}st1..stj$) , and the determination of the "existence" or "non-existence" of the particular phenotypic property ['phi (o)' or '¬phi (o)']stn are different, then due to the (above generalized) SROCS' embedded 'materialistic-reductionsitic' causal assumption wherein the determination of the "existence" or "non-existence" of the particular phenotypic property is assumed to be determined strictly- and solely- based on the direct (or indirect) physical interaction between the Genetic Factors and that phenotypic property, then this (generalized) SROCS computational structure inevitably leads to both logical inconsistency and computational indeterminacy – which (in turn) point at the Duality Principle's (abovementioned) computational constraint...

We are thus forced to accept the Duality Principle's conceptual computational constraint imposed upon the 'Genetic Encoding - Phenotypic Property' (secondary) SROCS structure wherein the determination of the "existence" or "non-existence" of any particular phenotypic property (within any given organism) cannot (e.g., in principle) be determined from within any direct- or indirect- physical interaction between any exhaustive set of Genetic Factors and any hypothetical phenotypic property, or through any hypothetical single- or multiple- computational levels associated with these direct or indirect physical interaction/s or based on the same or different (single- or multiple-) spatial-temporal points (or intervals) at which these Genetic Factors may interact with any particular phenotypic property:

$$PR\{G_{\{1...n\}}, phi\ (o)\}_{st1..stj} \nrightarrow ['phi\ (o)'\ or\ '\neg phi\ (o)']_{stn}/\ dih1...dihz.$$

As stated above, the conceptual computational proof for the Duality Principle's assertion arises from the inevitably 'logical inconsistency and 'computational indeterminacy' implications of the SRONCS computational structure wherein the particular phenotypic property seems to both "exist" and "not exist" at the same computational level (which not only constitutes an explicit 'logical inconsistency' but also produces an inevitable 'computational indeterminacy' that is contradicted by empirical findings indicating the capacity of genetic-biological computational systems to determine the "existence" or "non-existence" of any particular phenotypic property);

Instead, the Duality Principle asserts that there must exist a conceptually higher-ordered 'D2' computational framework which is capable of computing the "co-occurrences" of any hypothetical pair/s of Genetic Factor/s and any phenotypic property (e.g., existing at any spatial-temporal point/s):

D2: [{G{1...n}, 'ph$i$ $(o)'$ }st1; {G{1...n}, 'ph$j$ $(o)'$ }sti; ...{G{1...n}, 'ph$n(o)'$ }stn].

Therefore, the application of the computational Duality Principle to both Darwin's 'Natural Selection' (primary) SROCS computational paradigm, as well as to its (secondary) 'Genetic Encoding – Phenotypic Property' SROCS paradigm (e.g., which is assumed to serve as a contingency for the primary Natural Selection SROCS paradigm) has proven that it is not possible to determine the "existence" or "non-existence" of any 'organism'- or organism related 'phenotypic property' based on any direct- or indirect- physical interaction between any organism- and an exhaustive set of Environmental Factors or between any (organism's) phenotypic property and any exhaustive set of Genetic Factors e.g., including as carried out by single- or multiple- computational level/s, or taking place at any spatial-temporal point/s etc. Instead, the (novel) computational Duality Principle asserts that there exists a conceptually higher-ordered D2 computational level which computes the "co-occurrences" of any single or multiple hypothetical pairs of any exhaustive set of 'Environmental Factors' and any given 'organism' or of any exhaustive set of 'Genetic Factors' and any organism's 'phenotypic property', which may be represented in this manner:

D2: [{E$_{[1...n]}$, o}st$_1$; {E$_{[1...n]}$, o}st$_2$ ... [{E$_{[1...n]}$, o}st$_n$].

D2: [{G{1...n}, 'ph$i$ $(o)'$ }st1; {G{1...n}, 'ph$j$ $(o)'$ }sti; ...{G{1...n}, 'ph$n(o)'$ }stn].

Finally, it is hypothesized that with the advent of modern genetics, RNA and mRNA scientific research one additional (hypothetical) SROCS computational paradigm has emerged which is the *'Genetic Encoding – Protein Synthesis' (tertiary) SROCS* paradigm; This is because the latest developments in genetics research (in general) and those related to the investigation of the relationships that exist between genetic encoding and protein synthesis (in particular) are based on the assumption wherein any biological synthesis of proteins comprising- and constructing- the biological organism are contingent upon a direct (or indirect) physical relationship between an exhaustive set of Genetic Factors and a certain protein synthesis agent, e.g., such as for instance a particular RNA or mRNA synthesis of a particular protein through their direct or indirect physical interaction with a given set of exhaustive Genetic Factors (Burgess, 1971; Geiduschek & Haselkorn, 1969; Khorana, 1965; Rich & Rajbhandary, 1976; Schweet, & Heintz, 1966).

Indeed, it is suggested that this hypothetical (direct or indirect) physical relationship between a certain exhaustive set of Genetic Factors and any (hypothetical) protein synthesis agent precisely reproduces the (above mentioned) tertiary 'Genetic Expression – Protein Synthesis' SROCS paradigm.

iii. **Protein Synthesis (phenotype) – Genetic Expression SROCS***:* It is therefore hypothesized that both Darwin's (above mentioned 'primary') Natural Selection SROCS paradigm as well as the (secondary above mentioned) 'Genetic Encoding – Phenotypic Property' SROCS paradigms are anchored in- and contingent upon- a (tertiary) 'Phenotypic Expression – Protein Synthesis' SROCS computational paradigm, which assumes that the determination of the "existence" or "non-existence" of any particular

Protein (phenotype) is strictly- and entirely- dependent upon its direct (or indirect) physical interaction with an exhaustive set of Genetic Expression ;

SROCS III {Genetic Expression – Protein Synthesis}: PR{G{1...n}, p-synth}
→ ['p-synth or '¬p-synth].

we therefore obtain the full hierarchical-dualistic computational structure underlying Darwin's evolutionary theory as comprising of:

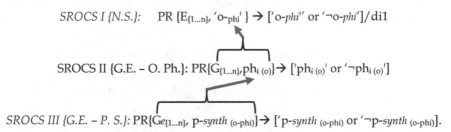

SROCS I {N.S.}:   PR {$E_{\{1...n\}}$, 'o-phi' } → ['o-*phi*'' or '¬o-*phi*']/di1

SROCS II {G.E. – O. Ph.}: PR{$G_{\{1...n\}}$, $ph_{i\ (o)}$} → ['$ph_{i\ (o)}$' or '¬$ph_{i\ (o)}$']

SROCS III {G.E. – P. S.}: PR{$Ge_{\{1...n\}}$, p-*synth* $_{(o\text{-}phi)}$} → ['p-*synth* $_{(o\text{-}phi)}$ or '¬p-*synth* $_{(o\text{-}phi)}$].

This (new) hypothetical hierarchical-dualistic computational structure underlying Darwin's evolutionary modeling is nevertheless constrained (i.e., at each and every one of its three layered SROCS scientific paradigms) by the *Duality Principle* which therefore forces us to replace each of these (three) SROCS computational levels with a conceptually higher-ordered singular 'D2' computation of the "co-occurrences" of multi-layered pairs of 'Environmental Factors – organism', 'Genetic Factors – (organism) Phenotype' and 'Genetic Expression - (organism-phenotype) Protein Synthesis'...

Based on the (above detailed) analysis of the Duality Principle's constraint of any (generalized) SROCS computational paradigm it is not necessary to repeat the details of the Duality Principle's conceptual computational proof for the inability of the (tertiary) 'Genetic Encoding – Protein Synthesis' SROCS to determine the "existence" or "non-existence" of any particular 'protein synthesis' based on its direct physical interaction with an exhaustive set of 'Genetic Expression'; Suffice to state that according to the (above generalized) conceptual computational proof of the Duality Principle, in the specific case of a SRONCS – i.e., in which any direct (or indirect) physical interaction/s between such (an exhaustive set of) Genetic Expression and any particular Protein Synthesis leads to the "non-existence" (e.g., 'non-synthesis') of any such particular protein, this produces the (abovementioned) 'logical inconsistency' and ensuing 'computational indeterminacy' which are contradicted by well-known empirical evidence indicating the *capacity* of biological-evolutionary systems to determine whether any particular protein is synthesized... As shown above, this leads to the Duality Principle's inevitable assertion regarding the existence of the conceptually higher-ordered 'D2' computational framework which computes the "co-occurrences" of any (hypothetical) series of 'Genetic Expression – Protein Synthesis' pairs occurring at any given spatial-temporal point/s in any given organism:

D2: [{Ge{1...n}, p*i-synth* (o-phi)}st1; Ge{1...n}, p*j-synth* (o-phi)}sti... ;
Ge{1...n}, p*n-synth* (o-phi)}stn]

Therefore, the Duality Principle's (abovementioned) constraint of the three ('Natural Selection', 'Genetic Encoding' and 'Protein Synthesis') SROCS computational paradigms (or levels) has proven the conceptual computational inability of each of these scientific paradigms (or computational levels) to determine the "existence" or "non-existence" of the particular 'y' element (e.g., particular 'organism', particular 'phenotypic property', or particular 'protein synthesis') – from within any direct or indirect physical interaction between the (given) 'x' factor and an exhaustive set of the (abovementioned) 'x' factor/s; Instead, the Duality Principle evinced the existence of a conceptually higher-ordered 'D2' computational level which (alone) can compute the "co-occurrences" of any of these (three-leveled) 'x' and 'y' factors (e.g., at any given hypothetical spatial-temporal point/s or computational level/s etc.), thus:

$$D2: [\{E_{\{1...n\}}, o\}st_1; \{E_{\{1...n\}}, o\}st_2 ... [\{E_{\{1...n\}}, o\}st_n].$$

$$D2: [\{G\{1...n\}, \text{'phi } (o)' \}st1; \{G\{1...n\}, \text{'ph}j (o)' \}sti; ...\{G\{1...n\}, \text{'ph}n(o)' \}stn].$$

$$D2: [\{Ge\{1...n\}, pi\text{-}synth (o\text{-}phi)\}st1; Ge\{1...n\}, pj\text{-}synth (o\text{-}phi)\}sti... ;$$
$$Ge\{1...n\}, pn\text{-}synth (o\text{-}phi)\}stn]$$

However, based on the previous (Bentwich, 2011c) conceptual proof for the *singularity* of the 'D2' computational framework forces us to accept the fact that there must be a (singular) simultaneous computation of all three-layered SROCS' "co-occurring" pairs (e.g., which according to the CUFT must comprise the same USCF frame/s):

$$D2: [\{E_{\{1...n\}}, o\}st_1; \{E_{\{1...n\}}, o\}st_2 ... [\{E_{\{1...n\}}, o\}st_n].$$

$$D2: [\{G\{1...n\}, \text{'phi } (o)' \}st1; \{G\{1...n\}, \text{'ph}j (o)' \}sti; ...\{G\{1...n\}, \text{'ph}n(o)' \}stn].$$

$$D2: [\{Ge\{1...n\}, pi\text{-}synth (o\text{-}phi)\}st1; Ge\{1...n\}, pj\text{-}synth (o\text{-}phi)\}sti... ;$$
$$Ge\{1...n\}, pn\text{-}synth (o\text{-}phi)\}stn]$$

Along these lines it is suggested that based on the Duality Principle's proof for the existence of a conceptually higher-ordered 'D2' computational framework for each of the two (Darwin's 'Natural Selection' and 'Genetic Factors – Phenotypic Property') SROCS paradigms, and a previous (Bentwich, 2011c) conceptual proof for the singularity of such higher-ordered 'D2' computational framework we are led to conclude that :

a. Darwin's evolutionary theory is based on a three-layered hierarchical-dualistic computational structure which consists of a primary 'Natural Selection' SROCS paradigm that is contingent upon a secondary 'Genetic Encoding – Phenotypic Property' SROCS paradigm that is (in turn) contingent upon a tertiary 'Genetic Expression – Protein Synthesis' SROCS computational paradigm...

b. Each of these SROCS computational paradigms is constrained by a (generalized) 'Duality Principle' which asserts that it is not possible to determine the "existence" or "non-existence" of any (hypothetical) 'y' element based on any direct or indirect physical interaction of that 'y' element with any (exhaustive set of) 'x' factor/s; Instead, the Duality Principle postulates that it is only possible to determine the "co-occurrences" of any series of (hypothetical) 'x-y' pairs taking place at different spatial-temporal point/s or interval/s as computed by a conceptually higher-ordered 'D2' computational framework that is (e.g., in principle) irreducible to any series of exhaustive hypothetical direct- or indirect- physical interaction/s, single- or multiple-computational level/s or any hypothetical series of spatial-temporal interactions or occurrences... and:

c. That there can exist only *one singular* such higher-ordered 'D2' computational framework (e.g., as proven by the application of the Duality Principle to each and every one of these hypothetical SROCS paradigms); (Later on, it will be shown that this (hypothetical) singular conceptually higher-ordered 'D2' computational framework must be equivalent to the previously indicated (Bentwich, 2011c) Computational Unified Field Theory's (CUFT) rapid series of 'Universal Simultaneous Computational Frames' (USCF's) which may underlie all microscopic (quantum) and macroscopic (relativistic) aspects of the physical reality.)

Note (however) that the full theoretical implications of accepting these conceptual computational constraints imposed by the Duality Principle upon any scientific SROCS paradigm (in general) and particularly which are set upon Darwin's three-layered must necessarily replace all material-causal (direct- or indirect- single- or multiple-) interaction/s with an a-causal (conceptually higher-ordered) singular computational framework (e.g., termed: 'D2') which alone can compute an exhaustive series of 'x-y' pairs that occur at different spatial-temporal point/s or level/s; Specifically, in the case of Darwin's biological-evolutionary theory the application of the computational Duality Principle to the (above-mentioned) three-layered (primary 'Natural Selection', secondary 'Genetic Encoding – Phenotypic Property' and tertiary 'Genetic Expression – Protein Synthesis') SROCS paradigms, may have potentially far reaching theoretical implications:

Essentially, the acceptance of the Duality Principle's postulated singular conceptually higher-ordered 'D2' computation of the "co-occurrences" of an exhaustive series of 'x-y' pairs implies that all three ('Natural Selection', 'Genetic Encoding –Phenotypic Property', and 'Genetic Expression – Protein Synthesis') 'material-causal' scientific SROCS paradigms must be replaced by a singular (conceptually higher-ordered) 'D2' computation of the "co-occurrences" of each of these (triple-layered) 'Environmental Factors - organism', 'Genetic Factors – phenotypic property' and 'Genetic Expression – Phenotypic Property' computational pairs simultaneously!

It is to be noted that the (above) detailed analysis of the three-layered SROCS computational structure points at two important (specific and more generalized) theoretical implications:

First, in the specific case of Darwin's (three-layered hierarchical-dualistic) SROCS computational structure, it becomes evident that not only is each one of the three constituent SROCS paradigms constrained by the computational Duality Principle – which therefore points at the existence of a singular (conceptually higher-ordered) 'D2' computational framework that computes the "co-occurrences" of each of the (above-mentioned) 'Environmental Factors – organism', 'Genetic Factors – phenotype' and 'Genetic Expression – protein synthesis' pairs (at any given spatial-temporal point/s); but also, an examination of the computational inter-relationships that exist between these (three-layered) SROCS paradigms reveals that each such (subsequent) computational SROCS layer in effect further fragments one of the components of the physical interaction/s in the (previous) layered SROCS structure:

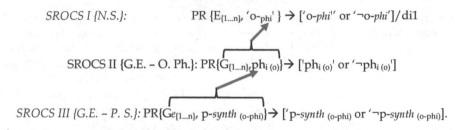

SROCS I {N.S.}:   PR $\{E_{(1...n)},$ 'o-phi' $\}$ → ['o-*phi*'' or '¬o-*phi*']/di1

SROCS II {G.E. – O. Ph.}: PR$\{G_{(1...n)},$ph$_{i (o)}\}$ → ['ph$_{i (o)}$' or '¬ph$_{i (o)}$']

SROCS III {G.E. – P. S.}: PR$\{Ge_{(1...n)},$ p-*synth* $_{(o-phi)}\}$ → ['p-*synth* $_{(o-phi)}$ or '¬p-*synth* $_{(o-phi)}$].

In fact, it is suggested that this hierarchical-dualistic computational structure underlying Darwin's evolutionary theory may point at a much more generalized 'Black-Box Hypothesis' (BBH) as underlying key materialistic-reductionistic (or "material causality" based) scientific paradigms; Indeed, before we attempt to further generalize this 'BBH' to other (key scientific) SROCS paradigms, we attempt to explicate the BBH in the case of these three-layered (above mentioned) SROCS computational structure: It was noted (above) that the inter-relationships between these three (layered) scientific SROCS paradigms is such that each subsequent computational leveled SROCS further fragments the previous level SROCS, i.e., further "de-composes" the previous level SROCS' 'y' element into two (or more) constituting factors; Thus, for instance, the 'y' element in Darwin's (primary) 'Natural Selection' SROCS which is the '*organism*' (e.g., which interacts directly or indirectly with the exhaustive set of Environmental Factors 'E{1...n}'- in order to determine whether such 'organism'shall exist/survive or not exist/gets extinct) – that 'organism' is further "de-composed" or 'fragmented' into the direct physical interaction between the exhaustive set of 'Genetic Factors' G{1...n} and a particular phenotypic property 'ph$i$ *(o)*' e.g., possessed by this particular organism; In other words, Darwin's (primary) Natural Selection SROCS' (direct or indirect) physical interaction between the organism and an exhaustive set of Environmental Factors is further decomposed in the secondary 'Genetic Encoding-Phenotype Property' SROCS computational structure into (two) sub-set fragments of the organism – i.e., which are assumed to consist of a (direct or indirect) physical interaction/s between the exhaustive set of Genetic Factors and (relevant) phenotype property (which is determined to "exist" or "not exist" based on this direct or indirect Genetic Factors – property interaction).

Hence, the secondary (Genetic Encoding – phenotype property) SROCS computational structure further decomposes one of the elements within the primary (Natural Selection) SROCS paradigm, i.e., the 'organism' ('y') element – into two interacting elements within the secondary (Genetic Encoding –phenotype property) SROCS paradigm, e.g., the exhaustive set of Genetic Encoding and a particular phenotypic property:

However, a closer application of the computational Duality Principle (in the case of this dual hierarchical-dualistic computational structure) indicates that not only is each one of these (inter-related) SROCS paradigms constrained by the Duality Principle; but it is also shown that the further fragmentation of the 'organism' element found in the primary (Natural Selection) SROCS paradigm – into the 'Genetic Encoding' (exhaustive set) and 'phenotype property' physical relationship in the secondary (Genetic Encoding – phenotype property) SROCS structure in effect does not alter the basic computational structure found in the primary SROCS paradigm: This is because both the Genetic Encoding exhaustive set and the (particular) phenotype property – are necessarily *included* within the organism (e.g., and its particular phenotype property expressed as: 'o-*phi*') within the primary ('Natural Selection') SROCS paradigm!

So, we can see that the initial ('generalized') SROCS computational structure: PR{E{1...n}, 'o-*phi*' } → ['o-*phi*'' or '¬o-*phi*']/di1 already contains within it any further (secondary) SROCS computational paradigms such as for instance the 'Genetic Encoding – phenotype property' SROCS paradigm; This is because the organism element within the primary SROCS paradigm (represented as: 'o-*phi*') already contains any further segmentation or fragmentation – i.e., as consisting of the Genetic Encoding and phenotype property (direct or indirect) physical interaction/s. Indeed, if we wish to represent the basic (generalized) SROCS computational structure as: PR{X{1...n),Y{1...n),} → ['y' or '¬y'] then any potential (further) breakdown or fragmentation of the 'Y{1...n) element is bound to be contained within the (original) generalized SROCS computational structure and therefore bound to be constrained by the computational Duality Principle.

In the specific case of Darwin's evolutionary theory – the generalized (above mentioned) SROCS computational structure may be represented by the (primary) 'Natural Selection' SROCS structure, thus:

SROCS I {N.S.}: PR{E{1...n}, 'o-*phi*' } → ['o-*phi*'' or '¬o-*phi*']/di1 which precisely replicates the (above mentioned) generalized SROCS structure of: PR{X{1...n),Y{1...n),} → ['y' or '¬y']; indeed, the further fragmentation of this basic (generalized-primary) SROCS computational structure into the secondary 'Genetic Encoding – Phenotype Property' and tertiary 'Genetic Factors - Protein Synthesis' SROCS computational does not alter the basic (generalized) SROCS computational structure (which is obviously constrained by the computational Duality Principle); This is because any further breakdown of the organism (Y{1...n) factor (e.g., within the basic SROCS generalized structure) – i.e., into the 'Genetic Factors' and 'Phenotype Property' [e.g., PR{G{1...n},phi *(o)*}] or into the 'Genetic Encoding' and 'Protein Synthesis' relationship [e.g., PR{G*e*{1...n}, p-*synth* (o-phi)} – obviously does not alter the basic (generalized) SROCS relationship between the organism (e.g., and all of its related phenotypic, genetic, protein… etc. factors) and its Environmental Factors!

More generally, we can see that any scientific SROCS paradigm which consists of the generalized format: PR{X{1...n),Y{1...n),} → [Y{1...n) or ' Y{1...n)']/di1 is not altered by any further breakdown (or fragmentation of the Y{1...n) element; Indeed, it is hypothesized that the BBH precisely constitutes such an explicit fragmentation of the basic SROCS computational structure (e.g., PR{X{1...n),Y{1...n),} → ['Y{1...n) or ' Y{1...n)']/di1) into further and further computational relationships – which are nevertheless comprised within the PR{X{1...n),Y{1...n),} basic SROCS computational structure which has already been shown to be constrained by the computational Duality Principle. Indeed, the abovementioned conceptual computational proof may point at the generalization of the Duality Principle which points at the fallacy of the 'Black Box Hypothesis' – i.e., wherein it becomes clear that the Duality Principle's basic computational constraint imposed upon any (generalized) SROCS paradigm remains unaltered regardless of how many further fragmentations, sub-divisions or computational levels (di1...din) the original 'y' element is comprised of- or divided into-…

Thus, it seems that the generalized form of the Duality Principle may point at the basic fallacy of the 'Black Box Hypothesis' (BBH) – i.e., proving that regardless of the number of factors- or computational levels- that any hypothetical SROCS is fragmented (or broken down into), any such (original) SROCS is necessarily (still) constrained by the Duality Principle; This means that the Duality Principle proves the conceptual computational inability of any such (single- or multiple- leveled) SROCS structure to determine the "existence" or "non-existence" of any hypothetical 'y' element from within its direct or indirect physical relationship/s with any exhaustive 'X-series' (e.g., at any 'di1...din computational level contained within this original SROCS computational structure):

$$\text{SROCS: PR\{X1...n, y\}} \rightarrow [\text{'y' or 'not y'}]$$

But, if indeed the generalized form of the Duality Principle can prove that any (single- or multiple- level) SROCS computational structure is constrained by the Duality Principle, then this means that for any such scientific SROCS paradigm (e.g., for which it is known that the given computational system *is capable* of determining whether a given 'y' element "exists" or "doesn't exist – there must exist a conceptually higher-ordered 'D2' computational level at which there is an 'a-causal' computation yielding the identification of (single- or multiple-) pairs of 'x' and 'y' factors (e.g., occurring at different spatial-temporal point/s, interval/s etc.). This is because the (generalized) Duality Principle has already proven that assuming that the determination of the "existence" or "non-existence" of any given 'y' element from *within* its direct physical interaction with another X(1...n) factor/s inevitably leads to both 'logical inconsistency' and 'computational indeterminacy' – which are (once again) contradicted by robust empirical findings. Moreover, it was shown earlier that the computational characteristics of such D2 level involves an 'a-causal' computation, which computes the "co-occurrences" of any (exhaustive hypothetical) series of 'x-y' pairs (occurring at any hypothetical spatial-temporal point/s or intervals etc.).

Indeed, in the above mentioned case of Darwin's tertiary SROCS computational structure e.g., (comprised of the primary 'Natural Selection Principle, which was further fragmented into the secondary 'Genetic Factors – Phenotypic Property' SROCS paradigm and finally further broken down into the third level 'Genetic Encoding – Protein Synthesis' SROCS paradigm) – the generalized Duality Principle proof pointed at the fallacy of the (tertiary leveled) 'BBH'; Instead, the (generalized) Duality Principle points at the existence of a conceptually higher-ordered 'D2' computational level which carries out computation yielding the (simultaneous) "co-occurrences" of all of the above mentioned three leveled 'x-y' pairs series: Specifically, it is suggested that in the case of Darwin's evolutionary theory an adoption of the Duality Principle's singular D2 computational level indicates that all (abovementioned) apparent tertiary SROCS computational paradigms need to be replaced by three (corresponding) series of 'x-y' pairs (e.g., Environmental Factors – organism; Genetic Factors – Phenotype Properties; Genetic Encoding – Protein Synthesis)...

This means that in the specific instance of Darwin's evolutionary theory instead of there existing multiple 'material-causal' interactions, i.e., between an exhaustive set of Environmental Factors and a single organism (e.g., which is assumed to determine whether that organism "survives" or "doesn't survive"); or between the organism's (deeper) 'Genetic Factors' and its 'Phenotypic Property' (e.g., which is supposed to determine whether particular phenotypic properties of that organism "exist" or "don't exist" – hence indirectly determining that organism likelihood of "surviving" or "not surviving"); or between the (still deeper) organism's 'Genetic Encoding' process and its expression of certain Protein Synthesis (e.g., which is once again assumed to determine the specific Phenotypic Property which determines the organism's "adaptability" or "compatibility" to the Environmental Factors, and hence determines whether that organism shall "survive" or be "extinct" etc.) – according to the computational Duality Principle there seems to exist only one singular conceptually higher-ordered computational level, 'D2' which is responsible for an "a-causal" computation of the existence of "co-occurring" pairs of 'organism-environment', genetic factors-phenotypic property, and genetic encoding process-protein synthesis etc...

Obviously, such conceptually higher-ordered "a-causal" D2 computation is quite "alien" to the basic Cartesian-causal conception wherein it is assumed that any naturally occurring phenomenon is necessarily caused by another material element/s (e.g., which are implicitly assumed to be caused by a series of ever more fine material-causal processes)... However, it is suggested that precisely through the above mentioned application of the Duality Principle analysis of any (single- or multiple- level) SROCS scientific paradigm it can be shown that such Cartesian-causal 'Black Box Hypothesis' is falsified and must necessarily point at the existence of a singular conceptually higher-ordered 'D2' computational framework which merely computes the "co-occurrences" of (single or multiple) computational 'x-y' pairs... Thus, in the case of Darwin's evolutionary tertiary SROCS structured computational paradigm it becomes clear that the material-causal (Cartesian) relationships must give way to a singular higher-ordered a-causal D2 computational framework which computes the "co-occurrences" of the above mentioned three pairs series, i.e., which "co-exist" rather than cause each other...

Indeed, it is suggested that precisely due to Cartesian science's (ingrained) material-causal working hypothesis, that the computational Duality Principle's conceptual proof for the principle inability to compute the "existence" or "non-existence" of any hypothetical 'di1...din' specific 'y' element – from *within* its direct or indirect physical relationship/s with any other (exhaustive) 'x-series' inevitably leads to both 'logical inconsistency' and (ensuing) 'computational indeterminacy' that are contradicted by robust empirical findings (e.g., in the case of each of the earlier mentioned SROCS scientific computational paradigms); Hence, the (generalized) Duality Principle has proven the conceptual computational fallacy of any such single- or multiple- 'Black Box Hypothesis' (BBH) based on an exhaustive analysis of any single or multiple SROCS computational level/s or factor/s – instead pointing at a singular conceptually higher-ordered D2 computational framework which can merely compute the "co-occurrences" of a series of 'x-y' pairs... Indeed, it is due to the generalized Duality Principle's conceptual proof for the principle inability of the multi-layered and (infinitely) complex BBH to determine any of its (single or multiple) SROCS x→y material-causal relationships that it is able to point at the conceptually higher-ordered singular D2 computational framework as the only viable means for determining the "co-occurrences" of any exhaustive series of 'x-y' pairs as underlying any such scientific SROCS paradigms! Finally, based on the earlier (Bentwich, 2011c) proof for the existence of only a singular such conceptually higher-ordered 'D2' Universal Computational Principle' which is responsible for computing a series of 'Universal Simultaneous Computational Frames' (USCF's) which give rise to all ('secondary') computational properties of 'space', 'time', 'energy', 'mass' (and 'causality'), it becomes clear that any such specific SROCS scientific paradigm can only be computed strictly based on this singular (higher-ordered) D2 USCF's series...

Hence, the next step is to prove in the case of each of the other scientific (key) scientific SROCS paradigms that their particular (single- or multiple-) computational (BBH) structure must necessarily be replaced by the singular D2 computational framework; Indeed, it is suggested that besides Darwin's (tertiary-leveled) SROCS evolutionary theory – there are two other (key) scientific paradigms that share the same (problematic) SROCS computational structure, and which therefore necessitate their reformalization based on the same singular conceptually higher-ordered D2 computational framework; These include: Genetics' fundamental 'genetic encoding' hypothesis and Neuroscience's basic 'psycho-physical' problem (e.g., and underlying 'materialistic-reductionistic' working hypothesis);

We've already seen that perhaps two out of three of Darwin's evolutionary theory SROCS paradigms, e.g., 'Genetic Factors – Phenotype Property' and 'Genetic Encoding – Protein

Synthesis' SROCS may be constrained by the computational 'Duality Principle' (and therefore call for their replacement by a corresponding higher-ordered singular 'D2 a-causal' computational framework); Indeed, when presented in the context of Darwin's evolutionary (tertiary) SROCS structure, it was shown that these specific 'Genetic Factors – Phenotype Property' and 'Genetic Encoding – Protein Synthesis' SROCS paradigms do not alter the basic constraint imposed by the (generalized) computational Duality Principle upon all SROCS scientific paradigms (as well as does not alter the need to replace all three-leveled Darwin's evolutionary theory SROCS with the singular higher-ordered D2 a-causal computational framework)... As such, the identification of these two genetics related computational SROCS paradigms (e.g., alongside Darwin's third evolutionary SROCS paradigm of 'Natural Selection') may indeed point at the (abovementioned) need to replace Darwin's tertiary SROCS computational structure by a singular conceptually higher-ordered 'D2 a-causal' computational framework...

But, given the fact that apart from the involvement of these two 'Genetic Factors – Phenotype Property' and 'Genetic Encoding – Protein Synthesis' SROCS paradigms within Darwin's (tertiary) evolutionary theory, these two SROCS computational paradigms also stand at the basis of the central scientific field of Genetics (e.g., in particular and Biology more generally), it is important to scrutinize these two basic genetics SROCS computational paradigms in terms of their fundamental definition of Genetics (and Biology)...

Indeed, it is suggested that the entire field of Genetics (and Biology more generally) may be founded upon these two basic 'Genetic Factors – Phenotype Property' and 'Genetic Encoding – Protein Synthesis' scientific SROCS paradigms; As such, their (above sown) constraint by the computational Duality Principle may call for a rather basic transformation of the scientific fields of Genetics (and Biology) based on the Duality Principle's proof for the need to base these SROCS computational paradigms upon the singular higher-ordered D2 a-causal computational framework;

In a nutshell, it is suggested that the entire field of Genetics is anchored in- and (completely) based upon- these two basic 'Genetic Factors – Phenotype Property' and 'Genetic Encoding – Protein Synthesis' SROCS paradigms... This is because the basic tenet of modern Genetics research (and understanding) is that any genetic process or phenomenon is anchored in and entirely based upon the (direct or indirect) physical relationship/s between certain Genetic Factors and particular Phenotypic Properties which are further mediated (or fragmented into) a secondary (direct or indirect) physical relationship between specific Genetic Encoding processes and particular 'Genetic Encoding' and 'Protein Synthesis' factors... Even more generally, it is suggested that the whole domain of modern Biological research (and scientific body of knowledge) is based upon the basic working assumption that the fundamental 'building-blocks' of all biological organisms is guided by- and based upon- these dual processes of 'Genetic Factors – Phenotype Property' and 'Genetic Encoding – Protein Synthesis' SROCS paradigms; Indeed, one may say that in much the same manner that Physics serves as the most basic building block for all other scientific domains (e.g., because it tells us what are the basic 'building blocks' of nature), these two genetics SROCS paradigms inform all the rest of Genetics and Biology in terms of the fundamental processes by which all biological phenomena, processes or organism/s are produced (and operate through etc.)

Thus, it is suggested that the whole domain of Genetics is based upon the basic working hypothesis wherein any characteristic/s- function/s- organ- tissue/s- or cellular structure/s etc. of any biological organism etc. is entirely dependent upon a series of (direct or indirect) physical interactions between an exhaustive set of Genetic Encoding factors and the production of specific Protein Synthesis, which in return are (solely) responsible for the production of an organism's particular Phenotypic Property; Hence, the production of any (possible) protein found within an organism is assumed to be solely determined through its (direct or indirect) physical interaction/s with an exhaustive set of Genetic Encoding processes, which are governed (solely and strictly) by an exhaustive set of Genetic Factors (e.g., responsible for the production of the specific Protein Synthesis processes). Therefore, we also obtain a (slightly similar) dual leveled SROCS computational structure of this form:

$$PR\{G(1...n), \text{P-synth}\} \rightarrow [\text{'P-synth' or 'not P-synth'}] \quad (7)$$

$$PR\{\text{P-synth}(1...n), \text{Phenotype}i\} \rightarrow [\text{'Phenotype}i \text{' or 'not Phenotype}i\text{'}] \quad (8)$$

Indeed, it is suggested that all genetic-originated biological processes and functions arise (e.g., in one form or another) from this dual-leveled SROCS paradigm: Thus, whether it is the genetic encoding of certain RNA proteins, mRNA activation of specific protein synthesis, the translation of any genetic (single or multiple) factor/s into three-dimensional protein structure/s or their translation into any (simple or complex) organism phenotype, trait or characteristics – all of these genetic encoding, transcription, synthesis and production/interface with any organism's phenotypic property must necessarily rely on the basic assumed (above mentioned) dual-leveled SROCS computational structure.

However, as shown (earlier) the composition of this dual-level Genetics SROCS computational structure is necessarily constrained by the (generalized) Duality Principle; This is due to the fact that each of the constituent SROCS paradigms is necessarily constrained by the Duality Principle (e.g., pointing at the existence of a conceptually higher-ordered D2 a-causal computational framework); Even beyond that the (abovementioned) fallacy of the BBH indicates that regardless of the number of intervening- or mediating- or complex- fragmentation (or makeup) of the basic Genetics SROCS computational structure of the form:

$$PR\{ G(1...n), \text{P-}phenotype(1...n)\} \rightarrow [\text{'P-}phenotype(1...n) \text{ ' or 'not P-}phenotype(1...n)\text{'}]/di1...din$$

the Duality Principle necessarily constrains any such (single or multiple) computational levels (di1...din) or any (single or multiple) mediating factor/s P-$phenotype$(1...n), and points at the existence of a conceptually higher-ordered D2 a-causal computational framework;

Indeed, in much the same manner in which the (generalized) Duality Principle has shown that all of Darwin's evolutionary (tertiary) SROCS computational levels must give way to (three) levels of simultaneously "co-occurring" ('x-y') pairs, so in the case of the (above mentioned) Genetics dual-level SROCS structure it is suggested that an application of the (generalized) Duality Principle points at the existence of the (same) conceptually higher-ordered singular 'D2 a-causal' computational framework which computes (simultaneously) the "co-occurrences" of dual levels of 'Genetic Factors – Protein Synthesis' and 'Protein Synthesis – Phenotype Property' computational pairs.

In other words, it is shown that an (embedded) part of the (above mentioned) tertiary computational structure of Darwin's evolutionary theory is the generalized (dual) 'Genetic Computation' SROCS structure; Therefore, since Darwin's (broader) evolutionary theory (tertiary) SROCS was shown to be based on a (triple strict) '$material$-$causal$' physical relationships between an organism's 'Genetic Factors $\rightarrow$ Protein Synthesis' ; which is assumed to also cause any specific (e.g., single- or multiple- relevant) Phenotypic Property, thus: 'Protein Synthesis→ Phenotypic Property'; which (in return) also caused the survival ("existence") or extinction ("non-existence") of any given organism: 'Phenotypic Property → Organism' ; hence, it is also shown that modern 'Genetic Computation' (dual) SROCS structure may be based on that organism's (direct or indirect) physical interaction/s between its 'Genetic Factors → Protein Synthesis '; and 'Protein Synthesis → Phenotypic Property'.

But, we've already seen that the discovery of the Duality Principle forced relinquishing any such strict –'materialistic-reductionistic' (generalized) SROCS computational structure, in favor of a conceptually higher-ordered 'D2 a-causal' computational framework which negates the existence of any such material-causal (tertiary) physical relationship. Instead, the (generalized) Duality Principle (format) has proven that *regardless of the number of computational levels or factors* that may be associated with the production of any given organism's phenotype or of the number of (direct or indirect) physical interactions between the organism and its environment, the only viable computation that determines any relationships between a given organism and its environment or any between constituent (genetic, protein synthesis or other) elements within the organism and its phenotypic property or properties is *a singular conceptually higher-ordered D2 computational framework which can only determine the simultaneous "co-occurrences"* of any such (single, multiple or exhaustive) pairs of *'Environmental Factors – Organism' ; 'Genetic Factors – Phenotypic Property'; or 'Genetic Encoding Factors – Protein Synthesis' pairs* series...

Therefore, it necessarily follows that the whole of Genetic Science (e.g., including all single- multiple- or exhaustive- factors, computational level/s, phenomena, processes etc. describing an organism's genetic, protein, biological etc. makeup, functioning, development or characteristics etc.) must be anchored in- and based upon- such singular (conceptually higher-ordered) D2 a-causal computational framework which can only compute the "co-occurrences" of any 'Genetic-Factors – Protein Synthesis'; and 'Protein Synthesis – Phenotypic Property' pairings (e.g., occurring simultaneously at any given spatial-temporal point/s)...

Hence, instead of the current 'materialistic-reductionistic' (dual) SROCS structure underlying all Genetic Science (research and theoretical body of knowledge), the (generalized) Duality Principle points at the existence of a singular (conceptually higher-ordered) 'D2 a-causal' computational framework which merely computes the "co-occurrences" of any given pairs of 'Genetic Factors – Protein Synthesis' and 'Protein Synthesis – Phenotype Property'. This means that instead of any exhaustive pool of Genetic Factors "causing" a given organism's resulting Phenotypic Property (or properties), the application of the (generalized) Duality Principle points at the existence higher ordered (singular) D2 computation which simultaneously computes the "co-occurrences" of all of the various aspects of an organism's genetic, protein synthesis, development, traits etc. (e.g., and in the broader scope of Darwin's tertiary evolutionary theory – also of all exhaustive series of any simultaneous 'Environmental Factors') taking place at any given spatial-temporal point/s or interval.

Indeed, it is suggested that such basic shift from the materialistic-reductionistic working assumption underlying current Genetic Science formulation towards a conceptually higher-ordered D2 a-causal computation may bear a few potentially significant theoretical ramifications: First, such conceptually higher-ordered 'D2 a-causal' computational framework necessarily replaces the currently assumed material-causal relationships between any exhaustive set of Genetic Factors which are assumed to cause particular

Protein Synthesis which (in turn) cause particular Phenotypic Properties to appear in a given organism (which may be further extended to include Darwin's Natural Selection SROCS' assumed causal relationship between the above 'Phenotypic Properties' which are assumed to directly interact with an exhaustive set of 'Environmental Factors', wherein it is assumed that the direct or indirect physical relationship of these Environmental Factors with the organism's Phenotypic Properties causes the determination of the "existence" or "non-existence" of any such given organism):

Instead, based on the (above mentioned) generalized Duality Principle's proof for the conceptual computational inability of any (single or multiple) SROCS structure to determine the "existence" or "non-existence" of any (SROCS') particular 'y' from within its direct (or indirect) physical interaction with any other exhaustive X series, the Duality Principle asserts the existence of a (singular) conceptually higher-ordered 'D2 a-causal' computational framework that computes (simultaneously) the "co-occurrences" of any (single or multiple levels) SROCS' 'x' and 'y' pairs series; Thus, the generalized Duality Principle points at the operation of a singular conceptually higher-ordered 'D2 a-causal' computational framework which computes (simultaneously) the "co-occurrences" of all of the abovementioned (dual or tertiary SROCS) computational pairs, thus:

D2 A-Causal Computation:

$$D2: [\{E_{\{1...n\}}, o\}st_1; \{E_{\{1...n\}}, o\}st_2 ... [\{E_{\{1...n\}}, o\}st_n].$$

$$D2: [\{G\{1...n\}, \text{'ph}i \ (o)\text{'}\}st1; \{G\{1...n\}, \text{'ph}j \ (o)\text{'}\}sti; ...\{G\{1...n\}, \text{'ph}n(o)\text{'}\}stn].$$

$$D2: [\{Ge\{1...n\}, p\textit{i-synth} \ (o\text{-phi})\}st1; Ge\{1...n\}, p\textit{j-synth} \ (o\text{-phi})\}sti... ;$$
$$Ge\{1...n\}, p\textit{n-synth} \ (o\text{-phi})\}stn]$$

Hence, the first (potentially significant) theoretical implication of the generalized Duality Principle (e.g., in the case of the currently existing Genetic Science dual SROCS computational paradigm) is that there cannot exist any real 'material-causal' relationships between any of the dual Genetic SROCS (or tertiary Darwin's evolutionary theory SROCS) particular 'x' and 'y' factors; In other words, based on the generalized Duality Principle conceptual computational proof it is asserted that neither the Genetic Factors can "cause" any real 'Protein Synthesis', not can such (particular) Protein Synthesis "cause" any real 'Phenotypic Property' in an organism; nor can any such 'Phenotypic Property' have any real physical interaction with an exhaustive set of 'Environmental Factors' – thereby "causing" the "existence" (survival) or "non-existence" (extinction) of any given (single or multiple) organism/s... Instead, the generalized Duality Principle asserts that there exist a singular conceptually higher-ordered D2 a-causal computational framework which computes *simultaneously* the "co-occurrences" of all of these 'Genetic Factors*st(i)*', 'Protein Synthesis *st(i)*', 'Phenotypic Property *st(i)*', or 'Environmental Factors *st(i)*'!

This means that as in the previous application of the computational Duality Principle in the case of the quantum and relativistic SROCS paradigms (Bentwich, 2011c) where it was shown that all of the physical properties of 'space', 'time', 'energy' and 'mass' cannot be computed based on any (quantum or relativistic) SROCS paradigms – but may only arise as secondary emerging (integrated) computational products of the singular conceptually higher-ordered 'D2 a-causal' series of 'Universal Simultaneous Computational Frames' (USCF's) computation; So also in the case of the Genetic model's dual level SROCS (or tertiary Darwin's evolutionary theory SROCS paradigm) we reach the inevitable conclusion that all of the above mentioned constituent biological elements of 'Genetic Factors$st(i)$', 'Protein Synthesis $st(i)$', 'Phenotypic Property $st(i)$', or 'Environmental Factors $st(i)$' can only exist as secondary emerging computational properties of a singular conceptually higher-ordered 'D2 a-causal' computational framework (e.g., which are therefore computed simultaneously as "co-occurring" at the D2 singular computational level). But, since it was earlier shown above (and also in Bentwich, 2011c) that there can only exist *one* such *singular* conceptually higher-ordered D2 computational framework – which has already been shown to consist of the CUFT's USCF's series that are computed by a Universal Computational Principle, thus:

$$\frac{c^2 x'}{h} = \frac{s \times e}{t \times m}$$

then it follows that the 'D2 a-causal' computation of the abovementioned multiple pairs series of 'Genetic Factors $st(i)$' - 'Protein Synthesis $st(i)$'; 'Protein Synthesis $st(i)$' - 'Phenotypic Property $st(i)$'; 'Phenotypic Property $st(i)$' - 'Environmental Factors $st(i)$ may only be carried out through the singular D2 a-causal computation of the series of USCF's! What's essential to understand is that given the Duality Principle's above mentioned conceptual computational proof for the principle inability of either of the Genetic (dual) SROCS paradigms (or Darwin's Natural Selection paradigm) to determine any 'material-causal' relationship/s between any of the (abovementioned) 'Genetic Factors $st(i)$' → 'Protein Synthesis $st(i)$'; 'Protein Synthesis $st(i)$' → 'Phenotypic Property $st(i)$'; 'Phenotypic Property $st(i)$' → 'Environmental Factors $st(i)$; but instead, the recognition that all of these 'x-y' pairs (series) are computed simultaneously as part of the same USCF's (e.g., at the conceptually higher-ordered singular D2 computational level)... Moreover, if (indeed) there cannot exist any real 'material-causal' physical relationship between any of these x→y (hypothesized particular SROCS) pairs, but only a conceptually higher-ordered (singular) D2 'a-causal' "co-occurrences" of all of these x-y pairs (series) as being computed simultaneously as part of the same (particular) USCF (frames), then it follows that the only computation responsible for such conceptually higher-ordered (singular) USCF's series (e.g., including all of its embedded particular 'x-y' pairs series) is the Universal Computational Principle which was hypothesized to be responsible for all USCF's series computation (i.e., including all of the "secondary computational integrated" physical properties of 'space', 'time', 'energy' and 'mass' etc.)

Note that despite the apparent "radical" theoretical conclusion that seems to stem from an application of the (generalized) Duality Principle in the case of the above (dual) Genetic Science SROCS computational structure (and its extended Darwin's Natural Selection assumed SROCS computational paradigm)- i.e., that there cannot exist any (real) "causal-material" physical relationship between any (exhaustive hypothetical) series of 'Genetic Factors $st(i)$' → 'Protein Synthesis $st(i)$'; 'Protein Synthesis $st(i)$' → 'Phenotypic Property $st(i)$'; 'Phenotypic Property $st(i)$' → 'Environmental Factors $st(i)$, but rather that there exists only one (singular) conceptually higher-ordered 'D2' a-causal' computational framework that computes simultaneously the series of USCF's (various) 'x-y' pairs, such conceptually higher-ordered D2/USCF's computational level is proven based precisely upon such a rigorous computational and empirical analysis (e.g., pertaining to any SROCS computational structure which inevitably proves the computational constraint imposed by the 'Duality Principle'). Furthermore, the adoption of such a conceptually higher-ordered

'D2 a-causal' computational mechanism – e.g., anchored in the USCF's series (computed by the singular 'Universal Computational Principle'), instead of the currently assumed 'materialistic-reductionistic' SROCS computational structure does not negate any of the (already known) empirical facts or body of knowledge pertaining to any biological intra-organism (genetic, protein synthesis, phenotypic etc.) or inter-organism (environmental or other evolutionary) empirical findings; Rather, the theoretical explanation (or construct) upon which these empirically well-validated facts are based is shifted (or even expanded) from the narrow constraints of any (hypothetical exhaustive) 'material-causal' (direct or indirect) physical relationship/s between any particular 'x→y' pair/s to a 'D2 a-causal' relationship/s between all potential 'x and 'y pairs (series) that are embedded within the exhaustive Universal Simultaneous Computational Frames (USCF's) series that are being computed by a singular Universal Computational Principle...

Finally, it should be noted that as shown previously (Bentwich, 2011b), the Computational Unified Field Theory's (CUFT) analysis of the production of the series of Universal Simultaneous Computational Frames (USCF's) is carried out by a *Universal Computational Principle* – which is the only computational (e.g., rather than "material" or "physical") element that exists "in-between" any two USCF's frames; This stemmed from the fact that it was shown that there can only exist one (singular) conceptually higher-ordered D2 computational level – which is (in principle) irreducible to any exhaustive-hypothetical 'x→y' (direct or indirect) physical relationship/s; Based on this conceptual computational constraint imposed by the 'Duality Principle' (e.g., negating the existence of any real 'x→y' physical relationship, but rather its replacement by a conceptually higher-ordered D2 computation of the "co-occurrences" of simultaneously occurring 'x-y' pairs embedded within the same USCF's) and empirical-computational postulate of the existence of these disparate USCF's (e.g., which coalesces well-validated quantum and relativistic empirical phenomena such as Planck's minimal inter-USCF's '*h*' constant as well as the hypothetical extremely rapid rate of USCF's computation given by $c^2/'h'$) it was hypothesized that there cannot exist any material element "in-between" two such postulated USCF's – except for the 'Universal Computational Principle' which computes each of these series of USCF's... Indeed, the hypothesized Universal Computational Formula:

$$\frac{c^2 x'}{h} = \frac{s \times e}{t \times m}$$

precisely outlines the fact that all computational features of 'space', 'time', 'energy', 'mass' (and 'causality') arise as secondary (integrated) physical properties of the conceptually higher-ordered D2 Universal Computational Principle's production of these series of USCF's frames). Therefore, when viewed from the conceptually higher-ordered perspective of the 'D2 a-causal' computational framework, all hypothetical (exhaustive) series of 'x-y' pairs may only be computed by the (singular) Universal Computational Principle as embedded within the series of USCF's (e.g., thereby replacing any of the currently assumed 'materialistic-reductionistic' direct or indirect "causal" relationship/s between any hypothetical exhaustive 'x→y' pair/s).

Indeed, perhaps a good mode of explaining the potential transformation from the contemporary purely 'materialistc-reductionistic' SROCS computational structure (e.g., underlying key scientific SROCS paradigms described in this article) to the conceptually higher-ordered 'D2 a-causal' computational framework – is to analyze the (metaphorically 'equivalent') case of the cinematic film sequence underlying any apparently "material-causal" relationships that may exist between any two 'x' and 'y' elements (e.g., within a given cinematic film); As hinted in a previous article (Bentwich, 2011c) it is suggested that underlying any such apparent "x$\rightarrow$y" physical relationship (within any given cinematic film sequence), there cannot be any "real" '*material*-causal' relationship within the film sequence; This is because it is shown based on the cinematic film metaphor that in order for any 'physical relationship' to exist through any (hypothetical) sequence of cinematic film frames, there must exist a certain pattern of "co-occurrences" of the given 'x' and 'y' elements – i.e., such as for instance that the 'x' factor appears to be located "spatial-temporally" closer and closer to the 'y' element (across a certain number of cinematic frames) which then leads to an alteration in the 'y' factor's (particular) condition (or spatial-temporal configuration etc.); In other words, for the appearance of any (hypothetical) "physical causality" to exist between the 'x' and 'y' factors within any film sequence there must be a (certain) series of film frames across which the "spatial-temporal" relationship between the 'x' and 'y' factors is transformed... To put it succinctly, it is suggested that it is not possible (e.g., in principle) to have any "causal-material" relationship between any two (hypothetical) 'x' and 'y' elements – that is not based on an alteration in the spatial-temporal (proximity and configuration) of any two such 'x' and 'y' elements across a number of cinematic film frames. But, once we realize that it is not possible to obtain any "material-causal" relationship between any two (hypothetical) 'x' and 'y' elements – which is not based on a change in the their "co-occurring" pattern *across a few cinematic film frames* the door is open to evince that there cannot in fact exist any "real material" element that can "pass" in-between any two such (hypothetical) cinematic film frames!? But since we know that there does not exist any "material" element that exists "in-between" any two such hypothetical cinematic film frames (e.g., '*f-i*' and '*f-i+n*'), then we must conclude that the only viable means for producing any such apparent "material-causal" relationship/s is based on the alteration in the spatial-temporal configuration of the 'x' and 'y' elements (across a series of cinematic frames)... In other words, since there is not "material" element that can pass "in-between" two such hypothetical cinematic film frames (e.g., '*f-i*' and '*f-i+n*') and since the existence of any hypothetical material-causal" physical relationship between any two hypothetical 'x' and 'y' elements is contingent upon a certain pattern of change in the 'x-y' spatial-temporal configuration across such (hypothetical) cinematic film sequence then it follows that the only means for producing any "causal" relationship between the 'x' and 'y' elements is only based on their "co-occurring" spatial-temporal across a certain number of cinematic frames... Finally, precisely based on this keen (computational) analysis wherein it is shown that any hypothetical "causal-material" x$\rightarrow$y relationship can only evolve based on their particular spatial-temporal configuration (across a series of cinematic film frames), and since there cannot be any "material" element that can pass "in between" any two subsequent cinematic film frames, then we are also led to conclude that the only means for arranging the particular "co-occurrence" of any apparently spatial-temporal "causal" pattern of change in the 'x' and 'y' configuration across a series of cinematic frames is based on a conceptually higher-ordered computation (or arrangement) of the 'x' and 'y' "spatial-temporal" sequencing across these series of frames... Ultimately, since there is no "material" element that can pass "in-between" any two subsequent (hypothetical) film frames and since the perception of any apparent "causal-material" physical relationship between the 'x' and 'y elements is contingent upon a particular pattern of change in the "spatial-temporal" configuration of the 'x' and 'y' elements across a series of cinematic film frames – then this points at the existence of a conceptually higher-ordered "non-material" computational element that is responsible for this particular spatial-temporal pattern of change across the film frames...

Indeed, it is hypothesized that the above metaphor of the cinematic film sequence may be entirely analogous to the Computational Unified Field Theory's (CUFT) (Bentwich, 2011c) account – not only in terms of the secondary (integrated) emerging physical features of "space", "time", "energy", "mass", but may also pertain to the basic (implicit) concept of "causality"; Previously, the cinematic film metaphor has been used as a 'pointer' to some of the hypothetical features of the CUFT including its delineation of the emerging (secondary) computational properties of 'space', 'time', 'energy' and 'mass' (e.g., wherein it was shown that the apparently physical properties of 'space' and 'energy', 'mass' and 'time' may arise as secondary computational combinations of a 'consistent' vs. 'inconsistent' computations of whole 'frame' presentations of the same object or event or of only partial segments of the whole frame entitled: 'object' - 'consistent' or 'inconsistent' presentations). The abovementioned postulated Computational Unified Field Theory's account of the four basic physical features (of 'space', 'time', 'energy' and 'mass') was also based on the existence of a hypothetical conceptually higher-ordered (D2) 'Universal Computational Principle' ("ר") which may carry out extremely rapid ('c^2/h') computational process giving rise to a series of 'Universal Simultaneous Computational Frames' (USCF's). The essential point to be noted is that based on the earlier outlined Duality Principle which proved that there can only exist one singular conceptually higher-ordered 'D2' computational framework that can (solely) determine all exhaustive hypothetical (quantum, relativistic or any other) 'x-y' "co-occurrences" across the series of (hypothesized) USCF's the CUFT was capable of replicating all known quantum and relativistic phenomena (as well as potentially harmonize all existing apparent contradictions between these two major pillars of modern Physics). But, if indeed the entire corpus of (all possible hypothetical) quantum and relativistic features, phenomena, laws and theoretical explanations can only be derived from such a Duality Principle based conceptually higher-ordered D2 (e.g., 'Universal Computational Principle' ' ר ') computation of a series of (extremely rapid) USCF's (Bentwich, 2011c), then it also necessarily follows that the CUFT's account of any (apparently) "material-causality" must also be transformed; Indeed, somewhat alike the cinematic film metaphor's demonstration that there cannot exist any real "material-causal" relationship between any hypothetical 'x' and 'y' factors – but only a conceptually higher-ordered ('D2') arrangement of the "co-occurrences" of a specific spatial-temporal configuration of the 'x' and 'y' factors (as discussed above), it is suggested that the CUFT's portrayal of a series of extremely rapid USCF's does not allow for any "material" element/s to pass "in-between" any two (hypothetical) USCF's except for the conceptually higher-ordered (immaterial) 'Universal Computational Principle' ('ר') which alone can compute the particular "co-occurrences" of a series of 'x-y' pairs that can give rise to the apparent existence of a "causal" relationship between the 'x' and 'y' elements...

Hence, we arrive at the inevitable conclusion wherein any apparent "material-causal" relationship/s between any hypothetical 'x' and 'y' factors (e.g., embedded within one of the key SROCS scientific paradigms) – must necessarily arise as secondary emerging computational property associated with a particular 'spatial-temporal' "co-occurrences" of the particular 'x' and 'y' factors' configuration across a series of USCF's... To follow the cinematic film metaphor, there does not exit any "real material-causality" between any two hypothetical 'x' and 'y' elements, but only the "co-occurrence" of the particular 'x' and 'y' factors across a series of USCF's (e.g., as computed by a conceptually higher ordered D2 computational principle – which in the case of the CUFT is the 'Universal Computational Principle'). Therefore, it may be said that perhaps underlying all scientific SROCS paradigms there cannot exist any (real) "material-causal" relationship/s between any two hypothetical 'x' and 'y' elements, but only the computation of their "co-occurrences" (e.g., in a particular spatial-temporal sequence as explained above) across a series of USCF's (as computed by the conceptually higher-ordered D2 'Universal Computational Principle' 'ל')...

This means that in the two (abovementioned) cases of Darwin's (tertiary) evolutionary theory SROCS computational structure as well as in the case of the (dual) Genetic Science SROCS computational structure an application of the (generalized) Duality Principle and its broader development within the CUFT has pointed at the existence of a series of USCF's that are computed by the conceptually higher-ordered ('D2') 'Universal Computational Principle' ('ל') and which give rise to any SROCS apparent "material-causal" ('x→y') relationships that are underlie by a particular series of "co-occurring" x-y pairs in which the 'spatial-temporal' relationships (e.g., as embedded within a series of corresponding USCF's, as explained above).

This means that both in the case of Darwin's (tertiary) SROCS computational structure as well as in the case of Genetic Science (dual) SROCS computational structure we must replace the currently assumed direct (or indirect) 'material-causal' relationship between any two particular 'x' and 'y' elements by the conceptually higher-ordered D2 computation of the "co-occurrences" of the corresponding (triple or dual) SROCS series of 'x-y' pairs that give rise to the appearance of any "material-causal" relationship; As discussed above, in both cases there exists a (hypothetical) conceptually higher-ordered D2 computational level which carries out the simultaneous computation of the "co-occurrences" of Darwin's SROCS paradigm's alternate 'Environmental Factors $st(i)$ and 'Phenotypic Property $st(i)$' pairs series, as well the two other (Genetic SROCS dual pairs of) 'Genetic Factors $st(i)$' and 'Protein Synthesis $st(i)$', and the 'Protein Synthesis $st(i)$' and 'Phenotypic Property $st(i)$' series. Indeed, according to the CUFT's (broadened application of the Duality Principle) such conceptually higher-ordered D2 simultaneous computation of each of these evolutionary and genetic encoding computational pairs constitutes the (extremely rapid hypothetical) series of USCF's that are carried out by the singular 'Universal Computational Principle' ('ל'). Thus, instead of the existence of any 'real' "material-causal" relationship/s between any of these (evolutionary or genetic) SROCS' particular 'x' and 'y' factors – all that truly exists is the conceptually higher-ordered (singular) Universal Computational Principle's ('ל') simultaneous computation of a series of (extremely rapid) USCF's in which there is an embedded series of 'Environmental Factors $st(i)$ and 'Phenotypic Property $st(i)$' ; 'Genetic Factors $st(i)$' and 'Protein Synthesis $st(i)$'; and the 'Protein Synthesis $st(i)$' and 'Phenotypic Property $st(i)$' pairs series (which give rise to the appearance of 'real' interactions within seemingly "material-causal" SROCS 'x→y' relationships)...

Finally, it is suggested that the application of the Duality Principle's asserted conceptually higher-ordered 'D2' (Universal Computational Principle's) computation of the series of USCF's which also embed all (exhaustive-hypothetical) 'x-y' pairs e.g., as replacing all scientific SROCS paradigms' apparent ('x→y') "material-causal" relationships should be implemented; Hence, the next step in the application of the computational Duality Principle to various other scientific SROCS paradigms consists of a (triple) demonstration that each of these (remainder) scientific SROCS paradigms is constrained by the (generalized) Duality Principle, may contain the (abovementioned) 'Black-Box-Hypothesis' (BBH) (e.g., which we've already seen cannot alter the basic computational constraint imposed by the generalized Duality Principle format), and therefore inevitably calls for the CUFT's assertion regarding the need to replace the currently assumed SROCS (particular) "material-causal" ('x→y') relationship with a conceptually higher ordered (Universal Computational Principle's '"') computed series of "co-occurring" 'x-y' pairs (as embedded within a rapid series of USCF's being computed by this hypothetical Universal Computational Principle).

3.2 The Duality Principle: Constraint of the 'Psycho-Physical Problem' (PPP) SROCS

It is hypothesized that another key scientific SROCS paradigm consists of Neuroscience's Psycho-Physical Problem (PPP); This is because the PPP which is defined as the question regarding how it may be possible for any given physical stimulus (or stimuli) to be translated into a neurochemical signal within the Central Nervous System (in humans) – is currently assumed to be resolved through Neuroscience's basic 'materialistic-reductionistic' (generalized) 'Psycho-Physical SROCS' computational structure: Essentially, Neuroscience's basic (generalized) 'Psycho-Physical SROCS' assumes that the determination of the "existence" or "non-existence" of any hypothetical (exhaustive) Psycho-Physical Stimulus or stimuli (e.g., 'PPs-i' - including all physical stimulation or any of its derived or associated physical features, properties, representations etc.) is determined solely based on its direct or indirect physical interactions with an exhaustive set of 'Neural Activation/s' (e.g., 'Na$_{(1...n)}$' – an exhaustive hypothetical series of neurons, neural connections, neural activation/s neurophysiological activity or pattern/s etc. which may take place at different single or multiple spatial-temporal points in the human Nervous System);

$$\text{SROCS: PR\{ PPs-}i\text{ , Na}_{(1...n)},\text{ \} → ['PPs-}i\text{' or 'not PPs-}i\text{']/di1...din}$$

Thus, for instance, it is currently assumed that the computation of the "existence" or "non-existence" of any such Psychophysical Psycho-Physical Stimulus, e.g., human consciousness or awareness to the existence of any given physical stimulus intensity (termed: termed: PPs-pp) – is strictly caused by the direct (or indirect) physical interaction of such 'Consciousness Psychophysical Stimulus' (Cs-pp) with an exhaustive hypothetical series of 'Neural Activation/s' (e.g., including any exhaustive hypothetical activity or activation of any neuron/s, neural activation, neuronal pattern/s etc. in the human brain):

$$\text{SROCS: PR\{N}_{(1...n)},\text{ Cs-}pp\text{\} → ['Cs-}pp\text{' or 'not Cs-}pp\text{']/di1...din}$$

But, such SROCS computational structure was previously shown (Bentwich, 2006a) to produce an inevitable SRONCS (e.g., 'Self-Referential Ontological Negative Computational System', as described earlier) in the case of *sub-threshold psychophysical stimulation*: SROCS: PR{ N$_{(1...n)}$, Cs-i} → 'not Cs-i '/di1...din

Indeed, such SRONCS was shown to produce both 'logical inconsistency' and ensuing 'computational indeterminacy' that are contradicted by robust empirical findings indicating the *capacity* of such psychophysical computational systems to determine the "existence" or "non-existence" of any given psychophysical stimulation (e.g., including in the case of sub-threshold psychophysical stimulus); therefore, the Duality Principle pointed at the existence of a conceptually higher-ordered 'D2 a-causal' computational framework which is capable of computing the existence of any series of pairs of any given Consciousness-Stimuli and an exhaustive hypothetical series of all possible 'Neural Activation' hypothetical), thus:

$$\text{D2: } [\{N_{(1\ldots n)\ st\text{-}i},\ C_{s\text{-}pp\ st\text{-}i}\};\ \ldots\ \{N_{(1\ldots n)\ st\text{-}i+n},\ C_{s\text{-}pp\ st\text{-}i+n}\}]$$

As proven previously (and represented in the generalized SROCS computational structure encompassing any single or multiple computational elements, factors etc., di1...din), the computational constraint imposed on the above Psychophysical SROCS structure is *conceptual* in nature – i.e., it holds true regardless of the number of neurons, neuronal interactions or spatial temporal point/s at which any direct or indirect physical interaction may take place between the given Consciousness Psychophysical Stimulus and any exhaustive hypothetical series of Neural Activations; This is because the formalization of this (primary) Psychophysical-Consciousness Stimulation SROCS already encompasses all direct or indirect physical interactions between any given Psychophysical Stimulation and an exhaustive set of all possible Neural Activations (occurring at any potential spatial-temporal point/s or interval/s etc.), and indicates that as such it inevitably leads to both 'logical inconsistency' and subsequent 'computational indeterminacy' (e.g., in the case of sub-threshold Psychophysical Stimulation SRONCS system) that are contradicted by well validated empirical findings...

Next, it is hereby hypothesized that the abovementioned Psychophysical Consciousness Stimulation SROCS may serve as a primary SROCS level within a multi-layered PPP SROCS computational structure, which may be generally divided into (at least) four separate SROCS computational levels including:

1. **Psycho-Physical Consciousness SROCS**: $PR\{Cs\text{-}pp\ ,\ Na(s\text{-}pp)\} \rightarrow$ ['Cs-pp ' or 'not Cs-pp ']/di1...din
2. **Functional Consciousness SROCS**: $PR\{Cs(pp)\text{-}fi,\ Na(spp\text{-}fi)\} \rightarrow$ [' C$s(pp)$- fi ' or 'not C$s(pp)$- fi '].
3. **Phenomenological Consciousness SROCS**: $PR\{Cs(pp\text{-}fi)\text{-}Ph\ ,\ Na(spp\text{-}fi)\text{-}Ph\ )\} \rightarrow$ [' C$s(pp$-fi$)$-Ph ' or 'not C$s(pp$- fi$)$-Ph ']/di1...din
4. **Self-Consciousness SROCS**: $PR\{$ C$s(pp$- fi-Ph$)$-S, Na(pp- fi-Ph)-S$\} \rightarrow$ [' C$s(pp$- fi-Ph$)$-S ' or 'not C$s(pp$- fi-Ph$)$-S '].

Below is a delineation of the various hierarchical-dualistic computational levels currently assumed by Neuroscience's materialistic-reductionistic working hypothesis;

1. **Psycho-Physical Consciousness SROCS**: $PR\{Cs(pp)\text{-}fi,\ Na(spp\text{-}fi)\} \rightarrow$ [' C$s(pp)$- fi ' or 'not C$s(pp)$- fi ']: wherein it is currently assumed that the (primary) Psychophysical Stimulation Consciousness SROCS' resulting output (e.g., ['Cs-pp ' or 'not Cs-pp ']/di1...din) undergoes a secondary SROCS computational structure in which the "existence" or "non-existence" of the (primary SROCS) 'Psychophysical Stimulation Consciousness' is analyzed in terms of the "existence" or "non-existence" of any particular 'Psychophysical Stimulation Functional Consciousness' (i.e., such as any given physical property, attribute, phenomenon etc., represented by: 'C$s(pp)$- fi '); It is hypothesized that this secondary 'Functional Consciousness' SROCS computational structure is comprised of: any direct or indirect physical interaction between a (given) Psychophysical Stimulation Functional Consciousness input (e.g., 'C$s(pp)$- fi ' or 'not C$s(pp)$- fi' which is equivalent to the above primary SROCS's: 'Cs-pp ' or 'not Cs-pp ' output), and another exhaustive set of Neural Activation/s responsible for computing "existence" or "non-existence" of that particular given Psychophysical Stimulation Consciousness Function; However, as shown earlier, this secondary SROCS paradigm also shares the same SROCS computational structure and as such is constrained by the same (generalized) Duality Principle;

2. **Functional Consciousness SROCS:** PR{Cs(pp)- fi, Na(spp-fi)}→[' Cs(pp)- fi ' or 'not Cs(pp)- fi ']/di1...din.
This is because this (secondary) Functional Consciousness SROCS computational structure also inevitably leads to both 'logical inconsistency' and ensuing 'computational indeterminacy' in the case of a SRONCS:
PR{Cs(pp)- fi, Na(spp-fi)}→ 'not Cs(pp)- fi '/di1...din . Once again, the generalized Duality Principle asserts that this last 'computational indeterminacy' is contradicted by validated empirical findings indicating the capacity of the human Central Nervous System (CNS) to determine any given particular functional properties of any given Psychophysical Stimulation. Therefore, the generalized Duality Principle points at the necessary existence of a conceptually higher-ordered 'D2' computational framework which computes simultaneously any series of "co-occurring" pairs of Functional Consciousness (attributes of a given psychophysical stimulus) alongside its Neural Activation correlate (e.g., at any given spatial-temporal point).

$$D2: [\{Cs(pp)f_i, Na_{(spp)}fi\}_{st\text{-}i} ; ... \{Cs(pp)f_{(i+n)}, Na_{(spp)}f_{(i+n)}\}_{st(i+n)}]/di1...din$$

Likewise, it is suggested that a further (subsequent third) potential SROCS computational paradigm level is that of 'Phenomenological Consciousness SROCS':

3. **Phenomenological Consciousness SROCS:** PR{Cs(pp- fi)-Ph , Na(spp-fi)-Ph)}→[' Cs(pp- fi)-Ph ' or 'not Cs(pp- fi)-Ph ']/di1...din wherein the previous (secondary Functional Consciousness) SROCS output of ' Cs(pp)- fi ' or 'not Cs(pp)- fi ' serves as the basis for the input to the third level Phenomenological Consciousness SROCS in the form of the phenomenological experience of any such particular Consciousness Function (i.e., Cs(pp-fi)-Ph) which directly interacts with an exhaustive set of Neural Activations which are assumed to be responsible for carrying out this processing; Hence, this third Phenomenological Consciousness SROCS assumes that the determination of the "existence" or "non-existence" of any particular 'phenomenological experience of any particular psychophysical stimulation function' (Cs(pp- fi)-Ph) is solely based on direct or indirect physical interactions between such given 'phenomenological experience of any particular psychophysical stimulation function' (Cs(pp- fi)-Ph) and an exhaustive set of Neural Activation/s (e.g., Na(spp-fi)-Ph) that are assumed to be responsible for carrying out such processing...
However, as in the two preceding SROCS computational structures it is clear that such (third-level Phenomenological Consciousness) SROCS must also be constrained by the generalized Duality Principle and therefore also inevitably leads to both 'logical inconsistency' and 'computational indeterminacy' in the case of the SRONCS:

$$PR\{Cs(pp\text{-} fi)\text{-}Ph , Na(spp\text{-}fi)\text{-}Ph)\}→ \text{'not } Cs(pp\text{-} fi)\text{-}Ph \text{'}/di1...din$$

wherein the specific phenomenological experience is asserted to both "exist" and "not exist" at the same (single or multiple) computational level/s (di1...din); But, since there exists ample empirical evidence indicating the capacity of human beings to determine (for each stimulus or stimuli) whether or not a certain phenomenological feature of function "exists" or "doesn't exist", then we must accept the (generalized) Duality Principle's assertion regarding the existence of a conceptually higher-ordered 'D2' computational level; Such conceptually higher-ordered 'D2' computational framework can compute the "co-occurrences" of any hypothetical series of such particular 'phenomenological experience of any particular psychophysical stimulation function' ($Cs_{(pp\text{-}fi)\text{-}Ph}$) and a corresponding exhaustive set of Neural Activations ($Na_{(spp\text{-}fi)\text{-}Ph}$):

D2: [{Cs(pp- fi)-Ph$_i$, Na(spp-fi)-Ph$_i$} $_{st\text{-}i}$; ...{Cs(pp- fi)-Ph$_{(i+n)}$, Na(spp-fi)-Ph} $_{st\text{-}(i+n)}$]

4.   **Self-Consciousness SROCS:** PR{Cs(pp- fi)Ph-S, Na(pp- fi-Ph)-S}$\rightarrow$[' Cs(pp- fi)Ph-S ' or 'not Cs(pp- fi)Ph-S ']/di1...din.

It is finally hypothesized that there exists one further (fourth and final) SROCS computational level of 'Self-Consciousness' which combines between all (third-level) Phenomenological Consciousness SROCS outputs of the "existence" or "non-existence" of any given phenomenological experience (e.g., of a particular psychophysical stimulus function) as the basis for its integrated input stimulus of a 'Phenomenological Self Stimuli' – which is assumed to directly (or indirectly) physically interact with an exhaustive hypothetical set of Neural Activation/s (e.g., comprised of all potential neuron/s, neural connection, neural activation/s etc. responsible to determine whether there "exists" or "doesn't exist" any such 'Phenomenal Self Stimuli' at any given computational level, 'di1...din').

However, as in all previous computational level SROCS since this (final) 'Self-Consciousness SROCS' is necessarily constrained by the (generalized) Duality Principle, then it also must be replaced by the conceptually higher-ordered 'D2' computational framework which computes the "co-occurrences" of any series of pairs of 'Phenomenal Self Stimuli' (e.g., comprised of the sum total of all phenomenal functional psychophysical stimuli – at any given spatial-temporal point/s) and any simultaneously occurring (exhaustive hypothetical) Neural Activation/s, thus:

D2: [{Cs(pp- fi)Ph-Si, Na(pp- fi)Ph-S $_i$} $_{st\text{-}i}$; ...{Cs(pp- fi)Ph-S(i+n), Na(pp- fi)Ph-S(i+n)} $_{st\text{-}(i+n)}$]

Therefore, it seems that the Psychophysical Problem of human Consciousness (PPP) is currently formalized as a (four-layered) computational SROCS structure which can be represented in the general format:

SROCS: PR{Cs-i , Na$_{(1...n)}$, }$\rightarrow$ [' Cs-i' or 'not Cs-i ']/di1...din

wherein it is assumed that an hypothetical series of direct or indirect physical interactions between any possible ("external") psychophysical or ("internal") 'functional', 'phenomenological' or 'self' stimuli and an exhaustive set of Neural Activations (e.g., as described above comprised of any single or multiple spatial-temporal neural activations, patterns, interactions, neurons or neural connections or neural networks etc.) is solely responsible for determining whether any such Psychophysical, Functional, Phenomenological or Self stimulus "exists" or "doesn't exist". But, it was shown (above and previously) that the generalized 'Duality Principle' constrains any such SROCS computational structure – by proving that any SROCS structure inevitably leads to both 'logical inconsistency' and 'computational indeterminacy' which are contradicted by known empirical findings indicating the capacity of the human Nervous System to determine whether or not any given 'psychophysical', 'functional', 'phenomenological' or 'self' stimuli "exists" or "doesn't exist"; Therefore, the generalized Duality Principle proves that there must exist a conceptually higher-ordered 'D2' computational framework which can compute the "co-occurrences" of any hypothetical series of corresponding pairs of:

D2:

1. Psychophysical: $[\{N(1...n)$ st-i, Cs-pp st-i$\}$; ... $\{N(1...n)$ st-i+n, Cs-pp st-i+n $\}]$
2. Functional: $[\{Cs(pp)f_i, Na(spp)fi\}_{st-i}$; ... $\{Cs(pp)f_{(i+n)}, Na(spp)f_{(i+n)}\}_{st(i+n)}]$
3. Phen.:$[\{Cs(pp-fi)-Ph_i, Na(spp-fi)-Ph_i\}_{st-i}$; ...$\{Cs(pp-fi)-Ph_{(i+n)}, Na(spp-fi)-Ph\}_{st-(i+n)}]$
4. Self: $[\{Cs(pp-fi)Ph-Si, Na(pp-fi)Ph-S_i\}_{st-i}$; ...$\{Cs(pp-fi)Ph-S(i+n), Na(pp-fi)Ph-S(i+n)\}_{st-(i+n)}]$

This means that instead of the currently assumed 'materialistic-reductionistic' SROCS paradigms – e.g., at the psychophysical- functional- phenomenological- and self- stimulus levels, the Duality Principle proves that there can only exist one (singular) conceptually higher-ordered 'D2' computational framework which computes the "co-occurrences" of each of the above (particular four level) PR$\{Cs$-i , $Na_{(1...n)}\}$ pairs... Moreover, instead of the currently assumed 'material-causal' physical relationships between the specific $\{Cs$-i , $Na_{(1...n)}\}$ pairs, and moreover between each of these four SROCS computational levels:

5. *Psychophysical:* $[\{N(1...n)_{st-i}$, Cs-pp $_{st-i}\}$; ... $\{N(1...n)_{st-i+n}$, Cs-pp $_{st-i+n}\}]$
6. Functional: $[\{Cs(pp)f_i, Na(spp)fi\}_{st-i}$; ... $\{Cs(pp)f_{(i+n)}, Na(spp)f_{(i+n)}\}_{st(i+n)}]$
7. Phen.:$[\{Cs(pp-fi)-Ph_i, Na(spp-fi)-Ph_i\}_{st-i}$; ...$\{Cs(pp-fi)-Ph_{(i+n)}, Na(spp-fi)-Ph\}_{st-(i+n)}]$
8. Self: $[\{Cs(pp-fi)Ph-Si, Na(pp-fi)Ph-S_i\}_{st-i}$; ...$\{Cs(pp-fi)Ph-S(i+n), Na(pp-fi)Ph-S(i+n)\}_{st(i+n)}]$

The Duality Principle conceptually proves that there cannot (e.g., in principle) exist any such direct or indirect material-causal relationship/s between any of these (assumed) four leveled scientific SROCS paradigms' particular $N(1...n)_{st-i}$ → Cs-$_{st-i}$ factors, or between any of these SROCS paradigms (themselves – as stipulated above);

Instead, the Duality Principle proves that at none of these (currently assumed) SROCS paradigms, or indeed at any other (exhaustive hypothetical) SROCS computational level/s – can there exist any real "material-causal" relationship between any Conscious stimulus (or stimuli – e.g., at any of the four above mentioned generalized computational levels or at any other exhaustive-hypothetical computational level/s) and any exhaustive hypothetical Neural Activation/s locus or loci etc. (e.g., at any hypothetical computational level 'di1...din'); Instead, the Duality Principle asserts that there can only exist the singular (conceptually higher-ordered) 'D2' computational framework which can compute simultaneously the "co-occurrences" of any of the four abovementioned psychophysical- functional- phenomenological- or self- pairs...

This means that instead of the currently assumed Neuroscientific 'materialistic-reductionistic' working hypothesis whereby all Conscious stimulus processing (e.g., whether involving an "external-psychophysical" or "internal- functional, phenomenological or self" stimulus types) – being reduced to a particular neurophysiological material (causal) interaction between the specific Conscious stimulus and the corresponding brain locus (or loci) regions responsible for processing that particular type of information; the Duality Principle conceptually proves that it is not possible (e.g., again in principle) to reduce any such Psycho-Physical Stimulus to any direct or indirect physical interaction/s between any such Psycho-Physical Stimulus and any exhaustive hypothetical Neural Activation/s. Instead, the Duality Principle asserts that the only viable means for determining which pairs of the psychophysical, functional, phenomenological or 'self' 'Consciousness' and corresponding 'Neural Activation/s "co-occur" – is given by the abovementioned singular higher-ordered 'D2' computational framework. But, since it was shown (earlier) that there can only exist *one singular* such conceptually higher-ordered (a-causal) D2 computational framework – which was also shown previously (Bentwich, 2011b) to be equivalent to the (hypothetical) Computational Unified Field Theory's (CUFT) rapid series of Universal Simultaneous Computational Frames (USCF's), then we must conclude that any (apparently) "external" (psychophysical) or "internal" (function- phenomenal- or self-) Psycho-Physical Stimulus (or stimuli) is necessarily computed simultaneously together with a corresponding Neural Activation/s locus as a series of pairs which are embedded- and computed- within the rapid series of USCF's... In other words, the current materialistic-reductionistic working hypothesis (underlying the key pillars of Neuroscience, Psychiatry Psychology and more fundamentally the Cartesian conception of all scientific inquiry) wherein the human brain is merely activated by- and can perceive- or interpret- "real-objective" psycho-physical stimulation and translate it (or reduce it) to specific Neural Activation/s patterns within specific loci in the brain – has to be abandoned in favor of the Duality Principle's proof for the non-existence of any such material-causal relationship between any (exhaustive hypothetical) computational level/s' (di1...din) SROCS Psycho-Physical Stimulus $\rightarrow$ Neural Activation/s; Instead, the existence of a singular conceptually higher-ordered D2 'Universal Computational Principle' must be recognized which can compute the rapid series of USCF's within which are embedded all hypothetical (exhaustive) 'a-causal' pairs (series) of all possible ("external" or "internal" 'psychophysical', 'functional', phenomenal', or 'self') Psycho-Physical Stimulus and corresponding Neural Activation/s!

Thus, instead of the currently assumed basic Cartesian 'split' that seems to exist between the "objective-material" 'psycho-physical' stimulus – which is assumed to materially "cause" an activation of a particular set of Neural Activations, e.g., which are assumed (in turn) to "cause" a series of 'Black Box Hypothesis' (BBH) material interactions within the CNS that give rise to all "subjective" phenomenological perceptions of the ("objective") physical Reality - the Duality Principle proves that all that truly exists is s series of ("external" psychophysical or "internal" functional, phenomenological or self) Conscious Stimulus – that are computed to "co-occur" simultaneously together with any exhaustive hypothetical Neural Activations within the CNS... Moreover, both the Psycho-Physical Stimulus and "co-occurring" Neural Activations pairs are computed simultaneously as embedded within a Universal Computational Principle's computed Universal Simultaneous Computational Frames (USCF's) rapid series...

But, since it was already shown (above and previously – Bentwich, 2011c) that it is the same USCF's series that give rise to all of the basic physical features of 'space', 'time', 'energy' or 'mass' (or 'causality'), then the recognition of the Duality Principle's asserted conceptually higher-ordered D2 Universal Computational Principle's computation of the series of USCF's in fact transforms Cartesian Science's fundamental conception of an "objective-physical" world that exists "externally" to our CNS' "internal-phenomenological" perception (and interpretation) of it! Instead, the discovery of the Duality Principle and the CUFT paves the way for a new (broader) understanding of both the "physical" universe alongside our "phenomenological" (CNS) conception of it – as mere integral pairs within the singular conceptually higher-ordered Universal Computational Principle computation of the rapid series of USCF's that embed all exhaustive hypothetical pairs of Psycho-Physical Stimulus and corresponding Neural Activations (within the CNS)...

4. Summary & potential theoretical implications

A previous publication (Bentwich, 2011c) hypothesized the existence of a novel 'Computational Unified Field Theory' (CUFT) which was shown to be capable of replicating the primary empirical findings and laws of both Quantum Mechanics and Relativity Theory based on a conceptually higher-ordered 'D2' rapid (e.g., c^2/h) series of 'Universal Simultaneous Computational Frames' (USCF's) which are computed by a singular 'Universal Computational Principle' (termed: 'ל'). Essentially, the CUFT is based on three fundamental theoretical postulates which consist of the computational 'Duality Principle', the existence of the rapid series of USCF's and the existence of three 'Computational Dimensions' associated with the dynamics of this rapid USCF's computation (e.g., by the singular Universal Computational Principle, 'ל'). Moreover, the CUFT was able to resolve the key theoretical inconsistencies (and contradictions) that seem to exist between quantum and relativistic models of physical reality.

The primary aim of the current chapter is to validate the Computational Unified Field Theory based on a dual approach which consists of contrasting the CUFT's identification of three particular empirical instances (or conditions) for which the critical predictions of the CUFT's may differ (significantly) from those offered by relativistic or quantum theories; and a broader application of one of the CUFT's three theoretical postulates, namely: the 'Duality Principle' towards key scientific 'Self-Referential Ontological Computational Systems' ('SROCS') (e.g., akin to the previously identified Quantum and Relativistic SROCS computational paradigms) in order to point at the need to reformulate these key scientific paradigms based on the Duality Principle's conceptually higher-ordered 'D2 a-causal computational framework' – which is no other than the CUFT's (singular) rapid series of 'Universal Simultaneous Computational Frames' (USCF's) (Bentwich, 2011c).

The CUFT's three critical predictions include: the 'CUFT's Universal Computational Formula's Relativistic & Quantum Derivatives', 'Differential USCF's Presentations of "Massive" vs. "Light" Objects', and the 'Reversibility of USCF's Spatial-Temporal Sequence'. Succinctly stated, the CUFT significantly differs from both relativistic and quantum theories in its complete integration of all four basic physical features (e.g., of 'space', 'time', 'energy' and 'mass') within a singular Universal Computational Formula. In contrast, Relativity Theory only unifies between 'space and time' (e.g., as a four-dimensional integrated continuum) and 'energy' and 'mass' ('E = mc^2') and describes the curvature of 'space-time' by massive objects etc., whereas Quantum Mechanics only constrains 'energy and space' or 'time and mass' as complimentary pairs whose simultaneous measurement accuracy cannot exceed Planck's constant ('h'). Therefore, by utilizing two specific (relativistic and quantum) derivatives of this Universal Computational Formula it is possible to critically contrast between the CUFT and existing relativistic and quantum predictions (e.g., regarding the relativistic 'energy-mass equivalence' or regarding the complete integration of the two quantum complimentary pairs – as embedded within the broader Universal Computational Formula).

The second empirical instance for which it seems that the critical predictions of the CUFT may differ (significantly) from those of quantum and relativistic theories is regarding the differential USCF's presentations of "massive" vs. "light" objects: Based on the CUFT's computational definition of "mass" as the number of 'object-consistent' presentations (across a given number of USCF's) (Bentwich, 2011c) it follows that when we measure the number of such 'object-consistent' presentations of a more "massive" compound (or atom/s) relative to a "lighter" compound (or atom/s, e.g., from the 'local framework' perspective - we should obtain that the "lighter" compound should appear on less USCF's, relative to the more "massive" compound)... In contrast, according to both quantum and relativistic theories the differences in masses (between relatively 'lighter' or 'more massive' compounds or atoms) is due to differences in the weight of their nucleuses but should not entail any differences in their number of consistent presentations across a series of USCF's.

The third critical prediction of the CUFT involves its capacity to reverse a given 'spatial-temporal' sequence of events (e.g., thereby de facto "reversing the flow of time" according to the CUFT); According to both relativistic and quantum theories the "flow of time" is assumed to be "uni-directional" and "un-altered" – due to the light speed limit set by Relativity theory on our capacity to reach any past relativistic event (object or phenomenon), or due to the probabilistic interpretation of quantum mechanics which assumes a strict 'SROCS' computational structure (Bentwich, 2011c) that is dependent on the "collapse" of the target's 'probability wave function' as a contingency for our capacity to determine (or even measure) any subatomic phenomenon, thereby negating the possibility of "un-collapsing" the target's probability wave function (e.g., which would be necessary if we wished to reverse the sequence of subatomic events such that the target's "collapsed" probability wave function would become "un-collapsed" as prior to its direct physical interaction with the 'probe' element). In contrast, the CUFT predicts that it may be possible to reverse a given object's spatial-temporal sequence by applying a certain electromagnetic field to the relevant series of that object's particular series of USCF's 'spatial-electromagnetic pixel/s value/s' – in such a manner which may allow to reverse its recorded series of USCF's 'spatial-electromagnetic pixel/s value/s'. It is suggested that in this manner it may be possible to "reverse the flow of time" of a given object/s, event/s or phenomenon (with other potentially associated phenomena that may allow for a "materialization" or "de-materialization" of objects or their modulation and their potential transference to other regions in space...)

The second segment of this chapter focused on attempting to apply one of the three theoretical postulates of the CUFT, namely: the computational 'Duality Principle' to key scientific 'Self-Referential Ontological Computational Systems' (SROCS) computational paradigms including: Darwin's Natural Selection Principle and associated Genetic Encoding hypothesis and Neuroscience's Psycho-Physical-Problem; The aim of applying the computational Duality Principle to such key ('materialistic-reductionistic') SROCS scientific paradigms was to demonstrate the broader potential applicability and construct validity of the Computational Unified Field Theory as a significant candidate for a 'Theory of Everything' (TOE) which therefore may possess a broader validity bearing on other (primary) scientific disciplines. Succinctly stated, this application of the computational Duality Principle to the abovementioned key scientific (SROCS) paradigms successfully demonstrated that each of these scientific paradigms does in fact constitute a SROCS computational structure and is therefore constrained by the Duality Principle; Specifically, the conceptual computational constraint imposed on each of these scientific SROCS paradigms by the Duality Principle pointed at the need to replace their current 'material-causal' working hypothesis by a conceptually higher-ordered 'D2 a-causal' computational framework which simultaneously computes the "co-occurrences" of an exhaustive series of (particular) spatial-temporal 'x-y' pairs, which are (in turn) embedded in the Computational Unified Field Theory's rapid series of USCF's (Bentwich, 2011c).

In terms of some of the potential theoretical implications of these (three) critical predictions differentiating the CUFT from the currently existing quantum and relativistic models of physical reality it is (first) suggested that a potential empirical validation of the CUFT (e.g., in contrast to the predictions of the existing quantum or relativistic theories) may indeed suggest that the CUFT may broaden the theoretical scope of our understanding of quantum and relativistic phenomena – as embedded within the more comprehensive (higher-ordered) rapid series of USCF's which are computed by the stipulated 'Universal Computational Principle' ('ל'), and which are delineated by the 'Universal Computational Formula'. Indeed, when taken together – the previous outline (Bentwich, 2011c) of the CUFT as being capable of both replicating all major quantum and relativistic phenomena (and laws) as well as bridging the apparent gap (and theoretical inconsistencies) between quantum and relativistic models of physical reality, together with the current chapter's identification of three critical predictions that may potentially validate the CUFT visa vis. the currently acceptable quantum and relativistic theories may point at the feasibility of the CUFT as a broader theoretical framework which may unify and embed the limiting cases of quantum and relativistic modeling within the higher-ordered ('D2') conceptualization of the rapid series of (a-causal) USCF's, which give rise to all known physical properties of 'space', 'time', 'energy', 'mass' (and 'causality') as secondary computational properties of the singular USCF's sequential process... Second, to the extent that the CUFT's critical predictions are validated empirically (and based on an acceptance of the CUFT's hypothetical computational structure, replication of quantum and relativistic findings and tentative resolution of any quantum-relativistic inconsistencies), a logical next step may also involve a closer analysis of the very "essence" of the 'Universal Computational Principle' ('ל') and its production of the rapid series of USCF's.

5. References

Bentwich, J. (2003a). From Cartesian Logical-Empiricism to the 'Cognitive Reality': A Paradigmatic Shift, *Proceedings of Inscriptions in the Sand, Sixth International Literature and Humanities Conference*, Cyprus

Bentwich, J. (2003b). The Duality Principle's resolution of the Evolutionary Natural SelectionPrinciple; The Cognitive 'Duality Principle': A resolution of the 'Liar Paradox' and 'Goedel's Incompleteness Theorem' conundrums; From Cartesian Logical-Empiricism to the 'Cognitive Reality: A paradigmatic shift, *Proceedings of 12th International Congress of Logic, Methodology and Philosophy of Science*, August Oviedo, Spain.

Bentwich, J. (2003c). The cognitive 'Duality Principle': a resolution of the 'Liar Paradox' and 'Gödel's Incompleteness Theorem' conundrums, *Proceedings of Logic Colloquium*, Helsinki, Finland, August 2003.

Bentwich, J. (2004). The Cognitive Duality Principle: A resolution of major scientific conundrums, *Proceedings of The international Interdisciplinary Conference*, Calcutta, January.

Bentwich, J. (2006a). The 'Duality Principle': Irreducibility of sub-threshold psychophysical computation to neuronal brain activation. *Synthese*, Vol. 153, No. 3, pp. (451-455)

Bentwich, J. (2011a). PCT/IL2010/000912 Computerized System or Device and Methods for Diagnosis and Treatment of Human, Physical and Planetary Conditions.

Bentwich, J., Dobronevsky E, Aichenbaum S, Shorer R, Peretz R, Khaigrekht M, Marton RG, Rabey JM. (2011b). Beneficial effect of repetitive transcranial magnetic stimulation combined with cognitive training for the treatment of Alzheimer's disease: a proof of concept study. J Neural Transm., Mar;118(3):463-71. Bentwich, J. (2011c) Quantum Mechanics / Book 1 (ISBN 979-953-307-377-3) Chapter title: The 'Computational Unified Field Theory' (CUFT): Harmonizing Quantum Mechanics and Relativity Theory. Born, M. (1954). The statistical interpretation of quantum mechanics, *Nobel Lecture, December 11, 1954*

Bentwich, J. (2011c) Quantum Mechanics / Book 1 (ISBN 979-953-307-377-3) Chapter title: The 'Computational Unified Field Theory' (CUFT): Harmonizing Quantum Mechanics and Relativity Theory

Brumfiel, G. (2006). Our Universe: Outrageous fortune. *Nature*, Vol. 439, pp. (10-12)

Burgess, R. R. (1971). "Rna Polymerase". *Annual Review of Biochemistry* 40: 711–740. doi:10.1146/annurev.bi.40.070171.003431. PMID 5001045

Darwin C (1859) *On the Origin of Species by Means of Natural Selection, or the Preservation of Favoured Races in the Struggle for Life* John Murray, London; modern reprint Charles Darwin, Julian Huxley (2003). *On The Origin of Species*. Signet Classics. ISBN 0-451-52906-5.

Ellis, J. (1986). The Superstring: Theory of Everything, or of Nothing? *Nature*, Vol. 323, No. 6089, pp. (595–598)

Geiduschek, E. P.; Haselkorn, R. (1969). "Messenger RNA". *Annual Review of Biochemistry* 38: 647. doi:10.1146/annurev.bi.38.070169.003243. PMID 4896247. edit

Greene, B. (2003). *The Elegant Universe*, Vintage Books, New York

Heisenberg, W. (1927). Über den anschaulichen Inhalt der quantentheoretischen Kinematik und Mechanik. *Zeitschrift für Physik*, Vol. 43 No. 3-4, pp. (172–198).

Khorana, H. G. (1965). "Polynucleotide synthesis and the genetic code". *Federation proceedings* 24 (6): 1473–1487. PMID 5322508

Rich, A.; Rajbhandary, U. L. (1976). "Transfer RNA: Molecular Structure, Sequence, and Properties". *Annual Review of Biochemistry* 45: 805–860. doi:10.1146/annurev.bi.45.070176.004105. PMID 60910. edit

Schweet, R.; Heintz, R. (1966). "Protein Synthesis". *Annual Review of Biochemistry* 35: 723 758. doi:10.1146/annurev.bi.35.070166.003451. PMID 5329473. edit

Permissions

The contributors of this book come from diverse backgrounds, making this book a truly international effort. This book will bring forth new frontiers with its revolutionizing research information and detailed analysis of the nascent developments around the world.

We would like to thank Mohammad Reza Pahlavani, for lending his expertise to make the book truly unique. He has played a crucial role in the development of this book. Without his invaluable contribution this book wouldn't have been possible. He has made vital efforts to compile up to date information on the varied aspects of this subject to make this book a valuable addition to the collection of many professionals and students.

This book was conceptualized with the vision of imparting up-to-date information and advanced data in this field. To ensure the same, a matchless editorial board was set up. Every individual on the board went through rigorous rounds of assessment to prove their worth. After which they invested a large part of their time researching and compiling the most relevant data for our readers. Conferences and sessions were held from time to time between the editorial board and the contributing authors to present the data in the most comprehensible form. The editorial team has worked tirelessly to provide valuable and valid information to help people across the globe.

Every chapter published in this book has been scrutinized by our experts. Their significance has been extensively debated. The topics covered herein carry significant findings which will fuel the growth of the discipline. They may even be implemented as practical applications or may be referred to as a beginning point for another development. Chapters in this book were first published by InTech; hereby published with permission under the Creative Commons Attribution License or equivalent.

The editorial board has been involved in producing this book since its inception. They have spent rigorous hours researching and exploring the diverse topics which have resulted in the successful publishing of this book. They have passed on their knowledge of decades through this book. To expedite this challenging task, the publisher supported the team at every step. A small team of assistant editors was also appointed to further simplify the editing procedure and attain best results for the readers.

Our editorial team has been hand-picked from every corner of the world. Their multi-ethnicity adds dynamic inputs to the discussions which result in innovative outcomes. These outcomes are then further discussed with the researchers and contributors who give their valuable feedback and opinion regarding the same. The feedback is then collaborated with the researches and they are edited in a comprehensive manner to aid the understanding of the subject.

Apart from the editorial board, the designing team has also invested a significant amount of their time in understanding the subject and creating the most relevant covers. They scrutinized every image to scout for the most suitable representation of the subject and create an appropriate cover for the book.

The publishing team has been involved in this book since its early stages. They were actively engaged in every process, be it collecting the data, connecting with the contributors or procuring relevant information. The team has been an ardent support to the editorial, designing and production team. Their endless efforts to recruit the best for this project, has resulted in the accomplishment of this book. They are a veteran in the field of academics and their pool of knowledge is as vast as their experience in printing. Their expertise and guidance has proved useful at every step. Their uncompromising quality standards have made this book an exceptional effort. Their encouragement from time to time has been an inspiration for everyone.

The publisher and the editorial board hope that this book will prove to be a valuable piece of knowledge for researchers, students, practitioners and scholars across the globe.

List of Contributors

Valery I. Sbitnev
B. P. Konstantionv St.-Petersburg Nuclear Physics Institute, Russ. Ac. Sci., Gatchina, Leningrad District, Russia

Fumihiko Sugino
Okayama Institute for Quantum Physics, Japan

S. P. Maydanyuk and V. S. Olkhovsky
Institute for Nuclear Research, National Academy of Sciences of Ukraine, Ukraine

A. Del Popolo
Istituto di Astronomia dell' Università di Catania, Catania, Italy
Dipartimento di Matematica, Università Statale di Bergamo, Bergamo, Italy

E. Recami
INFN-Sezione di Milano, Milan, Italy
Facoltà di Ingegneria, Università statale di Bergamo, Bergamo
DMO/FEEC, UNICAMP, Campinas, SP, Brazil

Călin Gh. Buzea
National Institute of Research and Development for Technical Physics, Romania

Maricel Agop
Department of Physics, Technical "Gh. Asachi" University, Romania

Carmen Nejneru
Materials and Engineering Science, Technical "Gh. Asachi" University, Romania

Kayode John Oyewumi
Theoretical Physics Section, Physics Department, University of Ilorin, Ilorin, Nigeria

Paul Bracken
Department of Mathematics, University of Texas, Edinburg, TX, USA

Jonathan Bentwich
Brain Perfection LTD, Israel

Francisco Bulnes
Technological Institute of High Studies of Chalco, Mexico

9 781632 381576